OXFORD CLASSIC TEXTS IN
THE PHYSICAL SCIENCES

THE STRUCTURE OF
PHYSICAL CHEMISTRY

BY

C. N. HINSHELWOOD

CLARENDON PRESS · OXFORD

*This book has been printed digitally and produced in a standard specification
in order to ensure its continuing availability*

OXFORD
UNIVERSITY PRESS

Great Clarendon Street, Oxford OX2 6DP

Oxford University Press is a department of the University of Oxford.
It furthers the University's objective of excellence in research, scholarship,
and education by publishing worldwide in

Oxford New York

Auckland   Cape Town   Dar es Salaam   Hong Kong   Karachi
Kuala Lumpur   Madrid   Melbourne   Mexico City   Nairobi
New Delhi   Shanghai   Taipei   Toronto
With offices in
Argentina   Austria   Brazil   Chile   Czech Republic   France   Greece
Guatemala   Hungary   Italy   Japan   South Korea   Poland   Portugal
Singapore   Switzerland   Thailand   Turkey   Ukraine   Vietnam

Oxford is a registered trade mark of Oxford University Press
in the UK and in certain other countries

Published in the United States
by Oxford University Press Inc., New York

ISBN 0-19-857025-2

PREFACE

PHYSICAL chemistry is a difficult and diversified subject. The difficulty can, of course, be overcome by a suitable intensity of application and the diversity dealt with in some measure, by judicious specialization. This is very well as far as it goes, but leaves something to be desired, because there is, it is to be hoped, still room for a liberal occupation with wide studies, and this in a manner which goes beyond the polite interest of the dilettante. In the light of a good long spell of University teaching, however, I have the impression that this aspect of the matter is in some danger of neglect. I thought, therefore, I would like to write a book of moderate compass which, in no way competing with more formal works, should lay emphasis on the structure and continuity of the whole subject and try to show the relation of its various parts to one another. Certain themes or, one might almost say, leitmotifs run through physical chemistry, and these would be used to unify the composition.

The treatment would be neither historical, nor formally deductive, but at each stage I would try to indicate the route by which an inquiring mind might most simply and naturally proceed in its attempt to understand that part of the nature of things included in physical chemistry. This approach I have ventured to designate humanistic. The proper study of mankind, no doubt, is man, but one of the greatest activities of man is to find things out.

Apart from the question of seeing the subject as a whole, there is that of seeing it with a sober judgement. It seems to me specially important in modern physical chemistry to be clear and honest about fundamentals. This is not so easy as it sounds. Some of the current working notions are expressed in words which easily become invested with a more literally descriptive character than they deserve, and many young chemists—this is my impression at least—are led to think they understand things which in fact they do not. Something simple and direct seems to be conveyed by words such as 'resonance' and 'activity', which is not legitimately conveyed at all. By certain descriptions, which it is easy to give, one is reminded of Alice: 'Somehow it seems to fill my head with ideas—only I don't exactly know what they are.' Many of the mathematical equations which serve important technical purposes in the modern forms of theoretical chemistry are of a highly abstract kind, but they have acquired a dangerous seductiveness in that they clothe themselves rather readily in metaphors. Occasionally it is salutary to regard this metaphorical apparel with the eyes of the child who surveyed the emperor's new clothes. I have done my best here and there to help the uninitiated reader keep in mind just what the content of theories amounts to.

On the other hand, I have not attempted any refined analysis of conceptions such as probability, or statistical equilibrium, since these are not difficult to have a working knowledge of, though purists could subject to harassing criticism almost any treatment save a very elaborate one. Excessively detailed analysis however, would have obscured the general plan which I was anxious to try and depict.

Evidently, therefore, my undertaking, although on a fairly modest scale, is a somewhat rash one involving various kinds of compromise, but as the book does not claim in the slightest degree to supersede or replace other sources of information (upon which, of course, it almost wholly depends itself) I hope the boldness is not quite unjustified.

Finally I should like to express indebtedness to various colleagues who have helped me in many ways; to Dr. R. F. Barrow for also reading through the proofs, and to the staff of the Clarendon Press who have given much more help than any author has the right to expect.

C. N. H.

OXFORD
January 1951

CONTENTS

CONTENTS

PART I

THE WORLD AS A MOLECULAR CHAOS

SYNOPSIS

SOME of the ancient philosophers conceived the world to be made up of primordial particles in random motion, but their theories were not very fruitful since they lacked the necessary empirical basis. The formation of this was a long, complex, and far from obvious process.

Quantitative relations between the masses, and in certain cases the volumes, of substances which combine chemically establish the atomic theory as a scientific doctrine, and rather subtle coherence arguments reveal the distinction between atoms and molecules. Developments in physics lead to the recognition of heat as the invisible motion of the molecules themselves, and the kinetic picture of matter emerges.

Even in its first primitive version, this picture gives a satisfying representation of nature in many of its broad aspects. Molecules are envisaged as microscopic masses following the laws which Newtonian mechanics prescribe for macroscopic bodies. They are believed to be in chaotic motion and also to exert forces upon one another at small distances. Their motions tend to scatter them through all space: the forces to agglomerate them into condensed phases. The conflict between these two tendencies governs the existence of material systems in their various states, determines the range of stability of gases, liquids, and solids, and regulates the extent to which various possible combinations of atoms into molecules occur.

Reasonable estimates can be made of the absolute sizes, masses, speeds, and modes of motion of molecules, and coherent explanations can be given of many of the physical properties of matter in its different forms.

At this level of interpretation nothing is yet postulated about the forces, except that they manifest themselves in the measurable energy changes which accompany almost every kind of physical and chemical transformation. It is necessary, therefore, to describe the state of a chaotic system of particles primarily in terms of its energy. Such a description is provided in statistical theory.

In the world of molecular chaos various energy states are regarded as so many boxes occupied by molecules at random, and everything tends to that condition which may be realized in the largest number of ways. This idea leads to the definition of entropy, a function which measures the probability of a molecular assemblage, and to the laws of thermodynamics, which prescribe the conditions of equilibrium and the direction of possible changes in all such processes as expansion and contraction, melting and evaporation, and in actual chemical transformations. These laws themselves are closely related to empirical observations about heat-changes which can be used to provide an independent basis for them.

Given only an empirical knowledge of the energy changes accompanying atomic and molecular regroupings, the dynamical and statistical laws (or the equivalent thermodynamical principles) predict the relations of solids, liquids, and gases for single substances and for mixtures, and also the dependence of chemical equilibrium upon concentration and temperature.

But the representation so made is limited in two ways. First, the forces remain unknown. The energy changes are measurable by experiment, but their nature remains unexplained. Secondly, the actual prescription of the condition of molecular chaos proves on closer inspection to present certain subtle problems which cannot be solved at this stage. If molecules are assigned to states according to the laws of probability, then what constitutes a state? In the first instance, equal ranges of momentum can be satisfactorily regarded as defining a series of states. But while this idea leads to valuable laws regarding changes in entropy, and the dependence of all kinds of equilibrium upon variables such as concentration and temperature, it leaves quite unsolved the question of the absolute position of chemical and physical equilibria.

Scientific explanations seek to describe the unknown in terms of the known. The first attempt at such an explanation of the material forms and relations of the world works in terms of particles which are themselves small-scale models of grosser objects. It goes a long way, but reaches a boundary beyond which it cannot pass.

I

ATOMS AND MOLECULES

Introductory observations

ADAM and Eve, tradition tells us, ate from a tree of knowledge and though driven forth from Eden were the forebears of a progeny which never lost their taste. Prometheus, we also hear, stole fire from heaven, yet the fate he met has not deterred posterity from emulation. These two old stories are the symbols of two deep desires: the one to understand the essence of the world, the other to achieve the power of dominating nature. It is not inappropriate that the allegories are taken from diverse mythologies, expressing as they do profoundly different attitudes of mind. To some men knowledge of the universe has been an end possessing in itself a value that is absolute: to others it has seemed a means to useful applications. These two divergent views are never reconciled by argument, yet by a strange entanglement the history of the sciences was made not by the one or by the other school of thought but from the interplay of both. The route to mastery has lain indeed through abstract knowledge, yet the path to knowledge by the ways of contemplation has seldom proved a practicable one. Direct solution of the problems which the useful arts present is not infrequently attainable by application of the principles already known and understood, but theories and hypotheses of other than a vague and misty kind are seldom based upon a panoramic survey of the general scene. They rest more often on a detailed searching into matters which a casual view might well dismiss as specialized, recondite, and obscure, and access to the realm of fruitful theories is usually by devious and unexpected ways, found out by men with very different ends in view.

As it is handed down, the outcome of past probings into nature presents a variegated and uneven picture, resembling in some ways those interlocking growths of crystals which have sprung from many centres. By the proper methods such tangles may be induced to recrystallize into uniformly oriented systems, and from the character of its origins any experimental science stands in need from time to time of the treatment which this analogy suggests.

The matter goes deeper even than the metaphor implies. Science is not the mere collection of facts, which are infinitely numerous and mostly uninteresting, but the attempt by the human mind to order these facts into satisfying patterns. Now a pattern or design is not a purely objective function but something imposed by the mind on what is presented to it, as is seen in those pictures of piled cubes which can be made at will to appear in advancing or receding order. The imposition of design on nature is in fact an act of artistic creation on the part of the man of science, though it is subject to a discipline more exacting than that of poetry or painting. Two painters may depict quite differently a given scene, though the can-vasses of both may present essential truth. Science has greater objectivity than painting, but the formulation of its laws cannot be rendered wholly independent of the individual mind. The limitation is the greater since knowledge is never absolute and its expression rarely perfect. Much of the content of a growing science is subsumed in working hypotheses, constructions which in the last analysis are not impersonal.

Upon the heterogeneity imparted to science by its multiple and fortuitous origins is thus superimposed another resembling that of a picture whose various parts are painted by artists of different schools. Judgements may often differ as to which of two conceptions is the more fundamental, so that related theories may sometimes start from very varying premisses. The unaesthetic element of variegation can never be wholly removed, but from time to time greater harmony can be introduced by rearrangement. To resort once more to metaphor: in a time of great expansion a museum might fill its galleries with the most disparate works of art, revealing to the scholar, stimulating to the craftsman, and to the man of taste a little disconcerting. But at intervals the acquisitions are regrouped according to a principle—in which the critics may not acquiesce—but which at least is consciously decided. In science, too, this rearrangement must from time to time be undertaken, and on the writer who describes a field of any breadth there falls the task of making what is practically an artistic judgement.

These more or less philosophical observations would a generation ago have seemed out of place as an introduction to an account of the field which we call physical chemistry. Today, however, they are relevant and serve to suggest what may be expected from the survey and what may not. To realize that the content of the subject is a series of accounts of how human minds have tried to represent their probings into nature, by diminishing unjustified expectation, increases admiration for what in fact has been achieved. What might appear as oddnesses, inconsistencies, and arbitrary assumptions fall into their perspective and no longer disconcert. The best understanding of the subject is attainable, to return to the metaphor already used, by proceeding from room to room of the gallery and studying the works of various schools with a due realization that values must be adjusted in transit.

It is in this spirit that we shall examine the scope and achievements of physical chemistry, and see what views about the nature of things it reflects. We shall attempt to show the subject in a continuous development which reveals its structure and displays the relation of its parts. We shall therefore not pay much attention to the accidents of history, but we shall be very much concerned with the methods by which an inquiring mind can penetrate the secrets of nature. In this sense the treatment may reasonably be called humanistic.

We shall find it necessary to keep before us what is meant by a scientific explanation: it is in effect the representation of the unknown in terms of the known, but we shall find that the idiom in which the representation is expressible has to suffer some remarkable transformations as we proceed. In the early stages, to employ yet again the metaphor of the picture gallery, we spend some profitable time in a school of primitives: presently we find that more abstract schools command our attention.

Atoms and molecules

Chemistry rests largely upon the theory of atoms and molecules. The idea of an atomic constitution of things was arrived at by the ancients, though the basis of their speculations differed considerably from that of modern chemistry. They thought that there must be a limit to the divisibility of matter, that living creatures must be reproduced from ultimate primordial bodies of the appropriate kind, that various natural phenomena such as the penetration of heat and cold depend upon the hidden motions of tiny particles, and so on. It is hardly conceivable that chemistry as we know it could have arisen directly in this way.

The theory, however, that the sensible qualities of objects could be interpreted in terms of the motion of minute particles of specific kinds was a very great achievement, and itself must have depended upon a long evolution of ideas. We need not enter into this history, but it is important to realize that it must have existed. To distinguish between the material substratum and the qualities for which it is responsible required enlightenment and profound thinking. As far as chemistry is concerned, the process cannot be said really to have been completed until the phlogiston theory disappeared.

The fundamental ideas of matter and motion only acquired their dominance after long processes of trial and error had shown how more and more could be described in terms of them. One accustomed to the ways of thought of science may now find it difficult to realize the elaboration of the analysis by which, for instance, colour came no longer to be regarded as something which a substance contains rather as a fabric contains a dye. Our notions about the scientifically describable substratum of the world are now so familiar that there is some danger of forgetting to what extent it is based upon a working hypothesis, and one which possesses natural limitations.

The basis of chemistry as it grew up in the nineteenth century is *Dalton's atomic theory*. In this the intuitive idea of ultimate particles is applied to explain definite quantitative laws of chemical composition, those, namely, of constant, multiple, and reciprocal proportions. These rules could only have emerged after a long empirical study of diverse and often obscure chemical substances, since many of the things most obviously attracting the attention of the contemplative philosopher—wood, rocks, plants, and the like—would have yielded singularly unfruitful evidence for an atomic theory in its infancy. When, however, the labours of many generations of alchemists, artificers, and craftsmen had provided the necessary facts, when the technique of weighing had been refined and the practice of quantitative measurement had become established, and when the various apt conjunctions of experiment and speculation had at length occurred, there crystallized out the notion that certain specific substances— and not principles such as fire—constituted the chemical elements, and that their union according to quantitative laws gave rise to all the other substances.

The same elements combine in constant and fixed proportions. This suggests that the union of macroscopic quantities is simply an n-fold repetition of the union of indivisible microscopic units of characteristic mass. Other mathematical descriptions of the phenomenon could doubtless be devised, but they could hardly possess the vividness of the atomic conception, which is further strengthened by the facts about multiple and reciprocal proportions. Two elements, which we may call A and B, not infrequently form several compounds with one another, and in these the masses of B which have combined with unit mass of A stand in the ratio of simple integers. This is the *law of multiple proportions*. To each element, further, there may be assigned an equivalent weight, such that it defines the relation of that element not to one other but to every other. The proportions in which any two elements unite are in the ratio of their equivalent weights or in one which is a simple integral multiple of it. This is the *law of reciprocal proportions*. By extending the idea of units to give a coherent set of relationships between all the elements it gives to the atomic theory a still higher status. Some kind of quantitative metric which would satisfy a mathematical physicist could conceivably be devised to embrace the combination of single pairs of elements, but that any other system as simple as that of Dalton could be found to describe the possible relations of every element to every other is unlikely.

A great advance became possible when the technique of making measurements with gaseous substances was introduced. *Gay Lussac's* famous rule states that when chemical combination occurs between gases the volumes of those consumed and of those produced, measured under standard conditions of temperature and pressure, stand in a simple numerical ratio. Since what applies to the masses according to the atomic theory itself now proves to apply to the gas volumes, the conclusion follows that equal volumes of substances in the gaseous state under the standard conditions actually contain, to within a small numerical multiple, equal numbers of the primordial combining units. After various trials and some groping in the dark it proved further that a coherent system requires the *hypothesis of Avogadro*, namely that the numerical multiple in question is unity, but that the primordial particles existing in the free state are molecules which may consist of more than one atom. For many common elements the number of atoms in the molecule was discovered to be two. According to the hypothesis equal volumes of gaseous compounds at equal temperature and pressure possess masses proportional to those of their constituent molecules, a principle of the greatest help in the establishment of chemical formulae and in the assignment of atomic weights to the elements.

The detailed arguments by which the individual atomic weights, after a long process of trial and error, were established need not be enlarged on here, but the principles of the method must be summarized. As many compounds as possible were analysed and the proportions of their elements determined. To a chosen element was arbitrarily given a conventional atomic weight (originally unity to hydrogen, later changed for practical convenience). As many molecular weights as possible were determined from the gaseous densities, and the least weight of a given element which ever appeared in the molecule of any compound came to be accepted as its atomic weight.

The system which was gradually built up, like the hypothesis of Avogadro upon which it was largely based, rested upon arguments of coherence, and could have involved errors of simple numerical multiples—for some time indeed carbon was given the atomic weight six. But the discovery of the periodic system of the elements, and the realization that Avogadro's hypothesis fitted in very well with the kinetic theory of gases, presently showed that the probability of error could be disregarded.

The numerical relations of Gay Lussac's law are only approximate and Avogadro's hypothesis is correspondingly inexact. For this fact the kinetic theory provides a ready explanation. It is, however, only intelligible in terms of developments which would probably never have occurred had the inexactitude not been in the first instance neglected. In science, as in everything else, there is need for courage to act upon the conviction that more significance may reside in an approximation to the truth than in the deviations which it for the moment disregards.

Before passing to consider the alliance of chemistry with physics we shall try to focus once more the way in which pure chemistry itself emerges from the complex whole which nature presents to contemplation. First, this complex is analysed in various tentative ways until substances are found to be enumerable separately from qualities, and we realize, for example, that when we say mercury oxide contains mercury, oxygen, and redness, the first two are differently significant from the third. Substances commonly existing in the world are found not infrequently to be separable into parts by simple means such as distillation. Some which resist this process of fractionation are classified as pure substances. But these sometimes change when brought together, and seem to disappear giving rise to others, often with evolution of heat or light. Such drastic changes can be reversed by various roundabout ways, and the original partici-pants in what is termed the chemical reaction can be regenerated. Most substances can be split up into others or made by the union of others, but some are incapable of further resolution or of synthesis and become recognized as the chemical elements. Gradually the list of them is filled, and presently coherent relations between the pro-perties of the individuals emerge. The structure of the periodic system is revealed and interlopers can be detected and ejected.

In chemical reactions, however spectacular, mass is found to be conserved within the limits of the sensitivity of chemical balances, and this is as it should be if the transformations are mere regroupings of the units of a material substratum. The atoms of ancient specula-tion fill the roles required. The quantitative laws of chemical com-bination then follow and permit the development of chemistry in the form in which it is known today.

It is hard to see how the evolution could have been other than long and painful, since the commonest objects are among the most unsuitable for elementary chemical investigation, and the major task for chemistry in one sense was the discovery of its own tools. Only by a fortunate disposition of providence has the story not been made more tangled still, since mass is not really conserved and the elements are only relatively unchanging. But the phenomena of nuclear physics are on such a scale that in chemical reactions they may be neglected, not only in a first but in much higher approximations.

Matter and motion

Chemistry concerns itself with the material substratum of the world. It was once said that the only thing you can do with matter is to move it, and though this statement would probably not sustain detailed analysis in the light of modern physics, it expresses an essential truth with a high degree of approximation. The least that can be said is that it brings into clear relief the need for a kinetic theory of molecules. The atoms of the ancient philosophers were in lively motion, and Lucretius tries to explain in detail how chaotic displacements of minute particles below the threshold of visibility can give the semblance of rest. Facts of common observation, such as diffusion, evaporation, and the like, are readily interpretable by a rudimentary kinetic theory which, however, only attains to the rank of a serious scientific hypothesis when the Newtonian laws are applied to the molecules conceived by chemistry. The laws of motion themselves are based upon the observation of massive bodies, and their application to the invisible, and indeed in the first instance hypothetical, particles of Dalton and Avogadro rests upon an assumption. On the face of it and in the absence of evidence to the contrary there is more reason for making this assumption than for not making it, and the taking of the risk proves to be abundantly justified. But there must be no surprise or concern when the utility of the picture so created turns out to have its limits.

One of the simplest routes to the understanding of the question is, as it happens, that which was followed historically. Observation of common phenomena such as the winds suggests the reasonableness of assuming a material substratum even for the tenuous and invisible parts of nature like the air. The next step consists in making experiments by confining gases in tubes, observing their 'spring', and measuring their pressures in terms of the heights of mercury columns which the spring supports. *Boyle's law* emerges in the well-known form

$$pv = \text{constant},$$

where p is the pressure exerted by a volume v of a gas at a constant temperature.

That the spring or pressure increases as the gas is compressed vividly suggests the picture of a crowd of flying particles seeking to escape and causing pressure on the surface of the containing vessel by their impacts. This idea leads immediately to the first quantitative result.

Gas pressure

Let there be n molecules of a gas in unit volume of an enclosure, and let their motion be random both in direction and in the magnitude of the speeds. Consider those which approach unit area of the surface at an angle of incidence within $d\theta$ of a given value θ and which possess velocities within du of a value u. When they impinge upon the surface they may be supposed to suffer an elastic reflection, the normal component of the momentum of each being changed from $mu\cos\theta$ to $-mu\cos\theta$. The change for each molecule is thus $2mu\cos\theta$. The number of particles which suffer reflection in each second is proportional to n and to u, so that the rate of change of momentum for the type of particle defined is αnmu^2, where α is a numerical multiplier independent of n, m, and u. Averaged over all possible directions of approach and over all permissible velocities, the result becomes $\alpha' nm\bar{u}^2$, where $\bar{u}^2$ is the average value of u^2 and α' is still a numerical multiplier. By Newton's second law the surface will be urged outward with the force necessary to maintain this rate of change of momentum, and the force on unit area constitutes the pressure, p.

Thus $$p = \alpha' nm\bar{u}^2.$$

A very elementary argument shows that α' is roughly $\frac{1}{3}$. If the speeds are fairly closely grouped about the average value all may be taken approximately as $\bar{u}$. Let all velocities be normal to one or other surface. Then in any given direction the number of molecules approaching unit area of the surface and reaching it in unit time is not far from one-sixth of those contained in a prism of unit base and height $\bar{u}$. Thus

$$p = \tfrac{1}{6}n\bar{u} \times 2m\bar{u} = \tfrac{1}{3}nm\bar{u}^2.$$

The more precise evaluation of the multiplier does not in fact make much difference, and for most purposes we shall accept the result

$$p = \tfrac{1}{3}nm\bar{u}^2.$$

It must be emphasized that the parts of the above argument which are stated without proof or sketched in only roughly relate to the value of the numerical multiplier only and not to the proportionality of p and $m\bar{u}^2$.

When a gas is expanded or compressed, n, and consequently p, varies inversely as the volume. This is Boyle's law. No real gases follow it exactly at higher pressures, but many do with a good degree of approximation, and all do at very low pressures.

Temperature

If a gas becomes hotter, the pressure which it exerts increases. In the first instance hotness is gauged simply by sensation. Since p is proportional to $m\bar{u}^2$, that is to the mean kinetic energy of the molecules, the hotness appears to be a function of the invisible translational motion. So rational does this interpretation of the origin of the sensation seem that it becomes expedient to introduce a scale of hotness and to define the degree of hotness, or temperature, as proportional to the pressure which a gas having that degree would exert. For a standard scale the gas must obey Boyle's law. This so-called perfect gas scale is simple theoretically and convenient practically, since the deviations of actual gases from the law are easily corrected for by extrapolation to low pressures.

To define the temperature of a gas we consider one gram molecule of it and write
$$T = pV/R,$$
or
$$pV = RT,$$

where R is a proportionality factor defined by the arbitrary condition that at 760 mm. pressure there shall be 100 degrees of temperature between the freezing-point and the boiling-point of water. What we mean by the temperature of the water, or of any non-gaseous substance, is simply the temperature which a gas would assume if left in contact with it for a long enough time. The idea of thermal equilibrium will require further discussion at a slightly later stage.

In terms of the primary definition of temperature which we have for the time being adopted, Charles's law, which states that gas pressure is proportional to the absolute temperature, would be a tautologous statement. It is not necessarily so, and was not so in its historical setting, since other scales of temperature, notably that based upon the expansion of a mercury column by heat, are possible. The mercury thermometer and the gas thermometer provide scales which in fact correspond rather closely, and it is the former which has been used in practice more than the latter. The approach to the theoretical aspect of temperature which is now being followed thus departs in an important way from the line of historical development.

The statements contained in the equations $p = \frac{1}{3}nm\bar{u}^2$ and $pV = RT$ constitute the simplest possible illustrations of the thesis, established in the course of the nineteenth century, that heat in general is a mode of motion. Since, as in the experiment where Joule warmed up water by the churning action transmitted from falling weights, mechanical energy of a mass is quantitatively convertible into heat, and since all attempts to account for such phenomena in terms of a special kind of caloric fluid prove to be sterile, the identification of heat with the invisible chaotic motion of the molecules is compelling. Mechanical energy becomes apparent when the motions of the invisible parts are so coordinated as to give rise to perceptible motion of the group: it is transformed into heat when the coordination is destroyed and the motion is no longer discernible by any of the senses save that which permits the appreciation of warmth. From the point of view of the individual molecule, nothing has occurred when mechanical energy has become heat, and the *quantitative equivalence of heat and energy* is simply a special example of the conservation of energy. This equivalence is asserted as an empirical principle in the *First Law of Thermodynamics*.

Chemistry gives reason to suppose that molecules, being groups of atoms, should possess shapes, and therefore be capable of rotations. Since, moreover, the union of atoms implies some kind of force to hold them together, there is the further possibility of internal vibrational motions. The existence of liquids and solids, as well as the departure of real gases from Boyle's law, shows that there are forces between molecules themselves, so that there must be potential energy stored up in any collection of them.

We may therefore expect that if we can find appropriate conditions in which to make the comparison, the total energy contents of grammolecular quantities of various substances should differ greatly according to the specific shapes and structures of the molecules. The possibility of a valid comparison depends upon the principle of thermal equilibrium.

Suppose a portion of substance to be placed in contact with a gas thermometer, or with a mercury thermometer calibrated in terms of one. The reading of the thermometer changes at first and then settles down to a steady value, say, T_1, which is taken as the common temperature of itself and of the substance with which it is now in thermal equilibrium. Suppose now that equal masses of two different substances at temperatures T_1 and T_2 respectively are placed in contact. T_1 and T_2 both alter until a common steady value T_3 is attained. T_3 is not in general the mean of T_1 and T_2. It may, however, be calculated, as elementary physics teaches, by the aid of the assumption that something called heat flows from one body to the other until a common temperature is reached, and that the quantity of heat required to raise a unit mass by one degree varies from substance to substance, each one possessing its own characteristic *specific heat*.

The interpretation which the molecular hypothesis gives of thermal equilibrium and of what constitutes equality of temperature between two bodies with different specific heats is one of the most important chapters of physical chemistry.

Let us begin by supposing that two gases, each of which obeys Boyle's law sufficiently well, are mixed. Initially their respective temperatures are T_1 and T_2. Their molecules interchange energy by collision, and the transfers in any given encounter may be in either direction, but on the average they will occur predominantly from the gas initially at the higher temperature to that at the lower. When the common temperature, T, is reached the interchanges still continue, but now they take place equally in both directions. What is called a *statistical equilibrium* is set up. In this state, for one gram molecule of each gas at a common temperature, pressure, and volume,

$$\tfrac{1}{3}n_1 m_1 \bar{u}_1^2 = \tfrac{1}{3}n_2 m_2 \bar{u}_2^2.$$

If, from the purely chemical evidence to which reference has already been made, we are prepared to believe Avogadro's hypothesis, we assume that $n_1 = n_2$ and infer that

$$m_1 \bar{u}_1^2 = m_2 \bar{u}_2^2.$$

The condition of thermal equilibrium is thus seen to be that the mean translational kinetic energy should be the same for each gas.

This latter result is derivable, as it happens, quite independently of Avogadro's principle and in a form which gives it even greater generality and importance.

The equipartition principle

The establishment of statistical equilibrium is not confined to translational energy. When two substances are in contact or mixed, whether they are gaseous, liquid, or solid, a state must be reached where the gains and losses of each kind of energy, translational, rotational, and vibrational, by the molecules of the two kinds balance. Such a condition corresponds to equality of temperature, and involves a quite definite relation between the different types of energy in the various kinds of molecule.

In a mixture of gases at a uniform temperature the average translational energies of the various kinds of molecule are the same. If some of the molecules present are of more complex structure than others, they will have more possibilities of motion, and the share of the total energy which they take might well be expected to be greater. A detailed investigation, which will be given later, shows in fact that the sharing out is governed by a law known as that of the *equipartition of energy*. The nature of this law is as follows.

Translational, rotational, and vibrational energies may be expressed in terms of vectorial quantities. Translational and rotational velocities may be represented by vectors, that is by lines with definite lengths and directions, which may be resolved into components along three spatial axes of coordinates. Similarly a vibration has an axis and an amplitude, and it is resolvable. Each component of each type of motion is called a *degree of freedom*. To a reasonable degree of approximation the energy may be expressed as a sum of terms in each of which a characteristic molecular constant multiplies the square of a suitable coordinate. Thus kinetic energy of translation is $\frac{1}{2}m(\dot{x}^2+\dot{y}^2+\dot{z}^2)$ where $\dot{x}$, $\dot{y}$, and $\dot{z}$ are the components of velocity along the three axes x, y, and z. Rotational energy may be expressed as a sum of terms of the type $\frac{1}{2}I\omega^2$, where I is the moment of inertia about the axis of rotation and ω is the angular velocity. When the vibrational motion is simple harmonic, which it may often be assumed to be, there is a term for the kinetic energy, $\frac{1}{2}m\dot{x}^2$, and also a term for the potential energy, $\frac{1}{2}\mu x^2$, where x is the displacement of the particle from its equilibrium position and μ is a constant. In so far as the total energy can be represented correctly by a sum of such *square terms*, the equipartition principle states that the average value of each for all types of molecule becomes the same when thermal equilibrium is attained.

The representation of the energy in this way is not really exact: nor are the different kinds of energy in the molecule strictly independent of one another. Thus, for example, as the amplitude of vibration increases the moment of inertia changes, so that the rotational energy is affected. While this complication limits the exactness of the equipartition principle (which is subject to even more important restrictions, as will appear later), it in no way affects the validity of the general conception of a definite functional relation defining the energy distribution in statistical equilibrium. It is the existence of such a relation which gives significance to the ideas of temperature and of heat flow.

The mathematical treatment of equipartition and the general consideration of energy distributions will presently occupy our attention a good deal. For the moment we shall turn to the way in which the simplest form of the principle has helped to deepen our knowledge about the existence of molecules.

Molecular reality and the determination of Avogadro's number

Considerations of the kind which have been so far advanced lay the foundations of a kinetic theory of molecules, which seeks to interpret physical and chemical phenomena in terms of particles and their motions. The conceptions of the ancient philosophers have gained in precision, and the way seems clear for further fruitful investigations. But the atoms and molecules remain, as far as any arguments so far considered go, inferential. They are creations of the mind, inaccessible to the ordinary senses. There was indeed a school of thought about the end of the last century which regarded them as fictions, deplored their excessive use in the theories of physical chemistry, and sought for the description of phenomena in terms of energy and of matter to which no minute structure was attributed. This plea, even in its day, was an extravagance, and it has since been put out of court by the detailed study of such phenomena as the emission of countable particles from radio-active substances and by the application of the techniques of X-ray and electron diffraction to reveal discrete structure in matter. The most direct way in which the challenge can be met is by the analysis of what is called the Brownian motion. This phenomenon provides a means, exploited by Perrin and others, of bridging the vast gulf between the invisible motions of molecules and the visible displacements of macroscopic bodies, of making molecular happenings sensible, and of determining the absolute masses of molecules. It consists simply in this: that certain minute particles still large enough to be directly discernible by the light which they scatter into a microscope are observed to be in rapid random movement.

According to the equipartition principle, the average translational energy of an oxygen molecule at a given temperature is the same as that of a molecule of hydrogen, benzene, or any other substance. No matter how many internal degrees of freedom there are, they do not affect the issue. In the derivation of the principle itself nothing is said about a limit to the size, mass, or complexity of the particle. Thus even a macroscopic lump of lead in thermal equilibrium with the ordinary air is buffeted on all sides by molecules of oxygen and of nitrogen, and acquires translational motion, now in one direction and now in another, such that the time-average of its energy of translation is the same as that for the gas molecules. Its mass being millions of times greater than theirs, its speed is of course imperceptibly small and its displacements quite below the threshold of detection. The question, now, is whether the smallest particles which are still accessible to direct observation may acquire a mean translational energy large enough for their motion also to be perceptible and measurable.

Fortunately this is so, as the existence of the Brownian motion shows. The particles in certain colloidal suspensions fulfil the required conditions. Observation with the ultra-microscope leaves no doubt about their particulate character, and they are seen to be in a state of random motion, darting hither and thither in a manner which gives a vivid impression of irregular impacts first from one side and then from another. These impacts must come from other particles which are themselves invisible, so that the Brownian motion (once the technical difficulty of showing that it is not due to convection currents is overcome) is justly claimed to provide evidence of molecular reality.

The mean kinetic energy of translation of a colloidal particle executing this kind of motion may be written $\frac{1}{2}M\bar{U}^2$ and is equal to that of any gas molecule, that is to $\frac{1}{2}m\bar{u}^2$. Now for the latter

$$p = \tfrac{1}{3}nm\bar{u}^2,$$

and if V is the molecular volume,

$$pV = \tfrac{1}{3}nVm\bar{u}^2 = RT.$$

nV is the total number of molecules in a gram molecule (Avogrado's number), N. Thus
$$\tfrac{1}{3}Nm\bar{u}^2 = RT.$$

Any means of determining $\tfrac{1}{2}M\bar{U}^2$ gives at the same time the value of $\tfrac{1}{2}m\bar{u}^2$ which is equal to it. Since RT is known, N becomes determinable, and thus the individual mass of any kind of molecule may be discovered.

It remains, therefore, to seek such a method of finding the kinetic energy of a particle accessible to observation in the ultra-microscope while it is executing the Brownian motion. There are several such methods. The simplest depends upon a study of the sedimentation equilibrium. Particles heavier than the solution in which they are suspended would if at rest sink to the bottom. Their motion, however, keeps them suspended, though more thickly in the lower layers of the medium than in the higher ones. This sedimentation equilibrium is analogous to the equilibrium of the Earth's atmosphere under gravity and may be similarly treated.

Consider a column of unit cross-section in which small particles are suspended in a medium of some kind. Suppose that the concentration of particles at height h is n and at height $h+dh$ is $n+dn$. In between two horizontal planes at these respective heights there are $n\,dh$ particles and these are urged downwards with a gravitational force $wn\,dh$, where w is the effective weight of each. (w is in fact $g \times$ volume of particle $\times$ difference in density of particle and medium.) This force produces a downward momentum of $wn\,dh$ units in each second. When the suspension is in equilibrium the communication of downward momentum must be balanced in accordance with Newton's second law. Now the motion of the particles renders them capable of communicating momentum and thereby of exercising what is dynamically analogous to a gas pressure. Its value is $\tfrac{1}{3}nM\bar{U}^2$, as in the equation derived on p. 16. Thus $dp = \tfrac{1}{3}M\bar{U}^2 dn$. $-dp$ must balance $wn\,dh$, and therefore

$$wn\,dh = -\tfrac{1}{3}M\bar{U}^2 dn,$$

$$\frac{dn}{n} = -\frac{w\,dh}{\tfrac{1}{3}M\bar{U}^2},$$

$$\ln\frac{n}{n_0} = \frac{w(h-h_0)}{-\tfrac{1}{3}M\bar{U}^2},$$

C

where n is the concentration at height h and n_0 that at height h_0.

$$n = n_0 e^{-\frac{w(h-h_0)}{\frac{1}{3}M\bar{U}^2}}.$$

The scale of the phenomenon is very small and the concentration falls off rapidly as the height increases. It so happens that with suitable colloidal suspensions, by the focusing of a microscope at different depths, $(h-h_0)$ being measured by a micrometer arrangement, counts of the particles (observed by scattered light according to the principle of the ultra-microscope) can be be made, and numbers proportional to n and n_0 conveniently obtained.

The particles being accessible to direct observation by counting and by coagulation, filtration, and weighing, w can be found. The only unknown is $M\bar{U}^2$ which is thus calculable.

Avogadro's number, N, is $RT/\frac{1}{3}M\bar{U}^2$, and the absolute mass of a molecule is the gram-molecular weight divided by N. The value of N is 6×10^{23} (the best measurements being in fact made by a quite different method).

Another method of finding the kinetic energy of the particles in Brownian motion depends upon the observation of their mean displacements: this need not be described, as from the present point of view it introduces no important new principle.

That, in the manner which has just been indicated, a limit to the fine-grained character of matter is determinable by the observation of visible phenomena directly and qualitatively explicable by the molecular hypothesis provides a strong argument for the reality of the entities with which this hypothesis deals.

Some other molecular magnitudes

The speeds of molecules are calculable very simply from the formula for the gas pressure. Since $p = \frac{1}{3}nm\bar{u}^2$, it follows that

$$\bar{u} = (3p/nm)^{\frac{1}{2}}.$$

But $nm = \rho$, the density, so that

$$\bar{u} = (3p/\rho)^{\frac{1}{2}}.$$

If p is expressed in dynes/cm.^{2} and ρ in gram/c.c., then $\bar{u}$ is given in cm./sec. It comes out to be of the order of magnitude of a kilometre a second for simple molecules, and varies inversely as the square root of the molecular weight.

With the reasonable presumption that molecules have a real

existence and ascertainable mass and translational energy, the next step is clearly to seek some means of obtaining information about their size. What precisely is meant by the size of a molecule raises some quite profound questions, the consideration of which, however, is better deferred until the nature of the possible experimental approaches has been surveyed. For a first attempt, a molecule may be envisaged as a small solid elastic sphere obeying the rules of Newtonian mechanics.

With a large number of mass particles in rapid motion the most obvious effect of their finite dimensions will be to cause collisions which interfere with their motion. Instead of pursuing an unbroken trajectory each particle will suffer abrupt changes of direction at each encounter, and will describe a zigzag path. The average length of an uninterrupted straight portion of this zigzag is called the *mean free path* (*l*). A smaller mean free path implies a greater interference of the particles with one another. Thus to find an experimental approach one must turn to a phenomenon which seems to depend upon the mutual obstruction of the moving molecules.

Such a phenomenon is that of viscosity, and the study of this property in gases does in fact provide clear-cut results, though not perhaps in quite the way which might have been expected. The bodily jostling of the molecules, which is more pronounced the bulkier they are, plays the major part in determining the viscosity of liquids, but the treatment of this effect proves to be rather complicated. The viscosity of gases depends not upon a crowding tendency but upon a *transfer of momentum* by molecules as they move from one volume element to another, and the collisions determine it in so far as they regulate the sort of momentum which is available for transfer. The matter may be rendered clearer by a simple calculation.

Suppose that a stream of gas moves in the direction of the x-axis and that there is a velocity gradient dv/dz in the direction z perpendicular to that of the flow. Consider a reference plane of unit area parallel to the flow and thus perpendicular to the velocity gradient. Let the velocity of the stream in this plane be v. In so far as the gas is treated as a continuous fluid one may say that between the faster moving and the slower moving layers there is a frictional force tending to slow down the former and to accelerate the latter until the velocity gradient is destroyed. To maintain this gradient an external force must be applied, and it is the magnitude of this force which measures the viscosity of the medium. The coefficient of viscosity is defined by the simple hydrodynamic relation:

$$\text{force per unit area of reference plane} = \eta \, dv/dz,$$

where η is the coefficient.

In the kinetic interpretation of this effect a distinction is drawn between the streaming velocity, v, the direction of which is common to all the molecules in a given volume element, and the thermal velocity $\bar{u}$, the direction of which for the various molecules is completely random. Molecules move across the reference plane from the faster to the slower layers and equally well in the reverse direction. There is thus a net transport of streaming momentum of a kind which would destroy the velocity gradient unless a force equal to the rate of transport across the reference plane were continuously applied.

The mass of gas moving in one direction perpendicular to unit area of the reference plane in one second is $\frac{1}{6}\rho\bar{u}$, by an argument exactly analogous to that used in the derivation of the expression for the gas pressure (p. 10). The streaming momentum carried through the plane in this direction in one second is $\frac{1}{6}\rho\bar{u}v_1$, where v_1 is the average value of the streaming velocity for those molecules passing across in virtue of their thermal motion. Now it is the value of v_1 which is determined by the mean free path. This latter represents the distance from the reference plane at which the molecules are last brought into mechanical equilibrium with the stream by mutual collisions. Thus v_1 will be $(v+l\,dv/dz)$. Similarly momentum is carried in the reverse direction and is equal to $\frac{1}{6}\rho\bar{u}$ multiplied by $(v-l\,dv/dz)$. The excess of that taken in one direction over that taken in the other is thus

$$\tfrac{1}{6}\rho\bar{u}(v+l\,dv/dz) - \tfrac{1}{6}\rho\bar{u}(v-l\,dv/dz) = \tfrac{1}{3}\rho\bar{u}l\,dv/dz.$$

This represents the viscous force, so that

$$\tfrac{1}{3}\rho\bar{u}l\,dv/dz = \eta \, dv/dz,$$

or
$$\eta = \tfrac{1}{3}\rho\bar{u}l.$$

A simple and obvious modification of the above argument yields an expression for the thermal conductivity of a gas

$$K = \tfrac{1}{3}\rho\bar{u}lc_v,$$

where c_v is the specific heat at constant volume.

According to the above formula, the viscosity of a gas increases with the mean free path. This result may seem a little paradoxical at first sight. The interpretation, however, is that the streaming momentum can be transferred from farther up or down the velocity gradient the remoter from the reference plane is the place at which the molecules last come into equilibrium with the general current. The viscosity does, however, depend upon mutual collisions directly, since without them there would be no hydrodynamic equilibrium at all.

The two equations just derived predict certain quite characteristic, and indeed slightly surprising, phenomena, the experimental verification of which affords good evidence for the essential correctness of the underlying ideas. Thus the viscosity of a gas is independent of the pressure, since ρ and l vary in a compensating way: it increases with rise in temperature, the momentum transfer becoming more lively: and the ratio of viscosity to thermal conductivity is constant and calculable from the specific heat. These results could hardly be explained in any other simple way.

The viscosity coefficient of gases may be determined from measurements of the rate of flow through a capillary tube of known radius under a given pressure difference, and from the result the mean free path, l, may be calculated. For simple gases at atmospheric pressure it is of the order 10^{-5} cm. It varies inversely as the pressure, and long before the highest vacuum given by a modern pump is reached it exceeds the dimensions of ordinary small-scale laboratory apparatus.

From the mean free path the size of molecules may be inferred. $\bar{u}/l$ gives the number of times in a second that the trajectory of a particular molecule has been interrupted and thus measures the number of collisions which it has suffered in unit time. Thus we have

$$Z_1 = \bar{u}/l.$$

The number of collisions is fairly simply related to the size. Let the diameter of a molecule be σ, which may be regarded as an effective value only, since we are not assuming much about the nature of a collision. With this convention two molecules may be regarded as entering into collision whenever their centres approach to within a distance σ of one another.

Let all the molecules in the system save one be imagined frozen into immobility and this selected one be thought to move about among the others with an effective velocity which we will denote by r. r is evidently proportional to $\bar{u}$ and may be written $\alpha\bar{u}$, where α is a numerical factor. If in reality all the molecules possessed the same speed, the directions alone being random, r would be the average value of the relative velocity of two molecules, the average being taken over all possible angles in space between the two directions. α, as a simple geometrical calculation shows, would then have the value 4/3. Correction, however, has also to be made for the fact that the velocities themselves vary according to the distribution law, and when the necessary calculations are made they show that α should be $\sqrt{2}$. Let the stationary molecules further be reduced to points and the selected molecule which moves among them swell so that its radius becomes equal to σ. This process will not affect the number of encounters.

As the selected specimen moves it sweeps out in the course of a second a cylindrical space of cross-section $\pi\sigma^2$ and length $\sqrt{2}\bar{u}$. The volume is $\sqrt{2}\pi\sigma^2\bar{u}$ and is inappreciably affected by the changes in direction which occur at each encounter, the mean free path being, as it turns out, normally very much greater than σ. The number of the point molecules contained in this cylinder is equal to $\sqrt{2}\pi\sigma^2\bar{u}$ $\times$ the number per unit volume, that is, to $\sqrt{2}\pi\sigma^2\bar{u}n$, which must represent the number of collisions made by any given molecule in unit time. Since the same applies to every molecule, the total number entering into collision in a second in one cubic centimetre is $\sqrt{2}\pi\sigma^2\bar{u}n^2$. Each collision involves two molecules, so that the number of collisions is half this. Thus we have

$$Z = \tfrac{1}{2}\sqrt{2}\pi\sigma^2\bar{u}n^2.$$

The viscosity of the gas is comparatively easy to measure, and from it l may be calculated and thus Z may be estimated. For all ordinary molecules, that is to say for those other than the macromolecules formed by complex polymerization processes, σ is of the order of magnitude 10^{-8} cm. Some typical values are given below.

	$\sigma \times 10^8$ cm.
Helium	2·18
Argon	3·66
Hydrogen	2·72
Oxygen	3·62
Water	4·66
Carbon dioxide	4·66

If the two molecules entering into collision are not of the same species, the above calculation has to be modified, and the result found is of the following form:

$$Z = n_A n_B \sigma^2_{AB} \{8\pi RT(1/M_A + 1/M_B)\}^{\frac{1}{2}},$$

where n_A and n_B are the respective numbers in unit volume, M_A and M_B the gram-molecular weights, and σ_{AB} the average of the two diameters. A derivation of this formula, by a method which is more precise than that sketched above for the molecules of the same species, will be given at a later stage (p. 384).

Although in the foregoing discussion the molecule was pictured as an elastic sphere, the result retains a good deal of its value even if this simple representation is entirely abandoned and the molecule is regarded, for example, as a mere centre of force which repels similar centres when they approach too closely. What the above calculations have really yielded is information about the average distance between the molecular centres at which transfers of momentum occur. For many purposes this constitutes quite a reasonable definition of the diameter. Ambiguity is, however, avoided if σ is termed the *collision diameter*. In actual fact it corresponds fairly closely to the molecular magnitudes determined by the methods of X-ray and electron diffraction.

The elementary treatment of the mean free path envisages, as it were, a balancing of the molecular accounts at a distance l from any reference plane. It is very useful as far as it goes, but it leaves certain important effects unaccounted for. Its limitations are well illustrated by the phenomenon of *thermal diffusion* (upon which one of the most efficient separations of isotopes depends). If a long vertical tube is heated axially by a wire while the walls remain cold, the heavier constituent of a gas mixture becomes concentrated towards the circumference. Here it sinks towards the bottom of the tube while fresh gas rises in the warm central region. When a steady state is established the lighter constituent of the mixture predominates at the top of the tube and the heavier at the bottom. The convection effect itself is easy enough to understand, but the initial enrichment of the heavier molecules in the colder region is not explicable in terms of the elementary theory. If pressure is uniform throughout, then, since $p = \frac{1}{3}nm\bar{u}^2$, n is inversely proportional to $m\bar{u}^2$. That is to say that the density is inversely proportional to the absolute temperature, whatever the value of m, since $m\bar{u}^2$ is the same for all types of molecule. Thus no separation of the gases in a temperature gradient seems explicable solely in virtue of the different molecular masses.

This consideration neglects the circumstance that a temperature gradient is only maintainable by a continual transport of heat from one region to another, and furthermore that the assumption of a true equilibrium state established at the end of each free path is only roughly correct. Molecules with a large component of velocity in a given direction tend in general to retain some fraction of it after collision, and the degree of this so-called persistence of velocity is a function both of the mass and of the law of force between the colliding particles. Normally it is such that the lighter molecules outstrip the heavier in penetrating into the warmer regions, so that the concentration of molecules in the cooler regions becomes more pronounced for the heavier than for the lighter. This is a second-order effect which is magnified by the continuous action of the vertical convection currents. With a different law for the inter-molecular forces the thermal diffusion could work in the opposite sense and lead to an accumulation of lighter molecules in the cooler parts.

In spite of the importance in certain special circumstances of these subtler phenomena, the achievements even of the rather naïve forms of the kinetic theory are undeniable. They amount to this: that in some major respects the properties of matter are interpretable in terms of the behaviour of sensible objects. Atoms and molecules seem not only to exist, but also to be enumerable and measurable, and representable in some degree as small-scale models of the objects which they themselves build up. This takes us a long step forward in describing the unknown in terms of the known, and this is what really constitutes a scientific explanation. The process can be carried even farther, but we shall soon find that a fresh turn has to be given to it, and that the kinds of known things by which the unknown have to be described become less homely.

II

MOLECULAR CHAOS AND ENTROPY

Molecular chaos

THE quantitative laws of chemical combination provide clear pointers to the molecular theory of matter, which increases progressively in vividness and realism with the application of Newton's laws to the motions of the particles. The interpretation of phenomena such as the pressure and viscosity of gases and the Brownian motion, and the assignment of definite magnitudes to molecular speeds, masses, and diameters render it clear that a continual interchange of energies must occur between the molecules of a material system, a circumstance which lies at the basis of temperature equilibrium and determines what in ordinary experience is called the flow of heat. It is responsible indeed for far more than this, and a large part of physical chemistry follows from the conception of the chaotic motion of the molecules. This matter must now be examined more deeply.

In a sufficiently numerous collection of molecules left to themselves a statistical equilibrium is established, usually, as it proves, with great rapidity. In this state the total quantities of energy in the various forms, translational, vibrational, and rotational, bear definite steady ratios to one another. Each kind of energy in every individual molecule fluctuates with time about a mean value, every kind of motion waxing and waning throughout the system. At a given instant the statistical equilibrium only becomes susceptible of complete definition as the number of molecules under consideration is indefinitely increased, but, even for a few molecules, the average state taken over an extended period of time corresponds to that of the true equilibrium for very large numbers at a single instant.

The state of such a molecular system is obviously chaotic in the sense that some molecules move fast and others slowly: some are in violent vibration or rotation while others are almost quiescent; and the condition of individuals is continually changing. A rough idea of the rapid and irregular variations of motion would be provided by the behaviour of a number of billiard balls propelled in random directions on a table, their translations and spins fluctuating according to the hazards of their mutual encounters.

For an elastic collision of two smooth spheres the laws of mechanics allow the final speeds and directions to be estimated from the initial values, and similar calculations are possible, in principle, for rough spheres, for non-spherical masses, and even for masses which do not collide in the ordinary sense but exercise mutual repulsive forces on close approach. Given the initial state of a mechanical system consisting of any number of particles, the final state is in principle calculable. In the initial state the distribution of motions may not have been random, but may in fact have been governed by definite conditions. For example, in a lead bullet cooled to a very low temperature and possessing a very high translational speed the motions of the molecules are largely ordered in a parallel way. If this bullet is stopped by a stone wall, it grows hot and the movements of the particles become random in the sense that all mass motion disappears and only the irregular invisible motion of the molecules remains. It requires, however, very subtle consideration to decide whether or not the state of motion after the impact can be strictly described as absolutely chaotic. The following reflection shows this problem in a clearer light. If the final motion of every particle in the mass were reversed, then a series of collisions would ensue such that the original initial state would be regained. Now the reversal of a chaotic set of velocities might be said still to leave them chaotic; yet the reversed set in this example would soon be replaced by a highly ordered system.

The original state of all the matter in the world is not known; nor can we predict what would happen if all molecular speeds were reversed. As far as the laws of mechanics go, we cannot assert that existing conditions are unrelated to an earlier condition of order. Whether, therefore, the complete randomness of all microscopic motion can be logically related to the Newtonian laws has in fact been a subject of controversy and no wholly satisfactory answer emerges.

What is undoubtedly true is that for practical purposes the chaos can be regarded at least as very nearly complete. That this is so irrespective of the origin of the present order of things is attested by such simple experiences as the shuffling of a pack of cards, which show how rapidly all vestiges of order become undetectable. While forgoing any attempt at a rigid application of the laws of dynamics to the question, we may therefore introduce as a *specific postulate* the *assumption of molecular chaos* for ordinary systems endowed with thermal energy. In any portion of matter in statistical equilibrium the distribution of motions is taken to be calculable from the laws of probability in conjunction with the appropriate rules of mechanics without any reference to the remote history of the system.

This postulate is broad, clear, and intuitive and quite worthy to serve as the basis of a theory of matter. With its aid wide tracts of physics and chemistry can be illuminated. Its consequences are much more positive than might have been suspected for what sounds at first like a somewhat negative principle.

The distribution law

The first task is to derive some of the important rules of statistical equilibrium among molecules. The state of a given individual at any instant is described by its position coordinates and by other coordinates which define its translational, rotational, and vibrational energy. For the present purpose the different contributions to the total energy of the individual will be taken as independent and expressible by square terms in the way previously outlined (p. 14).

The problem is to define how many molecules out of a very large number are at a particular instant in any given state. Here the definition of what we are to understand by a *state* requires detailed consideration. If the component $\dot{x}$ of the velocity along the axis x is denoted by u, the natural way of formulating the distribution question is to inquire about the number of molecules with velocities in the range from u to $u+du$, with corresponding inquiries about the other coordinates in terms of which the energy is expressed. With the idea, however, of introducing a convenient approximation we might agree to envisage a small arbitrary range in the neighbourhood of u and to consider all the molecules possessing velocities in that range as having the same definite energy ϵ in respect of that coordinate. In this way we should define a discrete series of energy states corresponding to each variable. Thus if the coordinates are p, p', p'',..., there are series corresponding to $p_1, p_2, p_3,..., p'_1, p'_2,..., p''_1, p''_2,...,$ and so on. As will appear, it is best to let the p's represent *momenta*. The contributions to the total energy of the molecule, which are proportional to the squares of these, may be written $\eta_1, \eta_2, \eta_3,..., \eta'_1, \eta'_2,..., \eta''_1, \eta''_2,...,$ etc.

The total energy of a molecule is then $\eta_i+\eta'_j+\eta''_k+...$, where $i, j, k,...$ can be taken in any combination. The multitudinous values of the total energy which a molecule might have as a result of these different combinations can themselves then be arranged in an ordinal series ϵ_1, ϵ_2, ϵ_3,....

It will remain to be decided whether subsequently the intervals in the various series of coordinates or energies are to be made vanishingly small. Later developments will show that this procedure is unnecessary and indeed incorrect, and that a discrete series of states, properly defined by what will be called quantum rules, is what corresponds to nature. For the present, however, the assumption of the series of numerous and fairly closely defined states may be regarded as a convenient simplification.

We envisage, then, a series of energy states like so many compartments into which molecules can be placed, and the distribution among which is to be investigated. The postulate of molecular chaos deliberately assimilates the problem to that of the random partition of a large number of objects among a number of boxes.

Let there be N molecules to be distributed among the states corresponding to the series of energies ϵ_1, ϵ_2, ϵ_3,.... N may conveniently be taken as the number in a gram molecule. Let there be, in a given distribution, N_1 in state 1, N_2 in state 2, and so on. The number of ways, W, in which such a distribution can be achieved is given by the formula

$$W = \frac{N!}{N_1! \, N_2! \, N_3! \, ...},$$

which follows in an elementary way from the principles of permutations and combinations.

The expression may be simplified by the use of Stirling's approximation for the factorials of large numbers, namely that $\ln N!$ tends to the value $N \ln N - N$ when N is large enough. Taking logarithms and making the substitutions for the factorials we obtain

$$\ln W = N \ln N - \sum N_1 \ln N_1. \tag{1}$$

The principle now to be applied is that in statistical equilibrium W, and thus $\ln W$, will attain a maximum value. N_1, N_2,..., can be subjected to small tentative changes δN_1, δN_2,..., and if the value of $\ln W$ is a maximum, then $\delta \ln W$ will be zero for such processes.

Thus we have as the major condition of our problem

$$\delta \ln W = 0,$$

and from (1), since N is constant,

$$\sum \delta(N_1 \ln N_1) = \sum (1 + \ln N_1) \delta N_1 = 0. \qquad (2)$$

Since the total energy of the whole collection of molecules remains constant and since the total of the numbers in all the states must always add up to N, we have the two auxiliary conditions

$$\sum N_1 = N,$$

so that

$$\sum \delta N_1 = 0, \qquad (3)$$

and

$$N_1 \epsilon_1 + N_2 \epsilon_2 + \ldots = E,$$

so that

$$\delta \sum N_1 \epsilon_1 = \sum \epsilon_1 \delta N_1 = 0. \qquad (4)$$

The solution of (2) subject to the conditions (3) and (4) is a standard problem in conditioned maxima, and what follows down to the result in equation (7) is purely mathematical.

Before we proceed with the solution, however, a word should be said about the use of the Stirling approximation. It has been objected that, although N is large, the number of states is also large, so that some of the numbers N_1, N_2,..., may not be great enough to justify the use of the approximation. If this is so, then one must take N to be much larger still until all the states really do contain enough molecules. The result might apply then only to the time average of any real system of finite size. This is quite all right, because in actual fact there would be a definite distribution law for such a system only if we averaged its condition over a period of time.

We proceed therefore to the solution of (2). (3) and (4) are multiplied by arbitrary constants, α and β, and added to (2) with the result

$$\sum (1 + \ln N_1) \delta N_1 + \alpha \sum \delta N_1 + \beta \sum \epsilon_1 \delta N_1 = 0. \qquad (5)$$

α and β may have any values required by other conditions of the problem, since (5) is an identity based upon three independent equations. Rearrangement of (5) gives

$$\sum (1 + \ln N_1 + \alpha + \beta \epsilon_1) \delta N_1 = 0. \qquad (6)$$

δN_1, δN_2,..., it must be remembered, are small arbitrary transfers made to test whether the condition for a maximum is fulfilled. They are variable and subject only to condition (3). Let α and β, being assignable at will, be given such values that

$$1 + \ln N_1 + \alpha + \beta \epsilon_1 = 0 \quad \text{and} \quad 1 + \ln N_2 + \alpha + \beta \epsilon_2 = 0.$$

Then in (6) the sum of all the terms from δN_3 upwards equals zero. From the nature of δN_4, δN_5,..., they may be chosen, if we so wish, to be zero. Let them be so chosen, but let δN_3 be given a value distinct from zero. Then we have

$$(1 + \ln N_3 + \alpha + \beta \epsilon_3) \delta N_3 = 0,$$

and since δN_3 itself is not zero its coefficient must be. Thus

$$(1 + \ln N_3 + \alpha + \beta \epsilon_3) = 0.$$

A repetition of the argument with appropriate modifications shows that in general

$$1 + \ln N_j + \alpha + \beta \epsilon_j = 0.$$

Therefore $\qquad N_j = e^{-1-\alpha} e^{-\beta \epsilon_j}.$

Since $\qquad \sum N_j = N,$

$$N = e^{-1-\alpha} \sum e^{-\beta \epsilon_j},$$

whence $\qquad N_j = N e^{-\beta \epsilon_j} / \sum e^{-\beta \epsilon_j}, \qquad (7)$

the sum in the denominator being taken over all possible states.

This equation is known as the *Maxwell–Boltzmann distribution law*. It rests, as has been seen, on the assumption that the distributions occurring in nature are those which can be achieved by the maximum number of permutations. All possible distributions are, as it were, explored in the course of the blind wanderings of the molecules from state to state, but the condition of real systems for most of their time corresponds closely to that of maximum probability.

The equipartition law

The Maxwell–Boltzmann law is of fundamental importance. We shall begin by applying it to the derivation of the equipartition law, which plays so prominent a part in determining the equilibrium of material systems. ϵ, the energy of a molecule, is the sum of terms η, η', η'',... corresponding to the different coordinates p, p', p'',... which describe the motions. Let attention be fixed upon one particular type of coordinate, p, and for this purpose let (7) be rewritten in the form

$$N_j = \frac{N e^{-\beta(\eta + \eta' + \eta'' + ...)}}{\sum e^{-\beta(\eta + \eta' + \eta'' + ...)}},$$

$$= \frac{N e^{-\beta(\eta' + \eta'' + ...)} e^{-\beta \eta}}{\sum e^{-\beta(\eta + \eta' + \eta'' + ...)}},$$

$$= \frac{N e^{-\beta(\eta' + \eta'' + ...)} e^{-\beta \eta}}{\sum e^{-\beta(\eta' + \eta'' + ...)} \sum e^{-\beta \eta}},$$

the factorization of the denominator being possible since every value of η (that is η_1, η_2,...) is combined in the sum with every value of η', η'',... (that is η_1', η_2',...; η_1'', η_2'',...).

Now let N_j be summed over all possible values of η', η'',..., and the result, which may be written $N_{(j)}$, gives the number of molecules which possess the energy η in the mode of motion corresponding to p, irrespective of the other components.

$$N_{(j)} = \frac{N\{\sum e^{-\beta(\eta'+\eta''+\cdots)}\}e^{-\beta\eta}}{\sum e^{-\beta(\eta'+\eta''+\cdots)} \sum e^{-\beta\eta}}$$

$$= \frac{Ne^{-\beta\eta}}{\sum e^{-\beta\eta}}. \tag{8}$$

Equation (8) might have been obtained intuitively from (7) by imagining the derivation of the latter for the series of η states only, without any consideration of the others.

If $N_{(j)}$ is now summed over the whole series of η states, $\sum N_{(j)}$ will clearly embrace all the molecules, so that it is equal to N.

Consider now the sum $\sum e^{-\beta\eta}$. As it stands it consists of a number of discrete terms, and the law of the series is not defined. One might assume from one term to the next equal increments of the energy itself or equal increments of the coordinate p which determines it. The best assumption, which makes a special appeal to those deeply versed in the science of dynamics, is that the series should be defined by equal increments in *momentum*, that is in general by equal increments of the variables corresponding to p. The basis of this idea is in fact wide experience of the way in which dynamical laws assume their simplest form when expressed in terms of momentum and space coordinates (Hamiltonian coordinates) as the fundamental variables. The successive energy states are then

$$f(p), \quad f(p+\Delta p), \quad f(p+2\Delta p), \quad \dots .$$

Equation (8) becomes

$$N_{(j)} = \frac{Ne^{-\beta f(p)}}{\sum e^{-\beta f(p)}}.$$

This may be multiplied top and bottom by Δp:

$$N_{(j)} = \frac{Ne^{-\beta f(p)}\Delta p}{\sum e^{-\beta f(p)}\Delta p}.$$

If Δp is made small enough we have for the number of molecules in the state corresponding to $f(p)$, or in the range Δp about it,

$$N_{(j)} = \frac{Ne^{-\beta f(p)}\,dp}{\int e^{-\beta f(p)}\,dp},\tag{9}$$

i.e.

$$\phi(p)\,dp = \frac{Ne^{-\beta\eta}\,dp}{\int e^{-\beta\eta}\,dp},\tag{9a}$$

$\phi(p)\,dp$ being for the continuous distribution the equivalent of $N_{(j)}$ for the discontinuous one.

Consider further the integral in the denominator of $(9a)$. It is taken over the whole range of coordinates. A partial integration with respect to p gives the result

$$\int e^{-\beta\eta}\,dp = \left[pe^{-\beta\eta}\right]_{\text{lower limit}}^{\text{upper limit}} + \beta\int pe^{-\beta\eta}\frac{d\eta}{dp}\,dp.$$

The first term on the right is zero at both limits: at the lower limit since $p = 0$ and at the upper limit since $e^{-\beta\eta}$ is a zero of higher order than the infinity of p. Rearrangement of the terms gives

$$\frac{1}{\beta} = \frac{\int \{p(d\eta/dp)\}e^{-\beta\eta}\,dp}{\int e^{-\beta\eta}\,dp}.$$

From $(9a)$,

$$\frac{1}{\beta} = \frac{\int \phi(p)\{p(d\eta/dp)\}\,dp}{N}.$$

The expression on the right is the average value of $p(d\eta/dp)$, so that the equation may be written

$$\overline{\left(p\frac{d\eta}{dp}\right)} = \frac{1}{\beta}.\tag{10}$$

It is clear from the derivation that the value of $1/\beta$ is the same for all possible momentum coordinates.

Now if the energy is expressible as a square term

$$\eta = ap^2, \quad \text{where } a \text{ is a constant.}$$

$$\frac{d\eta}{dp} = 2ap$$

and

$$p\frac{d\eta}{dp} = 2ap^2 = 2\eta.$$

Thus for all the square terms in the energy the average value is such that
$$2\bar{\eta} = 1/\beta,$$
or
$$\bar{\eta} = 1/(2\beta),$$
the bar signifying as usual an average value.

From the equations for the gas pressure,
$$p = \tfrac{1}{3}nm\bar{u}^2 \quad \text{and} \quad pV = RT,$$
whence $\qquad pV = \tfrac{1}{3}Nm\bar{u}^2 \quad \text{since} \quad nV = N.$

Therefore $\qquad \tfrac{1}{3}m(\dot{x}^2 + \dot{y}^2 + \dot{z}^2) = RT/N = kT,$

where $\qquad\qquad k = R/N.$

The average values of $\dot{x}^2$, $\dot{y}^2$, and $\dot{z}^2$ will be equal and thus
$$\overline{m\dot{x}^2} = kT,$$
or $\qquad\qquad \overline{\tfrac{1}{2}m\dot{x}^2} = \tfrac{1}{2}kT.$

But $\overline{\tfrac{1}{2}m\dot{x}^2}$ is $1/(2\beta)$ by the result just obtained, and thus
$$\beta = 1/kT, \tag{11}$$
and $\qquad\qquad \bar{\eta} = \tfrac{1}{2}kT.$

The discussion so far has shown that the average energy in each square term is $\tfrac{1}{2}kT$ for a given type of molecule. It remains to show that this result holds good in a mixture of different molecular species.

Let there be N_a molecules of type a and N_b of type b in the same enclosure, and let them be capable of exchanging energy. Let their individual energy states be $\epsilon_{a1}, \epsilon_{a2},...,$ and $\epsilon_{b1}, \epsilon_{b2},...,$ and the numbers in these states be respectively $N_{a1}, N_{a2},..., N_{b1}, N_{b2},....$

The number of ways of achieving this distribution is
$$W = \frac{N_a!}{N_{a1}!\,N_{a2}!\,...} \times \frac{N_b!}{N_{b1}!\,N_{b2}!\,...}, \tag{12}$$
and this must be a maximum. The auxiliary conditions to be observed are
$$\sum N_{a1} = N_a \tag{13}$$
and $\qquad\qquad \sum N_{b1} = N_b, \tag{14}$
and $\qquad\qquad \sum N_{a1}\epsilon_{a1} + \sum N_{b1}\epsilon_{b1} = E. \tag{15}$

The problem is solved as before: the logarithm of (12) is transformed by Stirling's formula, differentiated, and equated to zero. (13), (14), and (15) are also differentiated and equated to zero.

$$\delta \ln W = 0, \tag{16}$$

$$\sum \delta N_{a1} = 0, \tag{17}$$

$$\sum \delta N_{b1} = 0, \tag{18}$$

$$\sum \epsilon_{a1} \delta N_{a1} + \sum \epsilon_{b1} \delta N_{b1} = 0. \tag{19}$$

(17) and (18) are multiplied by arbitrary constants α_1 and α_2 and (19) by β. The results are added and the rest of the calculation proceeds as for the case of the single kind of molecule.

There is no need to carry it through in order to obtain the result required at the present juncture. The important thing to notice is that whereas there are now *two* separate constants α_1 and α_2 relating to the total numbers of each species of molecule, there is still only *one* single constant β relating to the constant total energy of the whole system. This expresses the obvious fact that whereas molecules of the types a and b are not interconvertible, energy, on the other hand, is both transformable and transferable from any one type of molecule to any other. Thus β, which plays exactly the same role as in the previous calculation, is the same for both species. The mean energies for a square term are therefore independent of the masses and other physical constants, such as the moments of inertia of the individual molecules.

As will have been seen, the derivation of the equipartition law depends upon certain rather general, and perhaps somewhat abstract, assumptions about molecular dynamics. It is desirable, therefore, to bring the result as soon as possible into relation with experimentally measurable matters. The most striking method of doing this is afforded by the study of specific heats.

Energy content and specific heat

The most obvious test of the validity of the equipartition principle is the attempt to calculate the energy content and thus the specific heat of substances composed of molecules simple enough for a fair guess at their structure and mechanics to be made.

Helium, argon, and the other inert gases might be assumed to consist of small masses devoid of internal motions, and as a first trial one might assume the absence of rotations. They would then possess three translational degrees of freedom only, and the energy of each gram atom would be $N \times 3 \times \frac{1}{2}kT$, or $(3/2)RT$. The increase of this for each degree of temperature is $(3/2)R$, and R being 1·98 cal./degree, the atomic heat should be 2·97, which corresponds exactly to the measured value.

Very stable diatomic molecules such as oxygen, nitrogen, and hydrogen should possess a structure, and might be schematized as objects with a sort of dumb-bell shape, but they might be devoid of vibrational energy, since the two atoms might be imagined to constitute a practically rigid body. If these guesses were correct, there would be three translational degrees of freedom and some rotational degrees. The rotational energy would be referable to three axes, one of which would be chosen to coincide with the geometrical axis of the dumb-bell. Viewed along this axis the structure would be seen in projection in a monatomic form and thus, by analogy with helium, might be supposed to have no rotation about it. Two rotational degrees of freedom would, however, be expected for the two other rectangular axes, making five degrees of freedom in all. These would account for a specific heat of $(5/2)R$ calories per gram molecule. Once again the estimated value is exactly right.

An unstable diatomic molecule, such as iodine, should perhaps be capable of vibration along the axis joining the two atoms, and there would then be an addition to the specific heat of $\frac{1}{2}R$ for the kinetic energy of this motion and (according to the properties of simple harmonic motion) of an equal amount for the potential energy. Thus the total value should be about 2 calories greater than that for oxygen. This, too, is in agreement with experiment.

The application of this method of calculation to crystalline monatomic solids (a category to which many metallic elements would according to a simple view belong) is interesting. Assuming the absence of rotations and translations, and the existence only of three degrees of freedom of vibration, one calculates that the atomic heat of all such solids should be $3(\frac{1}{2}R + \frac{1}{2}R)$ or 6 calories, which is a fairly good approximation to the famous *law of Dulong and Petit*.

There can be little doubt, therefore, that in many respects the equipartition law gives a rather accurate account of what happens. Its successes leave little doubt that when translations, rotations, and vibrations do exist, they are reasonably well describable as sums of independent square terms.

Second thoughts about the whole problem, however, suggest not so much that the equipartition principle is in any very important way inaccurate as that it altogether neglects some factor belonging to the essential nature of things.

The guesses which were made about the modes of motion of various types of molecule seem plausible enough at first sight, but they are strictly speaking inadmissible. There is, in fact, no reason in Newtonian mechanics why helium atoms should not rotate, why oxygen molecules should not have three rotational degrees of freedom, or why all diatomic molecules should not be in vibration. Smooth symmetrical molecules, it is true, are difficult to make rotate, but they are difficult to stop again when once they have been set in rotation. The equipartition law should, in fact, apply whether or not particular modes of motion happen to be difficult to excite. If they are difficult to excite they may acquire energy but slowly: they will also lose energy slowly, and in the equilibrium state they should possess the normal quota. All that can be said is that the dynamical constants characterizing some of the degrees of freedom which do not appear are quantitatively considerably different from those of the modes which do appear. But in the derivation of the law no assumption at all about these magnitudes is made.

Furthermore, detailed experiment shows that all specific heats are functions of temperature, at least over certain ranges, and this is completely inexplicable in terms of the equipartition principle, which makes the energy proportional to T, so that the specific heat, the differential coefficient of the energy with respect to T, should be constant.

The actual mode of variation of the specific heat is such as to suggest that in certain ranges of temperature some of the degrees of freedom pass entirely out of action. The principles so far introduced, then, need amplification by some quite fundamental new rules which provide reasons why sometimes degrees of freedom should be operative and sometimes not. These rules cannot be derived in any way except by the introduction of the *quantum theory*. In the meantime it appears that the equipartition principle is an incomplete statement. If the results it predicted were merely inaccurate in a numerical sense, the discrepancies could be attributed to causes within the framework of ordinary mechanics, for example, to the non-independence of rotations and vibrations, which would spoil the formulation of the energy as a sum of square terms, but the difficulty lies deeper.

There is a complete absence of contributions from modes of motion which according to mechanical principles should exist, a clear-cut qualitative absence, rendered all the more conspicuous by the relative precision with which the contributions of the other modes can be calculated.

At this juncture it is well to remember that the attempt to describe the nature of matter by the application to atoms and molecules of the rules which Newton inferred from the behaviour of massive bodies is no more than an hypothesis to be tested. In the present direction the conclusion about its validity is: thus far and no farther.

But in other directions the simple kinetic theory, without the specific considerations of quantum phenomena, has eminent services to render.

Entropy

We return to the conception of molecular chaos, of statistical equilibrium, and of a partition law defining the thermal balance between the various parts of a material system and making intelligible the idea of a temperature. The obvious task seems now to be the formulation of some quantitative rules about equilibria in systems of many atoms and molecules.

In the solution of this problem a quantity called the *entropy* has come to play a dominant role. What transpires is this: that the conception of molecular chaos, with very little in the way of subsidiary hypotheses except a few very general dynamical laws, goes a long way in explaining many of the major phenomena of physics and chemistry. This being so, it is clearly worth while examining carefully various expressions for the probability of molecular distribution.

We refer once more to the result found for the distribution of the N molecules of a gram molecule among their various energy states. For the present the volume of the system will be taken as constant, and to mark this restriction, which will presently be removed, the number of assignments will be written W_e instead of W. The equation which was derived on p. 28 takes the form

$$\ln W_e = N \ln N - \sum N_1 \ln N_1.$$

W_e measures the number of ways in which the molecules can be assigned to the various energy states for the distribution in question.

The basic principle employed in developing the theory was that W_e, and thus $\ln W_e$, tends to a maximum value.

It is convenient to work generally with $\ln W_e$ rather than with W_e itself, since, if there are two systems characterized by values W_1 and W_2 separately, the number of assignments for the combined system being $W_1 W_2$, the corresponding logarithm is $\ln W_1 + \ln W_2$, and the function describing the state is then formed additively from the values characterizing the individual components. This proves to be a great practical convenience.

$\ln W_e$ defines the degree of probability of the system and gives a quantitative measure of the state of chaos which prevails. Its properties will now be examined. It will be found to be closely related to important thermal magnitudes, and will prove to be one of the most important functions in physical chemistry.

By the calculation already given

$$N_1 = \frac{N e^{-\epsilon_1/kT}}{\sum e^{-\epsilon_1/kT}},$$

where the sum $\sum e^{-\epsilon_1/kT}$ will be written f_e.

For convenience we multiply $\ln W_e$ by k (the expediency of this trivial operation will appear later) and denote $k \ln W_e$ by the symbol S_e.

Thus
$$S_e = kN \ln N - \sum k N_1 \ln N_1.$$

The value of N_1 is now substituted in this expression for S_e, the result being
$$S_e = kN \ln f_e + kNT d\ln f_e/dT,$$
or
$$S_e = R \ln f_e + RT d\ln f_e/dT.$$

The appearance of the differential coefficient of f_e depends upon the relation
$$f_e = \sum e^{-\epsilon_1/kT},$$
$$\frac{df_e}{dT} = \frac{\sum \epsilon_1 e^{-\epsilon_1/kT}}{kT^2},$$
whence
$$\sum \epsilon_1 e^{-\epsilon_1/kT} = kT^2 df_e/dT.$$

Now the total energy, E, of the system is clearly related to f_e. We have
$$E = \sum \epsilon_1 N_1 = \frac{\sum \epsilon_1 N e^{-\epsilon_1/kT}}{f_e} = \frac{N}{f_e} kT^2 \frac{df_e}{dT},$$
whence
$$E = RT^2 \frac{d\ln f_e}{dT},$$
and thus
$$S_e = R \ln f_e + \frac{E}{T}.$$

This relation of the *entropy* to the total energy is extremely important. We shall now throw it into a simple form, in which it will prove to play a quite dominant role. We take the differential coefficient of the entropy, S_e, with respect to the temperature, that is we examine the function which tells us how the degree of disorder changes as the temperature rises and molecules pass into generally higher energy states.

$$\frac{dS_e}{dT} = R\frac{d\ln f_e}{dT} + \frac{1}{T}\frac{dE}{dT} - \frac{E}{T^2}$$

$$= R\frac{d\ln f_e}{dT} + \frac{1}{T}\frac{dE}{dT} - R\frac{d\ln f_e}{dT} = \frac{1}{T}\frac{dE}{dT}.$$

We have supposed the volume of the system to remain constant, so that the last equation is more accurately written

$$\left(\frac{\partial S_e}{\partial T}\right)_v = \frac{1}{T}\left(\frac{\partial E}{\partial T}\right)_v.$$

Now the rate of change of energy with temperature is the specific heat. We thus arrive at the fundamental result that the change of entropy due to an increase of temperature dT at constant volume is given by

$$\boxed{\mathbf{dS_e} = \left(\frac{\partial \mathbf{S_e}}{\partial \mathbf{T}}\right)_v \mathbf{dT} = \frac{\mathbf{C_v dT}}{\mathbf{T}}}$$

As the temperature rises S_e increases according to the relation

$$S_e = C_v \ln T + S_{e0},$$

where S_{e0} may be a function of the volume.

It is important to have an intuitive understanding of the reason why S_e increases with the temperature. It is that as the total energy of the system becomes greater, more and more of the higher energy states become accessible to more and more of the molecules, and that as the distribution so widens, the number of ways of realizing it correspondingly increases. In this sense it is correct to say that when a substance becomes hotter its state becomes more probable.

The question of an analogous treatment of the distribution of molecules in space now arises.

Entropy and volume

If chaotically moving particles are unconfined, they will wander in all directions and fill the whole of any volume accessible to them.

If the temperature is uniform and if the molecules exert no appreciable attractive forces upon one another, they will distribute themselves with a uniform density over all the volume elements, provided that these elements contain numbers great enough for the application of statistical averages. This result could be derived formally by a method similar to that employed in the calculation of the energy distribution, but is itself as obvious intuitively as would be the postulates made in a seemingly more rigorous treatment.

Suppose a gaseous system of total volume V contains N molecules which exert negligible forces upon one another and are of a size negligible in comparison with their average separations. Let the volume be divided into many small elements v, this being a constant standard reference magnitude, entirely arbitrary except in so far as it is supposed small compared with V while being still large enough to contain many molecules. In the equilibrium state the number in each element is Nv/V. The number of ways in which the N individuals can be distributed among the V/v compartments into which the gas is virtually divided so that there are equal numbers in each is given by

$$W_v = \frac{N!}{\{(Nv/V)!\}^{V/v}}.$$

Application of Stirling's formula, $\ln N! = N \ln N - N$, gives the result

$$\ln W_v = N \ln N - N - \frac{V}{v}\left(\frac{Nv}{V}\ln\frac{Nv}{V} - \frac{Nv}{V}\right)$$

$$= N \ln V - N \ln v.$$

Thus $\qquad\qquad k \ln W_v = R \ln V - R \ln v.$

As the volume increases, therefore, the state becomes more probable. By analogy with the corresponding value of $k \ln W_e$, we may write

$$S_v = R \ln V - R \ln v,$$

and thus, since v is a constant arbitrary magnitude,

$$\left(\frac{\partial S}{\partial V}\right)_T = R \frac{dV}{V}.$$

If we now consider systems in which the temperature and the volume are both variable, we can combine the two foregoing results. The number of ways in which the energy distribution can be realized is W_e: and the number in which the volume distribution is satisfied is W_v. The number of ways in which both can be satisfied together is expressed by

$$W = W_e \times W_v:$$

then we have

$$k \ln W = k \ln W_e + k \ln W$$
$$= C_v \ln T + R \ln V + \text{arbitrary constant}$$
$$= S.$$

S, which takes into account both the spatial positions of the molecules and their motions, is called the *statistical entropy of the system*.

At the present stage the equation is most useful in the differential form

$$\boxed{dS = C_v\, dT/T + R\, dV/V}$$

This expresses the increase in entropy and probability which accompanies rise in temperature or expansion of volume.

Imperfect gases and substances in general

When there is interaction between the molecules, as in an imperfect gas, and still more in solids and liquids, the form of the volume-dependence becomes too complex to be satisfactorily formulated. f_e, the sum of the terms like $e^{-\epsilon_1/kT}$, is itself a function of the molecular separations. The number of possible energy states becomes indefinitely larger because variable potential-energy terms, each depending upon the proximity of other molecules, are included in the energy of each molecule in any given state.

One relation, however, retains its simplicity: and this fact is of vital importance for the whole of the science of thermodynamics. *However complex the expression for $\partial S/\partial V$ may become, the value of $(\partial S/\partial T)_v$ remains C_v/T.* This crucial result may be seen by inspection of the derivation which was given earlier (p. 38) and where S_e was expressed in the form $R \ln f_e + E/T$, after E itself had been given in terms of f_e. No matter how complex the series of states may become, or how involved a function of volume f_e may become, the relation of E to f_e and of S_e to f_e remains, and hence the relation of S_e to E. Therefore, in a differentiation at constant volume the value of $\partial S/\partial T$ remains C_v/T. Put rather crudely, complex interactions have the same effect on C_v/T as they do on $\partial S/\partial T$. Thus the key relation

$$dS_e = \left(\frac{\partial S}{\partial T}\right)_v dT = C_v\, dT/T$$

holds good.

Conditions of equilibrium in molecular systems

At first sight it might appear that the conception of a molecular chaos, though picturesque and affording the opportunity for some ingenious calculations, could hardly bear fruit of a very substantial kind. Nothing could be farther from the truth. It leads directly to the treatment of physical and chemical equilibrium in a manner which is not only illuminating but very useful.

The basic principle to be applied is this: a chaotic collection of molecules will always tend to pass into the state of maximum probability. If it moves nearer to equilibrium from a state in which equilibrium has not yet been established, then the probability increases. If, on the other hand, a completely isolated system moves from one state of equilibrium to an alternative one (as it may on occasion do), the probability does not change.

Molecular systems, however, are not normally found in complete isolation. They are in contact with other systems, and, since the world as a whole is not in equilibrium, they are liable to gain or lose heat by exchange with their surroundings. Moreover, in their blunderings towards a state of maximum chaos the molecules of a given assembly tend to transgress their bounds, and if impeded by obstacles, exert pressure on these, pushing them back in appropriate circumstances and performing mechanical work.

The laws of these interactions determine in a remarkable way the form of many of the phenomena of chemistry and physics.

It is convenient to begin by considering the adiabatic expansion of a quantity of gas, that is, the increase in volume of an assemblage of molecules isolated entirely from their surroundings. The system is supposed to be held in equilibrium throughout the process, the external pressure being always adjusted to the exact value characteristic of the momentary temperature and volume of the gas itself. Thus the system is not to be regarded as moving spontaneously from a given initial state to one of greater probability, but simply as being led successively from one possible equilibrium state to a neighbouring one. In these circumstances $\ln W$ remains constant, and therefore dS is zero. Volume and temperature change, so that we have

$$dS = \left(\frac{\partial S}{\partial V}\right)_T dV + \left(\frac{\partial S}{\partial T}\right)_V dT = 0.$$

If the transition had been, not to the neighbouring state of equilibrium, but towards equilibrium from a less probable state, then dS would have been positive, and we should have had $dS > 0$.

Now we may consider an isothermal expansion, also to a neighbouring state of equilibrium. This may be imagined to occur in two stages. First, the change of volume, dV, takes place adiabatically, the change in entropy being zero, so that

$$\left(\frac{\partial S}{\partial V}\right)_V dV = -\left(\frac{\partial S}{\partial T}\right)_V dT.$$

In this part of the process the system cools. Next a source of heat is found, by contact with which the change of temperature dT may be neutralized, and T restored to its original value. In this stage a quantity of heat must be communicated to the system, the amount required being given by the equation $dQ = C_v dT$. Now by the general result obtained on p. 39, the value of $(\partial S/\partial T)_v dT$ is $C_v(dT/T)$, which is the same as dQ/T. This represents, in fact, the whole entropy change accompanying the isothermal expansion, since that occurring in the initial adiabatic part of the double elementary process was zero. Thus, for an isothermal expansion we have that the entropy change is given by dQ/T.

Next let us consider the general case where V and T both change, the conditions, however, not being adiabatic. Once again the process may be imagined to occur in elementary stages. In the first, which is adiabatic, there is a temperature change dT_1, such that

$$\left(\frac{\partial S}{\partial V}\right) dV + \left(\frac{\partial S}{\partial T}\right) dT_1 = 0.$$

An amount of heat $dQ_1 = C_v dT_1$, is then put in to neutralize dT_1, the change in entropy by the previous argument being dQ_1/T. Finally the temperature is further changed by the prescribed amount dT at constant volume, heat $dQ_2 = C_v dT$ being absorbed and the entropy changing by dQ_2/T. The total change of entropy has been for the various partial steps

$$dQ_1/T + dQ_2/T = dQ/T,$$

where dQ is the total heat taken in.

The argument can be generalized still farther. Let a system as complex as we please, including different chemical substances (which may even suffer interconversions), have a total entropy S, and let it move from a state of equilibrium to an adjacent one, being caused to do so by the change of certain variables, X_1, X_2,..., and T. X_1, X_2,..., may be pressure, chemical composition, or anything that has a physically intelligible meaning in this connexion. Once again let the change take place in stages: first an adiabatic one for which, since the probability remains a maximum,

$$dS = \left(\frac{\partial S}{\partial X_1}\right)dX_1 + \left(\frac{\partial S}{\partial X_2}\right)dX_2 + ... + \left(\frac{\partial S}{\partial T}\right)dT = 0.$$

The change of temperature, dT, is now neutralized by the communication of heat, the amount required being $\sum C_v dT$. The corresponding value of dQ/T represents the change in entropy expressible by

$$\left(\frac{\partial S}{\partial X_1}\right)dX_1 + \left(\frac{\partial S}{\partial X_2}\right)dX_2 ....$$

Thus we arrive at the fundamentally important result that whenever any system, however complex, or whatever the law of dependence of entropy upon volume or other variables may be, moves from one state of equilibrium to a neighbouring one,

$$dS = dQ/T,$$

where dQ is the heat taken in from outside the system itself.

This result, it should be emphasized once again, depends upon the fact that, however complex $\partial S/\partial X$ may be, $(\partial S/\partial T)_v$ remains C_v/T (or $\sum C_v/T$), and however complex a function of variables C_v itself may be, it is always the measure of the heat required to produce unit rise of temperature.

Since in practical chemistry and physics—not to mention engineering—measurements of heat quantities play a very important part, the equation $dS = dQ/T$ is of great value. It provides a mathematical formulation of a condition of equilibrium.

When a system moves from a condition in which equilibrium does not prevail to one in which it does, the probability increases. The change of entropy will therefore be greater than it would have been in the movement to a neighbouring state of equilibrium. Thus the applicability of the equation $dS = dQ/T$ to a small displacement is a condition that the initial state is one of equilibrium. The important applications of this will appear shortly.

Some of the basic principles of the calculus of molecular chaos may be illustrated in a more direct way by the detailed consideration of the expansion of a perfect gas.

Suppose the volume to increase from V_1 to V_2. No matter how the change is brought about, the last state may be regarded as more probable than the first, since it is that to which the aimless diffusion of the molecules would lead them if they were unconstrained Suppose that they could wander at random through the whole of the volume V_2, then the chance that a given one of them finds itself at a particular instant in a reserved part of the enclosure of volume V_1 is evidently V_1/V_2. The chance that the N molecules of a gram molecule should do so simultaneously is $(V_1/V_2)^N$. This may be regarded as the ratio of two probabilities and written W_1/W_2, where W_2 is an unknown absolute probability of the state of larger volume.

Then
$$W_1/W_2 = (V_1/V_2)^N,$$
$$\ln W_1 - \ln W_2 = N \ln V_1 - N \ln V_2,$$
thus
$$d(k \ln W) = k N d \ln V,$$
or
$$dS = R \, dV/V.$$

In an expansion by dV the entropy increases by $R \, dV/V$ if the temperature remains constant. This result holds good whether the expansion is to a neighbouring state of equilibrium or not.

Suppose first that it is, then the condition of equilibrium implies that the expansive pressure of the gas is held in check by an equal and opposite balancing pressure. This opposing resistance must be forced back as the gas expands and work is done equal in amount to $p \, dV$. If the temperature is to remain constant, this work must be provided for by a flow of heat from without, since in a perfect gas the molecules exert no attraction on one another and the internal energy is independent of the volume. Thus an amount of heat given by $dQ = p \, dV$ is taken in.

$$dQ = p \, dV = RT \, dV/V = T \, dS, \quad \text{since} \quad dS = R \, dV/V,$$
or
$$dS = dQ/T,$$

in accordance with the general law.

If the gas is not in equilibrium but expands, for example, into a vacuum, the change in entropy is still $R \, dV/V$, but no work is done and no heat need be taken in, so that $dQ = 0$, and therefore in this case
$$dS > dQ/T.$$

Now consider a case where the expansion is adiabatic and where it takes place against a balancing pressure, in overcoming which work is done. Since no heat is taken in, the work of expansion, $p\,dV$, must be balanced by a decrease in the internal energy of the gas, which does in fact cool in the process. Thus

$$p\,dV + C_v\,dT = 0,$$

or
$$RT\,dV/V = -C_v\,dT.$$

There is no change in the probability of state in this process: the spatial distribution of the molecules in the expanded gas is more probable, but the energy distribution at the lower temperature is correspondingly less so. The contribution to dS due to the volume change is $R\,dV/V$, so that the contribution due to the temperature drop must, according to the last equation, be $-C_v\,dT/T$. This again agrees with the general statistical result given above.

If the equilibrium conditions were not maintained and if the gas expanded against no opposing pressure, then the entropy would still increase by $R\,dV/V$ in virtue of the volume increase, but since no work would be done, there would be no cooling and no corresponding drop in entropy to compensate the increase. Thus we should have $dQ = 0$ and $dS = R\,dV/V$, so that $dS > dQ/T$.

The kind of transformation which occurs when a system passes from one state of equilibrium to an adjacent one is known as a *reversible change*: that which occurs when it passes from a state farther removed from equilibrium to one less far removed is called *irreversible*.

The condition for equilibrium is that all forces acting upon the system shall be exactly balanced by opposing forces. In a reversible change the maximum resistance is overcome and thus the maximum possible amount of work is done. In an irreversible change the work done is less and may fall to zero.

In a reversible change $dS = dQ/T$: in an irreversible change $dS > dQ/T$.

The condition, then, that a given small displacement should be reversible is also the condition that the system in question starts out from a state of equilibrium.

Thermodynamics and molecular statistics

The ideas outlined in the foregoing pages provide a transition from the notions of molecular statistics to those of heat and work, and thus to the science known as thermodynamics. They have just been illustrated by reference to gases which obey Boyle's law, but the more detailed statistical argument which preceded showed the generality of the law relating dS with dQ/T for equilibrium and non-equilibrium systems respectively.

The exploitation of the idea of molecular chaos, which in turn is based upon the view that molecules are small particles subject to mechanical laws, thus leads to a new point of departure from which wide regions of physical chemistry can be explored.

The entropy principle which has just been developed is susceptible of experimental verification and is equivalent to what is usually called *the Second Law of Thermodynamics*.

Before examining the scope of thermodynamical principles, it may be well to consider the degree of validity and the limitations of the statistical ideas so far developed. In the first place, it has been expedient for the calculations to assume a finite series of discrete energy states, among which the molecules are distributed (p. 27). The number of assignments is then definite, and a calculus of probabilities can be applied. If the spacing of the successive states in the series is made infinitesimally fine, then the number of assignments assumes a large and indefinite value. The absolute magnitude of the statistical entropy, as far as the present discussion has gone, is therefore devoid of significance. That ratios of numbers, both of which tend towards infinity may retain significance is, of course, perfectly possible, so that differences in entropy may be finite and non-arbitrary even when the absolute values are unknown, unknowable, or possibly meaningless. But to assume that this should be so involves something of a working hypothesis.

The positive fruitful results of the foregoing statistical calculations did not in any way depend upon the extrapolation to the case of infinitesimally small differences of energy. The calculations related to small finite differences. If there were a definite law specifying the spacing of discrete energy levels, and if these did not merge into a continuous series, then the number of assignments and the statistical entropy would possess exactly knowable absolute values, and a whole range of possible new applications of the entropy principle might open out.

The series is in fact a discrete one, the laws in question are those of the quantum theory, and the applications of the knowledge which they bring are indeed extensive, as will appear.

To obtain the equipartition law in the form which gave, apart

from the mystery of the missing degrees of freedom, the correct form
for the specific heat, the finite and discrete spacing of the energy
levels had to be given a particular character. The series corresponded
not to equal increments of energy but to equal increments of momentum. Even if an extrapolation to a continuous distribution had
been made, this hypothesis would have had to be retained. Thus
it again becomes apparent that something beyond the normal
mechanical laws is needed for a complete solution of the problem.

In the meantime the entropy principle itself has a host of applications which do not involve knowledge of what this something is.

III

THERMODYNAMIC PRINCIPLES

The laws of thermodynamics

THE entropy principle emerges from the conception of the world as a chaos of particles endowed with random motions. It states that when any system undergoes a small displacement from equilibrium the heat taken in, dQ, is related to the entropy change by the equation

$$dQ/T = dS.$$

Such a change is called reversible: for the irreversible change

$$dQ/T < dS.$$

These relations hold whether the molecules exert forces on one another or whether they are nearly independent masses as, approximately, are those of dilute gases. When mutual influences are not negligible, then the internal energy becomes a function of the volume. In general it may also be a function of other external variables.

The difference between dQ, the heat taken in by any system, and dE, the change in the internal energy, is accounted for according to the principle of energy conservation by the performance of external work as, for example, when a gas expands against an external pressure. This available balance is a maximum when dQ itself has its maximum of $T\,dS$ in the reversible transformation. It is then termed the change in *free energy* or the change in *thermodynamic potential*.

The question of the relation between the heat changes accompanying various kinds of natural process and the work which may be performed is one that has been of dominating importance. The conception of the statistical entropy is evidently one which penetrates more deeply into the nature of things, but the idea of deriving work from heat is one which has appealed more strongly to minds bent upon the useful arts such as the construction of steam-engines, and the laws of thermodynamics were discovered independently of the statistical ideas which interpret them.

The first law of thermodynamics states that there is an exact quantitative equivalence between heat consumed and mechanical work generated and vice versa. In the light of the notion that heat is the invisible energy of molecules, differing only from mechanical energy in being disordered rather than ordered, the first law appears as a simple corollary of the principle of the conservation of energy.

What is known as the second law follows from the entropy principle in a fairly direct way. Suppose there are two systems, one of which transfers heat to the other. Let the temperature of the one which loses the heat be T_1, and that of the one which gains it T_2. The former gives up an amount of heat dQ, and the drop in its entropy is dS_1: the latter receives an equal amount of heat, and its entropy increases by dS_2. If the whole transaction is one which takes place spontaneously, the probability increases on balance, so that $dS_2 - dS_1$ will be a positive quantity. In the limit when the transfer of heat is only just possible as a spontaneous process, $dS_2 - dS_1$ just approaches zero. We have

$$dS_1 = dQ/T_1,$$

$$dS_2 = dQ/T_2,$$

$$dQ(1/T_2 - 1/T_1) = \Delta.$$

Δ being positive (or in the limit just zero) and dQ by hypothesis a positive quantity we have

$1/T_2 - 1/T_1$ is positive (or in the limit zero).

Therefore, for a spontaneous transfer, $T_1 > T_2$, or in the limit when the flow just ceases to occur, $T_1 = T_2$.

This is in fact the second law of thermodynamics which states that it is impossible for heat to pass spontaneously from a body at a lower temperature to one at a higher.

An equivalent form of this statement, which relates it more obviously to practical considerations, is that work cannot be derived by the continuous utilization of the heat contained in a system at uniform temperature. An aeroplane, for example, cannot propel itself with the aid of some device which consumes the heat of the atmosphere in which it moves. If it did so, it would leave the surroundings cooler than itself, and the process could only continue if heat flowed against the adverse temperature gradient so created.

The second law in this form denies the possibility of what has sometimes been called perpetual motion of the second kind, just as the first law denies that of the first kind, which is generation of work from nothing at all.

These principles follow from the molecular nature of things, but they have in the past themselves played the part of fundamental laws and, since they are directly relatable to our experience of the sensible world, some people would still prefer to regard them as more satisfactory starting-points than the statistical considerations into which they may be translated. It has been argued that we can be sure of the validity of the two laws of thermodynamics in a way in which we cannot be sure of the molecular theory. Perhaps there is no answer to this, except that the molecular theory might claim greater intelligibility than the laws of thermodynamics by themselves possess.

One claim that is sometimes made is that the laws of thermodynamics are more objective and realistic than molecular-statistical considerations. This purist attitude looks somewhat artificial in the light of the fact that the second law of thermodynamics is really not rigidly true at all, except as a statistical average for large numbers of molecules. For systems small enough to show the Brownian motion it becomes patently false if advanced as a law which takes no cognizance of molecules.

Nevertheless, the adoption of the two formal laws as the starting-point of thermodynamic considerations has many conveniences. Various modes of development are possible, and each has some advantage in yielding a different point of view and clarifying a different set of relations.

Discussion of the second law of thermodynamics

In the first place, by a suitable reversal of the arguments already given, the entropy principle may be derived from the second law. The process is a little cumbrous as traditionally carried out. The following is perhaps the simplest way.

Gases exist which obey Boyle's law nearly enough for the properties of perfect gases to be inferred with some precision. The pressure, volume, and temperature of a perfect gas (for convenience we consider one gram molecule) are connected by the relation $pV = RT$. The temperature which this equation defines is taken as the standard scale. When a perfect gas expands into a vacuum there is no temperature change, so that the internal energy must be independent of the volume. When expansion takes place against an external pressure, work is done, its amount reaching a maximum of $p\,dV$ for an increase of volume dV against a pressure just equal to that exerted by the gas itself. If the temperature is to remain constant, heat, dQ, must flow in, the amount being given by

$$dQ = p\,dV = RT\,dV/V.$$

If the temperature changes by dT, there is an additional absorption of heat, $C_v\,dT$, so that

$$dQ = p\,dV + C_v\,dT = RT\,dV/V + C_v\,dT.$$

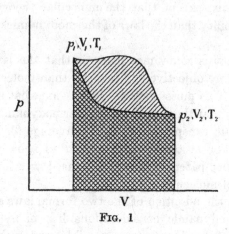

FIG. 1

Now suppose the reversible change is of finite magnitude. The work done is $\int p\,dV$, where p varies in a way which depends upon the course of the temperature change. The total heat taken in can be expressed by the integral of the last equation, namely

$$\int (RT\,dV/V + C_v\,dT),$$

a quantity which cannot be evaluated unless an auxiliary equation is provided giving T in terms of V at every stage. The value of the integral can vary widely according to the nature of this T, V relation, as may be seen in the diagram (Fig. 1). The two curved lines represent two possible paths from p_1, V_1, T_1 to p_2, V_2, T_2 and the two shaded areas correspond to the two different values of $\int p\,dV$, the work done.

On the other hand, the value of $\int dQ/T$ is perfectly definite and depends only upon V_2, T_2 and V_1, T_1.

$$\int dQ/T = \int (RT\,dV/VT + C_v\,dT/T) = \int R\,dV/V + \int C_v\,dT/T$$
$$= R\ln V_2/V_1 + C_v\ln T_2/T_1.$$

If the changes of pressure, temperature, and volume are cyclical, so that the system returns to its original state, the value of $\int dQ/T$ is clearly zero, $(\ln 1 + \ln 1)$, as the last equation shows, though the work done and the total heat absorbed, $\int dQ$, need not by any means be zero, as is seen in Fig. 2, where the area enclosed in the curve is proportional to the work done.

What has just been said is equivalent to the statement that there exists, for a perfect gas, a function, which may be called the thermo-

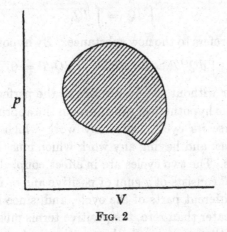

Fig. 2

dynamic entropy, defined by the equation $dS = dQ/T$ (reversible), and which depends only upon the instantaneous values of the variables describing the state. Had the changes involved in the foregoing argument been irreversible, then the work done in a given expansion would have been less, so that

$$dQ_{(\text{irrev})} < dQ_{(\text{rev})};$$

but $dQ_{(\text{rev})} = T\,dS$, by definition, and therefore

$$dQ_{(\text{irrev})} < T\,dS,$$

or

$$dS > dQ/T_{(\text{irrev})}.$$

So far all this has followed merely from the properties of the perfect gas. But the perfect gas has no special status except in so far as it is used to define the normal temperature scale. The important task, therefore, is to show that the entropy principle also applies to other substances.

This is done by a proof that unless the principle applied to all substances the second law of thermodynamics could not be true. The demonstration is as follows.

Suppose there were a substance which could be caused to go through a reversible series of operations and to return to its starting-point in such a way that the value of the function $\int dQ/T$ were greater than zero. One may obviously choose such an amount of this substance that the heat which it requires to take in while traversing some defined small cycle is equal to that taken in by a standard amount of a perfect gas going through a corresponding cycle. That is

$$\int dQ' = \int dQ,$$

where the dash refers to the new substance. By hypothesis, however,

$$\int dQ'/T > 0 \quad \text{while} \quad \int dQ/T = 0,$$

where the letter without the dash refers to the perfect gas.

Now we let the hypothetical substance, in the appropriately chosen amount, traverse its cycle, using any work which it performs to compress the gas, and having any work which must be done upon it done by the gas. The two cycles are in effect coupled together. The integral $\int dQ'/T$ consists of a sum of positive and of negative contributions from different parts of the cycle, and, since for the cycle as a whole it is greater than zero, the positive terms must predominate. In $\int dQ/T$, on the other hand, the positive and negative ones are equal. Thus for the joint system of gas and hypothetical substance the positive contributions must predominate over the negative. But for this joint system the positive and negative contributions to dQ and dQ' cancel, since we chose the amounts of material so that these two quantities were numerically equal, and, in the circumstances prevailing, they are of opposite sign, the new substance forcing the gas to traverse its cycle in reverse. Thus the positive contributions must be associated in general with lower values of T than the negative ones. In other words, the system as a whole must have taken in heat at lower temperatures and rejected it at higher temperatures. This possibility is denied by the second law, and we must therefore conclude that the assumption $\int dQ'/T > 0$ is inadmissible. A similar argument shows that an integral less than zero is equally inadmissible. Hence we must conclude that for all possible substances $dQ'/T_{(rev)} = 0$. It follows by the same argument which was used in connexion with the gas that dS or $dQ/T_{(rev)}$ measures a quantity which is a function of the variables of state only. For an irreversible change, moreover, $dS = dQ/T_{(rev)} > dQ/T_{(actual)}$.

The application of this result to the formulation of conditions of equilibrium follows in exactly the same way as for the statistical entropy.

The principle of Carathéodory

There is yet another method by which the laws of thermodynamics may be derived, and one which exemplifies a completely different attitude towards the interpretation of nature.

The quantity of heat introduced into a system may be expressed as a sum of the quantities going into various separate parts of it, that is,
$$dQ = dQ_1 + dQ_2 + \dots.$$
Each term in the sum is of the form
$$dQ_1 = dE_1 + p_1 \, dV_1,$$
where dE_1 itself may be expressed in terms of other variables
$$dE_1 = \left(\frac{\partial E_1}{\partial x_1}\right) dx_1 + \dots.$$
Thus dQ is of the general form
$$dQ = X_1 \, dx_1 + X_2 \, dx_2 + \dots,$$
where X_1, X_2,..., are functions of the different variables which describe the state of the whole system. For an adiabatic transformation
$$X_1 \, dx_1 + X_2 \, dx_2 + \dots = 0,$$
a differential equation of the type known as Pfaff's equation.

Expressions of this kind have of course geometrical interpretations in terms of lines and surfaces, and it is in the light of such ideas that Carathéodory develops the principles of thermodynamics.

First we enter into certain considerations regarding a single substance. Pressure, volume, and temperature are connected by an equation of state, any one of the variables p, V, and T being calculable in principle from the other two. ($pV = RT$ for a perfect gas is the simplest possible example of an equation of state.) It follows that dQ can be expressed as an equation in two variables, and such an equation is in principle always susceptible of integration. The example of a perfect gas will make this point clear.
$$dQ = C_v \, dT + p \, dV.$$
Since $pV = RT$, p can be eliminated from the equation for dQ, which becomes
$$dQ/T = C_v \, dT/T + R \, dV/V,$$

after division by T. The right-hand side is now immediately integrable and yields

$$S = \int dQ/T = f(V, T).$$

The factor $1/T$ which multiplies dQ is called an *integrating factor*. Now $dQ = T\,dS$, and therefore when $dQ = 0$, $S =$ constant. For an adiabatic change, then, $dQ = 0$ and $dS = 0$.

These results are in no way dependent upon the second law, and follow simply from the existence of an equation of state where two independent variables may be chosen and the third expressed in terms of them.

The following result is also obvious for a single substance. In any adiabatic change all attainable states are represented by points on a curve for which $S =$ constant. There are other points which do not lie on this curve, and which represent states which cannot be reached by an adiabatic transition.

Geometrically the condition that some states are reachable by an adiabatic transition and that others, as near to them as we please, are not, is intimately connected with the existence of an integrating factor for the equation $dQ = X_1\,dx_1 + X_2\,dx_2$. ($x_1$ and x_2 were V and T in the above example.) If there are adiabatically unreachable points, they fill an area bounded by the line of reachable points. This line must be described by an equation of the form

$$f(V, T) = \text{constant},$$

which in turn must be derivable from the equation $dQ = 0$ by a purely mathematical process. Hence the existence of the integrating factor.

The argument thus briefly outlined survives more rigid analysis and, what is important, can be generalized to the case of more than two independent variables. The result which emerges is this: that if in the immediate neighbourhood of any given point, corresponding to coordinates $x_1, x_2,...$, there are points not expressible by solutions of the equation $X_1\,dx_1 + X_2\,dx_2 + ... = 0$, then for the expression $X_1\,dx_1 + X_2\,dx_2 + ...$ itself there exists an integrating factor. When such a factor multiplies dQ, the product becomes a perfect differential, that is the differential of a function which depends only on the coordinates and not upon the shape of the path connecting two sets of values of these coordinates.

Now, as has been said, the application of this geometrical and analytical result tells us little or nothing about the behaviour of single substances which could not have been inferred directly from the mere existence of the equation of state. But it involves a special condition about the equilibrium between any two substances, and this condition can be used as a basis for a formulation of the second law of thermodynamics.

For two substances in contact

$$dQ = dQ_1 + dQ_2,$$

and to express this we need a three-variable equation of the form

$$dQ = \left(\frac{\partial U_1}{\partial V_1} + p_1\right)dV_1 + \left(\frac{\partial U_2}{\partial V_2} + p_2\right)dV_2 + (C_{1v} + C_{2v})\,dT.$$

The equations $dQ_1 = 0$ and $dQ_2 = 0$ individually have integrating factors in any case, and may be written $T_1\,dS_1 = 0$ and $T_2\,dS_2 = 0$ respectively. The equation $dQ = 0$ has an integrating factor only if the assertion is true that in the neighbourhood of any point representing a state of the system there are other points not accessible by adiabatic transformations. If this assertion is in fact true, then geometrical arguments, which are an elaboration of the simple considerations outlined above for the two-variable system, show the existence of an integrating factor for dQ, which can then be written as $T\,dS$.

When this last operation is permissible we have

$$T\,dS = T_1\,dS_1 + T_2\,dS_2$$

for the three-variable system.

Further detailed argument of a purely mathematical nature shows that in this case there is one universal factor, T, for the separate systems and for the joint system, and that $dS = d(S_1 + S_2)$, that is, that there is an absolute temperature and an entropy function for all substances. From this point the derivation of the principles of thermodynamics follows the same course as before.

In the orientation given to it by Carathéodory, thermodynamics is made to depend upon the postulate that in the immediate vicinity of any state of a system of more than one body there exist other states which cannot be reached by reversible adiabatic transitions.

That this should be analytically equivalent to the second law may perhaps be seen in a general way by reflecting that if there were no limitation upon the direction of heat transfers, such as the second law asserts, then, even subject to the condition $dQ = 0$, dQ_1 and dQ_2 could assume values corresponding to unnatural transitions, and states could thereby be reached which in reality are unattainable.

Comparison of various methods of formulation of the second law

There are thus three main avenues of approach to thermodynamics. The first is that which starts from the laws of probability, and in its origins is specially associated with the name of Boltzmann. The second is that of Carnot, of Clausius, and of Thomson, and sets out from experience of the flow of heat and of the convertibility of heat into work. The third is that of Carathéodory.

A comparison of these three approaches is of more than historical significance, in that it raises a question which cannot legitimately be evaded, that, namely, of what is meant by a scientific interpretation of the world.

It has been claimed that Carathéodory's method avoids what is alleged to be the unsatisfactory notion of the flow of heat, and dispenses with cumbrous conceptions such as cycles of operations in which heat is utilized and work performed: that it reduces the whole argument to a clean-cut mathematical consideration of the geometry of lines and surfaces, and that it makes only one simple postulate about the possibility of reaching certain states by adiabatic means.

As to the matter of the 'flow of heat', one might ask for a definition of the symbol dQ employed in the development, and it may be suspected that the answer would make a rather thinly veiled use of the notion to which exception is taken. This, however, is a rather unimportant point.

More seriously it can be said that the rigid mathematical development of Carathéodory's arguments is no less elaborate than the alternative mode of discussion, and the question could certainly be raised whether his fundamental postulate would really be of any interest or significance in itself were it not known to be going to lead to the second law of thermodynamics. From one point of view it might be regarded rather as an incidental corollary to the latter, and the derivation of the law from the principle might even be represented as putting the cart before the horse. Moreover, in its primitive form, Carathéodory's postulate is hardly open to confirmation by direct observation.

There is, however, no means of deciding whether the one or the other of these different views about the matter is the more correct one. Something that conforms to a mathematical pattern of an established type will always appeal strongly to some people and there are even parts of science, as will appear, where this mode of description seems to be the only one available. But it is well to remember that this conformity provides a means of description only, and that its significance is partly a matter of aesthetic judgement.

To some men of science such descriptions seem satisfying and to some they do not, just indeed as to certain other people any purely scientific description of the world lacks depth and cogency. Bearing this consideration in mind, we may turn for a moment to the other thermodynamic formulations.

It is in the forms which deal with the flow of heat and the possibilities of obtaining mechanical work that the second law is linked most closely with the everyday experience upon which science partly rests, and with the useful arts to which it owes so much. Here perhaps an objective criterion does exist for giving the traditional formulation precedence over that of Carathéodory. If a major object of science is discovery, then, to that extent, the most satisfactory presentation is that which proceeds by the smallest steps from the known to the unknown. This the methods of the nineteenth-century masters do, whereas those of Carathéodory are more of the nature of a sophisticated commentary on discoveries already made by other means.

The doctrine of molecular chaos, leading to the interpretation of entropy as probability, is in a somewhat different case again. It is based, though not upon direct experiment, upon the primary hypothesis of all chemistry, that of the existence of molecules, and upon the assumption, common to most of physics, that these particles are in motion. It is related very closely to such facts of common observation as diffusion and evaporation, and it takes its place among the major theories about the nature of things. In scope and significance it is of a different order from rather colourless assertions about the geometry of lines and surfaces constructed with the variables of state.

Yet while the geometrical method is self-consistent and, within its limitations, complete, the statistical method is beset with problems. One at least of these, the formulation of what constitutes a molecular state, has already appeared. But it is precisely the examination of this question which leads to fresh realms of discovery.

The geometrical method has qualities about it which might seem to appeal to purists, and as far as the setting forth of the arguments goes, with reason. Yet these qualities are to some extent illusory. No approximation like Stirling's formula is used, yet the very rigidity results in a derivation which ignores in principle the possibility of the Brownian motion, and does not take into account the fact that, in the last analysis, the most interesting thing about the second law of thermodynamics is that it is not absolutely true.

Mode of application of the second law

In its usual form the second law of thermodynamics appears as a negative principle, denying the possibility of certain kinds of change. This being so, it is remarkable how many positive and quantitative results, constituting landmarks throughout physics and chemistry, are built upon it. Some general consideration of the methods by which fruitful conclusions are drawn from the second law is therefore opportune.

All the spontaneous processes which occur in the world are movements from states where equilibrium does not prevail towards those where it does. In these transitions the probability increases. Detailed examination of any individual example reveals that a means can always be devised by which the change could be opposed and the system made to yield work in overcoming a resistance in its passage towards its stabler final state. The greater the opposing influence, the greater is the amount of work which can be obtained, up to a maximum when the system can overcome no greater resistance. When yielding the maximum work the system passes through what is really a series of states of equilibrium between the defined initial and the defined final state, the opposing force being adjusted at each moment to what can just be overcome. The whole process is then carried out in what is called a reversible or quasi-static manner.

These ideas may be illustrated by simple examples. If a concentrated solution is placed in contact with more solvent, molecules of solute diffuse spontaneously until the concentration is uniform throughout. No work is done (provided that there is no volume change on mixture) and the process is irreversible. But there exist what are called semi-permeable membranes, which retain solute while permitting the passage of solvent, and one may imagine a vessel provided with a movable piston permeable to solvent. If solution is placed on the inner side and solvent on the outer, the piston will be urged outwards with a pressure, Π known as the *osmotic pressure*, and can overcome any opposing pressure which does not exceed this in magnitude. If a volume dV of solvent enters the solution, the maximum work which can be performed is $\Pi\, dV$. This in fact is the work of reversible or quasi-static dilution.

This quantity is, of course, of no particular significance in itself and only becomes interesting when compared with the maximum work obtainable in other methods of reversible dilution. Such alternative methods do in fact exist.

Suppose the volume dV of liquid corresponds to dx gram molecules of the substance. Imagine this quantity to be evaporated under its own constant vapour pressure p_0. The volume of vapour generated is $RT\, dx/p_0$, and the work which is performed against the external pressure in the course of its generation is $RT\, dx\, p_0/p_0 = RT\, dx$. This vapour could be condensed into the solution by a compressing force which would expend work $RT\, dx$ in the process, this latter quantity cancelling that obtained in the evaporation. But the vapour pressure, p_1, of the solution is lower than that, p_0, of the pure solvent. Thus, before the condensation is caused to occur, a levy of work can be taken by expanding the vapour against a maximum opposing force as its own pressure drops from p_0 to p_1.

The work obtainable in this process is

$$dx\, RT \int_{V_0}^{V_1} dV/V = dx\, RT \ln V_1/V_0 = dx\, RT \ln p_0/p_1.$$

If this levy is not taken, there is a catastrophic and irreversible condensation of vapour when the solvent is placed in contact with the solution with its lower vapour pressure.

The operation described provides, then, an alternative method of reversible dilution. The maximum work obtainable by the first method is expressible in terms of the osmotic pressure: that obtainable by the second method is expressible in terms of the vapour-pressure ratio. These two quantities of work, as is easily shown, *must be equal*, provided that temperature is constant.

The proof follows in a quite general form from the second law of thermodynamics. Let the maximum amounts of work derivable from two alternative methods of carrying out a given transformation be

A_1 and A_2 respectively. If one of the processes were reversed, say the second, the work done by the system as it returned to the initial state would be $-A_2$. Now let the transformation take place in the forward direction by the first mode and subsequently be reversed by the second, the temperature being maintained constant throughout. The total work derived from this cycle of operations is $A_1 - A_2$. The system returns to its starting-point and its internal energy is therefore unchanged. The amount of work $A_1 - A_2$ must, therefore, have been done at the expense of heat absorbed from the surroundings. The second law denies that work can be obtained in this way, so that the value of $A_1 - A_2$ must be zero. It follows that $A_1 = A_2$.

In the above particular example the two quantities to be equated are the two amounts of work obtained by reversible dilution of the solution. We have therefore the expression

$$\Pi \, dV = dx \, RT \ln p_0/p_1.$$

Certain simple substitutions make this equation more useful. dx gram molecules are equivalent to $M_0 \, dx$ grams where M_0 is the molecular weight of the solvent. $M_0 \, dx$ grams occupy dV c.c., so that $M_0 \, dx/dV = \rho_0$, the density of the liquid solvent. Thus

$$\Pi = \frac{\rho_0 RT}{M_0} \ln \frac{p_0}{p},$$

an equation which proves to contain a major part of the theory of dilute solutions.

Free energy and related functions

When a solution is diluted reversibly, work, as has been seen, may be derived from the process. The total volume of liquid, solvent plus solute, may change only negligibly. When the volume can be regarded as constant, it is expedient to represent the maximum work as the decrease, $-\Delta F$, in a potential function, F, which is called the *constant volume free energy*. When, on the other hand, the pressure is constant and the volume changes, the maximum work may be represented as the decrease, $-\Delta G$, in another potential function, G, the *constant pressure free energy*. F and G are called *thermodynamic potentials*, and exist in general for all systems. The idea underlying their use is simply that if the maximum work is independent of the nature of the reversible process by which a system passes from a given initial to a given final state, it represents the decrease in some inherent capacity of the system to do work. Such a capacity is what is normally called a potential.

In the particular example which was quoted in the last section the work is derived partly, or, with a sufficiently dilute solution, entirely from the heat of the surroundings. This is known from the fact that there is no heat change on further irreversible addition of solvent to a solution already dilute, so that the internal energy must be unaffected by the mixing. The inherent improbability of a conversion of heat into work in the reversible dilution is exactly compensated by the increased probability of the spatial distribution of the molecules, which become more randomly scattered as the solute spreads through the larger volume of solvent.

The theorem of the equality of the maximum work derivable isothermally from alternative reversible transformations is applicable in many ways, since, as has been stated, all spontaneous processes can be made to yield work. That is to say, they all occur with diminution of free energy. The exercise of imagination provides means for expressing the maximum work in terms of very varied physical properties, and the equating of the various values leads to connexions between these properties in formulae which are often of great importance.

The scope of the application of these ideas is widened still farther when variation of temperature is brought into the picture. For this extension a preliminary codification of formulae and equations is desirable.

The following quantities are conveniently introduced:

$$U = \text{internal energy of a substance or system,}$$

$$H = U + pV = \text{heat content.}$$

Changes in U and H determine the heat taken in or evolved when a change occurs in a calorimeter, at constant volume or at constant pressure respectively, under such conditions that no work is done (other than that of simple expansion).

$$S = \text{entropy.}$$

If a quantity of heat dQ is added to a system, the balance-sheet is given by

$$dQ = dU + p\,dV,$$

that is to say, the heat supplies the increased internal energy and the work of expansion.

By the second law, if the process is reversible, that is, so long as equilibrium conditions prevail,

$$dQ = T\,dS.$$

Therefore

$$T\,dS = dU + p\,dV. \tag{1}$$

Since in general

$$d(TS) = T\,dS + S\,dT$$

and

$$d(pV) = p\,dV + V\,dp,$$

equation (1) can be written

$$dU - d(TS) = -p\,dV - S\,dT,$$

so that

$$d(U - TS) = -p\,dV - S\,dT.$$

For any small displacement from equilibrium in which dV and dT are zero

$$d(U - TS) = 0.$$

We write

$$\mathbf{U - TS = F},$$

where F is defined as the constant volume free energy, so that for the small displacement from equilibrium

$$dF = 0.$$

Since, for displacements of systems not in equilibrium, S increases, F must decrease. That is to say, in spontaneous changes at constant temperature and volume, the free energy decreases.

Equation (1) can also be written

$$dU + d(pV) - d(TS) = V\,dp - S\,dT,$$

so that

$$d(U + pV - TS) = V\,dp - S\,dT$$

or

$$d(H - TS) = V\,dp - S\,dT.$$

For small displacements from equilibrium at constant temperature and pressure dp and dT are zero, so that

$$d(H - TS) = 0.$$

We write

$$\mathbf{H - TS = G},$$

where G is defined as the constant pressure free energy, so that the equilibrium condition becomes $dG = 0$. In spontaneous changes G decreases.

The variations of F and G with temperature are specially important.

In general

$$dF = -p\,dV - S\,dT.$$

When F is expressed as a function of volume and temperature, its total variation is the sum of the partial variations which it would suffer by change of each of these separately. In the notation of partial differentials

$$dF = \left(\frac{\partial F}{\partial V}\right)_T dV + \left(\frac{\partial F}{\partial T}\right)_V dT,$$

where $(\partial F/\partial V)_T$ represents the rate of change of F with V at constant temperature and $(\partial F/\partial T)_V$ that of F with T at constant volume. Comparison of the last two equations shows that

$$\left(\frac{\partial F}{\partial T}\right)_V = -S,$$

but
$$F = U - TS,$$

and therefore
$$\left(\frac{\partial F}{\partial T}\right)_V = \frac{F-U}{T},$$

or
$$\mathbf{F-U} = \mathbf{T}\left(\frac{\partial F}{\partial T}\right)_V. \tag{2}$$

Similarly
$$dG = V\,dp - S\,dT$$

and
$$dG = \left(\frac{\partial G}{\partial p}\right)_T dp + \left(\frac{\partial G}{\partial T}\right)_p dT,$$

so that
$$\left(\frac{\partial G}{\partial T}\right)_p = -S,$$

but
$$G = H - TS.$$

Thus
$$\left(\frac{dG}{\partial T}\right)_p = \frac{G-H}{T}$$

or
$$\mathbf{G-H} = \mathbf{T}\left(\frac{\partial G}{\partial T}\right)_p. \tag{3}$$

The importance of equations (2) and (3) will presently appear in various ways.

Entropy and volume

Another useful equation which may be derived by rearrangement of the fundamental equation (1) is that giving the variation of entropy with volume.

Since
$$dF = -p\,dV - S\,dT,$$

and since in general

$$dF = \left(\frac{\partial F}{\partial V}\right)_T dV + \left(\frac{\partial F}{\partial T}\right)_V dT,$$

it follows that

$$\left(\frac{\partial F}{\partial V}\right)_T = -p \quad \text{and} \quad \left(\frac{\partial F}{\partial T}\right)_V = -S.$$

By the well-known property of partial differentials

$$\frac{\partial}{\partial T}\left(\frac{\partial F}{\partial V}\right) = \frac{\partial}{\partial V}\left(\frac{\partial F}{\partial T}\right),$$

so that

$$\left(\frac{\partial p}{\partial T}\right)_V = \left(\frac{\partial S}{\partial V}\right)_T.$$

Systems of several phases or components

Before embarking on a brief survey of what thermodynamics has to say about the general pattern of physico-chemical phenomena it will be convenient to deal with certain formalities and one or two general principles.

Normally the systems which come under consideration consist of several substances or phases. These may change one into the other by physical or chemical processes, and we are often concerned with the free energy or entropy differences accompanying such transformations as solid into liquid, or hydrogen and oxygen into water. Energies, entropies, and free energies are additive, so that in the following chemical reaction, for example:

$$2H_2 + O_2 = 2H_2O$$

the total energy change is:

(energy of two gram molecules of steam) minus (energy of two gram molecules of hydrogen plus energy of one gram molecule of oxygen). This may be written in the following conventional way:

$$2U_{H_2O} - (2U_{H_2} + U_{O_2}) = \sum nU = \Delta U,$$

the reaction products being given a positive sign and the initial substances a negative one. The total change ΔU has a positive sign when the energy of the system is greater after reaction than before, that is, when there is absorption of heat if the reaction takes place in a constant volume calorimeter without performance of any kind of work.

With this convention the equations (2) and (3) of p. 65 can be applied to express the changes of free energy accompanying physical or chemical transformations. They assume the forms:

$$\Delta F - \Delta U = T\frac{\partial(\Delta F)}{\partial T} \tag{4}$$

$$\Delta G - \Delta H = T\frac{(\partial \Delta G)}{\partial T}, \tag{5}$$

the Δ referring to the finite increase which occurs in the whole composite system as the result of the change.

Entropy of gas mixtures

Entropy is an additive function, but a special question arises in connexion with the entropies of mixtures of gases. Here, if the gases obey Boyle's law, the total entropy is the sum of the entropies which each gas separately would possess if it occupied the whole volume filled by the mixture.

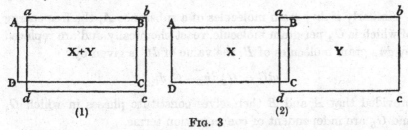

Fig. 3

This result follows at once from the fact that the probability of the state is the product of two other probabilities which, in so far as the molecules do not exert appreciable influences on one another, are independent. Macroscopically the result follows from considerations which can be more clearly envisaged if the idea of a semipermeable membrane is introduced (though of course the conclusion implies nothing at all about the real existence of such devices). In Fig. 3 $ABCD$ and $abcd$ are two containers which move with the aid of suitable seals and joints from position 1 to position 2 and back. The wall BC is semipermeable to the gas Y only, and ad to X only. Movement from 1 to 2 and back can occur without performance of work and without absorption or evolution of heat. Thus there is no entropy change in this type of mixing or unmixing. The entropy in condition 1 is the same as that in condition 2, and the principle stated above is evidently true. This case, it is to be noted, is quite different from that where two gases intermingle in such a way that each expands from its own original volume to one which is the sum of the two. In such conditions the entropy would of course increase.

Variation of masses and proportions of components

With a substance in one phase the changes which occur in S, F, G and other thermodynamic quantities are due to variations of temperature, pressure, or concentration. In a system consisting of more than one phase, or of more than one chemical compound, there are further important possibilities.

If a mass, dm, of a solid phase, for example, is transformed into liquid, then there is a gain of $G_2 dm$ units of free energy on account of the fact that dm of a phase not previously present and contributing G_2 per unit mass appears: and there is a corresponding loss of $G_1 dm$ units on account of the bodily disappearance of the mass dm of the solid phase with free energy G_1 per unit mass. Thus

$$dG = G_2\,dm - G_1\,dm.$$

Similarly if dn_A gram molecules of a substance A, the free energy of which is G_A per gram molecule, react chemically and are replaced by dn_B gram molecules of B, the value of dG is given by

$$dG = G_B\,dn_B - G_A\,dn_A,$$

provided that A and B themselves constitute phases in which G_A and G_B are independent of concentration terms.

Still more generally we may write

$$dG = \frac{\partial G}{\partial m_1}\,dm_1 + \frac{\partial G}{\partial m_2}\,dm_2 + \ldots,$$

where $\partial G/\partial m_1, \ldots$ may individually be functions of all the concentrations in any given phase.

Similar expressions of course apply to dF.

It should be observed that when G is written with a subscript, G_1, G_A,..., it refers to the free energy per unit amount of a substance or phase. If we have, for example, a pure solid phase, G_1 means just the same as $\partial G/\partial m_1$, m_1 being the mass of solid present. When concentrations change, $\partial G/\partial m_X$ can still be written G_X as long as one remembers that G_X is a function of concentration.

Activity

The free energy function, G, is defined as $H - TS$. For a perfect gas the entropy is of the form

$$S = C_v \ln T + R \ln V + \text{const.}$$

If the concentration, c, is introduced in the form $c = 1/V$, then at constant temperature S is of the form (const.$-R \ln c$). Therefore G is of the form $G = G_0 + RT \ln c$, where G_0 is a constant.

For substances other than perfect gases it is convenient to write

$$G = G_0 + RT \ln a,$$

where a is a function of the concentration which is called the activity. For a perfect gas $a = c$, and for other substances it is a function of c—in general quite a complicated one. Nevertheless, except in special circumstances, a at least varies in the same direction as c, increasing and decreasing with it, though in no simple manner.

Forces and ordered structures

The particles of which the world is assumed to be built up are conceived to be in chaotic motion. But this cannot be the whole story, or everything would consist of rarefied gas. Solids and liquids testify to the existence of ordering and agglomerating forces. At the present stage the nature of these is unknown. The union of atoms to give molecules depends upon attractions of another kind, also unknown. The inquiry into the origin of these mutual influences has to pursue a route other than that of statistics and thermodynamics, but for many purposes the forces are sufficiently characterized, both experimentally and theoretically, in terms of the energy changes which manifest themselves when they operate. As is now known, atoms are composed of electrical particles, molecules of atoms, solids and liquids of agglomerations of molecules (or occasionally atoms). At each level of this hierarchy of structures there is a balance between two tendencies: that, on the one hand, for the random motions to dissipate the constituent particles as widely as possible through the available space, and that, on the other hand, for the forces to order them into patterns possessing a minimum of potential energy.

These two opposing effects are reflected in the two terms of the free energy equations. For displacements from equilibrium, at constant pressure and constant volume respectively, we have, if $\alpha\tau = 0$,

$$dG = d(H - TS) = 0,$$
$$dF = d(U - TS) = 0,$$

so that, S being maximal at equilibrium, G and F are minimal. At a given temperature the influence of H and S or of U and S are opposed to one another.

From what has been said of the nature of the problem and from the character of these thermodynamic functions it is obvious that they will define important properties of physical and chemical equilibria. Among the major problems which confront us in the attempt to see how the world emerges from chaos are three: first, that of the equilibrium between gases, liquids, and solids, and in general of the equilibrium between any number of gaseous and condensed forms; secondly, that of the relations between systems of molecules which pass from states of segregation to states of admixture, in other words, the process of solution; and thirdly, that of the equilibrium between free elements and their compounds, and in general between the reacting substances and the products in any chemical change.

IV

AGGREGATION OF MOLECULES TO SOLIDS AND LIQUIDS

Matter in various states of aggregation

WHEN molecules pass from one phase to another, for example, from the solid to the gas, the intensity of the binding forces changes and the potential energy is altered. As the molecules become freer, energy is absorbed, and constitutes what is measured calorimetrically as latent heat.

For the purposes of a statistical or thermodynamic treatment of phase equilibria the magnitude of the forces themselves is sufficiently characterized by the value of ΔU or ΔH accompanying the transition from one phase to another.

The conditions under which condensed phases are formed from vapour, or one condensed phase is transformed into another, are subject to important general laws. Suppose two phases of a single pure substance coexist and suppose that the first is capable of passage into the second with absorption of energy; that is, ΔU and ΔH are positive. This would apply, for example, to the transition from solid to liquid, where the potential energy increases because the orderly orientation of molecules is destroyed, or to the passage from a condensed phase to the vapour state, where the potential energy increases because the molecules have been separated against the action of the attractive forces.

We may first examine the conditions under which two phases can coexist at constant pressure. Suppose dn gram molecules of substance pass from phase 1 to phase 2. The change in free energy is given by

$$dG = (G_2 - G_1)\,dn,$$

G_2 and G_1 being the respective free energies per gram molecule in the two phases. For equilibrium $dG = 0$, so that

$$G_2 = G_1.$$

$G_2 - G_1$, according to the convention explained above, is written ΔG, so that, under conditions of stable coexistence, $\Delta G = 0$.

From the definition of G itself

$$G_2 - G_1 = H_2 - H_1 - T(S_2 - S_1),$$

or
$$\Delta G = \Delta H - T\Delta S.$$

When $\Delta G = 0$, $T = T_{\text{equil}}$.

Thus
$$T_{\text{equil}} = \Delta H / \Delta S.$$

Now ΔH and ΔS are both single-valued functions of temperature, and, therefore, ΔH, on the one hand, and $T\Delta S$, on the other, when plotted against temperature are respectively represented by lines. Only at the intersection of these two curves can $T\Delta S = \Delta H$ and $\Delta G = 0$. Thus, for a given pressure, there exists a single equilibrium temperature at which $G_1 = G_2$. At any other temperature these two quantities are not equal and no equilibrium is possible.

If out of one gram molecule of a substance a fraction α exists in phase 1 and $(1-\alpha)$ in phase 2, then we have

$$G = \alpha G_1 + (1-\alpha)G_2,$$

and G reaches its minimum value corresponding to equilibrium either when $\alpha = 1$ or when $\alpha = 0$, according as G_1 or G_2 is the smaller, except of course when $G_1 = G_2$.

If, therefore, we consider a given pressure, molecules of a pure substance will pass all into one phase or all into the other, except at a single temperature where a given pair of phases can coexist, that is at the melting-point, condensation point, or transition point. At this equilibrium temperature, since $G_1 = G_2$, G remains at its minimum for all values of α. Thus the equilibrium does not depend upon the relative amounts of the two phases present.

This result corresponds, of course, with ordinary experience, the temperature of melting of a pure substance, for example, remaining constant so long as solid and liquid are both present.

Variation of pressure: vapour-pressure formula

The question now arises as to what happens when the constant pressure at which the phase transition occurs is altered from one value to another. The influence of variation of pressure may be treated as follows.

Let dn gram molecules of a substance pass from one phase in which the molecular volume is V_1 to another phase in which it is V_2. The volume change is given by

$$dV = (V_2 - V_1)\,dn,$$

and the heat absorbed by $\lambda\,dn$, where λ is the molecular latent heat. The change in entropy is dS, where

$$dS = \left(\frac{\partial S}{\partial T}\right)_V dT + \left(\frac{\partial S}{\partial V}\right)_T dV.$$

In this example the temperature stays constant and

$$dS = \left(\frac{\partial S}{\partial V}\right)_T dV.$$

If the system remains in equilibrium, then by the general law

$$dS = dQ/T.$$

Therefore $\left(\frac{\partial S}{\partial V}\right)_T dV = \frac{dQ}{T} = \frac{\lambda\,dn}{T} = \frac{\lambda\,dV}{(V_2-V_1)T}.$

By the result on p. 66,

$$\left(\frac{\partial S}{\partial V}\right)_T = \left(\frac{\partial p}{\partial T}\right)_V,$$

so that $\left(\frac{\partial p}{\partial T}\right)_V = \frac{\lambda}{(V_2-V_1)T}.$

Since at the equilibrium temperature the proportion of the phases is immaterial, we may write simply

$$\frac{dp}{dT} = \frac{\lambda}{(V_2-V_1)T},$$

or $\quad\dfrac{dp}{dT} = \dfrac{\Delta H}{T\Delta V},$

since $V_2 - V_1 = \Delta V$, and λ is a special value of what in general is represented by ΔH.

The equilibrium temperature will vary with pressure, and in a direction which depends upon the sign of ΔV.

When liquid or solid passes into vapour, ΔV is positive and p increases with T. With rising temperature the equilibrium pressure becomes higher: this is the well-known increase of vapour pressure with temperature.

The last equation can be thrown into a convenient approximate form with the help of two assumptions, namely that the volume of the condensed phase is negligible in comparison with that of the vapour, and that the vapour obeys the laws of perfect gases. In these circumstances $\Delta V = V_2$ and

$$\frac{dp}{dT} = \frac{p\lambda}{RT^2}; \quad \frac{1}{p}\frac{dp}{dT} = \frac{\lambda}{RT^2},$$

i.e.
$$\frac{d\ln p}{dT} = \frac{\lambda}{RT^2}.$$

If, further, the variation of the latent heat with temperature is assumed negligible in comparison with that of the vapour pressure itself, the last equation may be integrated in the simple approximate form

$$\ln p = C - \frac{\lambda}{RT},$$

where C is a constant. Thus the logarithm of the vapour pressure should be a linear function of the reciprocal temperature. Over moderate ranges this relation is in fact rather well satisfied.

For other phase transitions occurring in the sense for which ΔH is positive, the sign of ΔV may be positive or negative and, accordingly, the equilibrium pressure increases or decreases with rising temperature; or, if the pressure be regarded as arbitrarily controlled, the equilibrium temperature rises or falls with increasing pressure. In the transition from ice to water, for example, there is a contraction, ΔV is negative, and hence the melting-point falls as the pressure increases.

Coexistence of several phases

Most substances can assume the three forms: solid, liquid, and vapour, and we may inquire under what conditions the three can coexist. For any given pair of phases at a selected pressure there is in general a definite equilibrium temperature (with an exception which will be referred to later). For solid and liquid the equilibrium temperature may be plotted as a function of pressure, and yields a definite curve. For liquid and vapour there is a corresponding curve, neither identical with nor parallel to that for solid and liquid. The two lines will in general, therefore, cut at a definite point, representing a fixed temperature and a fixed pressure. Under these conditions the three phases can coexist stably. They can do so nowhere else. The point of coexistence is called the *triple point*.

When there are several components in the system, fresh possibilities open out. We may begin by considering two substances A and B which as solids are immiscible, but which in the liquid state are miscible in all proportions, a behaviour exemplified by many real pairs. For equilibrium of solid A with liquid we have

$$G_{2(A,B)} dn_A - G_{1(A)} dn_A = 0 \qquad (1)$$

and for the equilibrium of solid B with liquid

$$G_{2(B,A)} dn_B - G_{1(B)} dn_B = 0, \qquad (2)$$

where the subscript 2 refers to the liquid phase and the subscript 1 to the solid phase. dn_A and dn_B represent the amounts of A and B respectively which pass from one phase to the other in the small displacement from equilibrium by which the constancy of G is tested. $G_{2(A,B)}$ is the free energy per gram molecule of A in the mixed liquid at the prevailing composition, and $G_{2(B,A)}$ that per gram molecule of B. Both these latter quantities are functions of the composition of the liquid, but it is to be noted that the fixing of the value of one automatically fixes the value of the other. $G_{1(A)}$ and $G_{1(B)}$ contain no concentration terms and neither depends upon the ratio of A to B in the system as a whole, since each refers to a pure solid phase. Since $G_{2(A,B)}$ has a different value for each proportion of A to B, there will be a whole range of temperatures at which equation (1) can be satisfied. Thus solid A can exist in equilibrium with liquid at temperatures which vary with the composition of the system. Similar considerations apply to the solid B. If a fixed temperature is chosen (within a certain range), a composition of liquid can be found which will permit equilibrium with A, but this composition, though determining $G_{2(B,A)}$, will not in general permit equilibrium with B. For the simultaneous equilibrium of the two solids with the mixed liquid at a given pressure it is necessary to choose from the range of compositions possible for A and B separately that one which suits both together. This being fixed, the temperature is also fixed. Thus the two solids can only coexist with liquid at one particular temperature, which is known as the *eutectic* temperature.

The situation is again changed if the two substances A and B are miscible in the solid state. For equilibrium at constant pressure we now have

$$G_{2(A,B)} - G_{1(A,B)} = 0,$$

$$G_{2(B,A)} - G_{1(B,A)} = 0,$$

where $G_{1(A,B)}$ and $G_{1(B,A)}$ are now functions of composition as well as of temperature. For any given temperature (within a certain range) a composition can be found which corresponds to equilibrium, since now the concentration in the solid phase as well as that in the liquid can adjust itself in such a way that the two equations can be satisfied together. The equilibrium temperature will vary continuously between the melting-points of the separate substances.

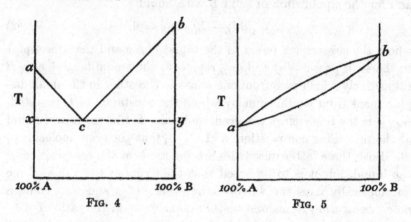

FIG. 4   FIG. 5

The relations between solid and liquid to which these principles lead are shown in Figs. 4 and 5. In Fig. 4 a is the melting-point of pure A, b that of pure B. If there is no miscibility in the solid state, the composition of solid must correspond either to A or to B, except at the eutectic, where it can correspond to any proportion of the two solid phases. Thus composition of solid follows the line $axyb$. That of liquid follows acb.

Where miscibility in the solid phase is complete, the relations are as shown in Fig. 5, where the upper line represents the composition of the liquid phase and the lower line that of the solid in equilibrium with it at a given temperature. Variants of this case are shown in Figs. 6 and 7, and an intermediate case where the miscibility in the solid state is partial is shown in Fig. 8.

Fig. 4 illustrates the fact of common observation that the melting-point of each pure component is lowered by the addition of a second substance. That it should be changed has already been shown: that it should specifically be *lowered* is a matter for more detailed consideration.

The condition for equilibrium of the pure substance is that $G_2 = G_1$.

Also, $\left(\dfrac{\partial G_2}{\partial T}\right)_p = -S_2$ and $\left(\dfrac{\partial G_1}{\partial T}\right)_p = -S_1$.

Now since heat is absorbed when solid changes reversibly into liquid, the entropy of the liquid is the greater, that is, $S_2 > S_1$. Therefore

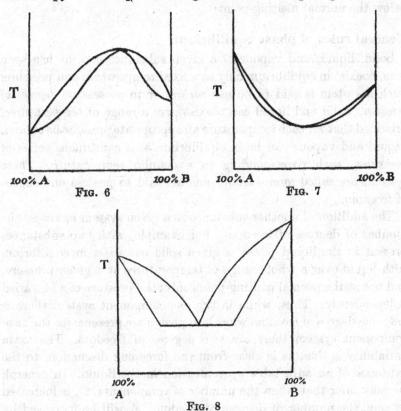

FIG. 6   FIG. 7

FIG. 8

G_2 decreases more rapidly with rise of temperature than G_1. If, then, the temperature *drops*, the free energy of the liquid *increases* relatively to that of the solid. The two phases are in equilibrium at the melting-point, so that as the temperature drops below it there develops a positive tendency for the liquid to change into solid with decrease of free energy. This tendency could be counteracted and the equilibrium could be preserved if the free energy per unit quantity of the liquid were reduced in some compensating way. A simple way is to add a foreign substance to the liquid. Free energy in general is expressible in the form $G = G_0 + RT \ln a$, where a is a function which, except in quite special circumstances, increases with concentration

(p. 69). If the pure liquid phase of A is diluted by a substance B the value of $G_{2(A)}$ is lowered. The addition, therefore, provides the compensating influence which, by decreasing the free energy of the liquid phase, allows equilibrium with the solid at a temperature below the normal melting-point.

General rules of phase equilibrium

Solid, liquid, and vapour of a given substance can, as has been seen, coexist in equilibrium only at a fixed temperature and pressure. Such a system is said to be *non-variant* or to possess no *degrees of freedom*. Solid and liquid can coexist over a range of temperatures, provided that for each temperature an appropriate pressure be chosen. Liquid and vapour can be in equilibrium at a continuous series of pressures, each corresponding to a definite temperature. These systems are called monovariant, and are said to possess one degree of freedom.

The addition of another substance to a given system increases the number of degrees of freedom. For example, with two substances present in the liquid phase, a given solid can exist in equilibrium with liquid over a whole range of temperatures at a given pressure, and not at the normal melting-point only: the pressure can be varied independently. Thus, while in the one-component system there is only one degree of freedom when two phases are present, for the two-component system there are two degrees of freedom. The extra variability is due, as is clear from the foregoing discussion, to the existence of an adjustable concentration in the liquid. In general, we may infer that when the number of components, C, is increased by one, the number of degrees of freedom, F, will be increased by one also. In other words, for a given number of phases in equilibrium we should suspect the relation, $F - C = $ constant.

Furthermore, as was seen above, if there is one solid phase and one liquid phase in the two-component system A, B, the equilibrium is possible over a range of temperatures even at one pressure, whereas, if the two solid phases are separately present, the equilibrium, for a given pressure, is possible at one temperature only. Thus, for a given number of components, the addition of a new phase reduces the number of degrees of freedom by one. This result, too, is general, and we may well suppose that $P + F = $ constant, where P is the number of phases.

Combining these two tentative rules, we obtain

$$P+F-C = \text{constant}.$$

Since for a single substance, three phases coexist in an invariant system,

$$3+0-1 = \text{constant},$$

whence the constant is evidently equal to 2, and the above induction takes the form

$$P+F = C+2.$$

This is known as the *Phase Rule*.

A more formal derivation is possible and runs as follows. For each pair of phases, as has been seen, there must be an equation expressing the change in free energy which accompanies the virtual transfer of any given component. If there are P phases, the number of independent pairs is $(P-1)$. With C components there will be C equations for each pair of phases, the total number being $C(P-1)$. To define the composition of the whole system the concentrations of $C-1$ components in each phase must be known (the remaining concentration always being calculable by difference). Thus there are $P(C-1)$ concentration variables. To these must be added temperature and pressure, making $P(C-1)+2$ in all. The $C(P-1)$ equations define a corresponding number of variables, leaving

$$P(C-1)+2-C(P-1)$$

to be assigned at will. These assignable variables constitute the degrees of freedom. Thus

$$P(C-1)+2-C(P-1) = F,$$

or

$$P+F = C+2.$$

The Boltzmann potential energy relation

So far the equilibria between phases have been considered in the light of thermodynamic principles. These, of course, are only a formal embodiment of the statistical laws governing the behaviour of large numbers of particles, and the major results which they predict should be interpretable directly in terms of the molecular-kinetic picture. Though sometimes more difficult to carry through in detail, the molecular interpretations are of considerable interest in themselves. All are variations on the familiar theme that the random motions of the molecules tend by themselves to dissipate matter into the state of rarefied gas, while attractive forces tend to collect it into condensed phases. Equilibria result from the opposition of these two tendencies.

The first departure from random distribution is expressed in the Boltzmann potential energy law. If in any region the potential energy of molecules differs by an amount ΔU from that in the rest of space, U_0, the molecules will cluster there more or less densely in the ratio n/n_0 such that

$$n = n_0 e^{-\Delta U/kT}.$$

If ΔU is negative, that is, if the potential energy is lower in the special region, they cluster more thickly.

The proof of the theorem is as follows. From the general statistical law, the relative numbers of molecules in two states of energy ϵ_j and ϵ_k respectively are

$$\frac{N_j}{N_k} = \frac{e^{-\epsilon_j/kT}}{e^{-\epsilon_k/kT}} = e^{-(\epsilon_j-\epsilon_k)/kT}.$$

If the state j differs from the state k only in the value of the potential energy, so that $\epsilon_k = \epsilon_i + U_0$ and $\epsilon_j = \epsilon_i + U_0 + \Delta U$,

$$\epsilon_j - \epsilon_k = \Delta U,$$

$$\frac{N_j}{N_k} = e^{-\Delta U/kT}.$$

The last expression is quite independent of the value of ϵ_i, so that N_j and N_k may, for all internal energy states of the molecules, be replaced by the general values n and n_0.

The existence of attractive forces leads to a certain condensing tendency, since potential energy runs down as attracting molecules approach one another. The average disturbance of density (or orientation) is insignificant as long as ΔU is small compared with kT. When temperature drops there comes a point at which complete condensation or orientation occurs and a phase change results.

The nature of the phase changes and the conditions of equilibrium between matter in different states is a subject for detailed examination.

Kinetic consideration of equilibrium between different phases of matter

In the consideration of actual phase changes evaporation is the simplest case with which to begin. Consider the surface which separates a liquid from its vapour. If there is a condition of equilibrium, the rate of condensation of molecules from the vapour equals the rate of evaporation from the liquid. Since the surface of separation of the two phases is unchanging, the rate of evaporation at a given temperature is constant.

The rate of condensation is, however, proportional to the pressure of the vapour. The latter increases until the rate of condensation equals the rate of evaporation, at which point

$$c_1 = c_2 p,$$

where c_1 and c_2 are functions of temperature only. Thus $p = $ constant at any given temperature. For small changes of temperature c_1 does not change very much, the rate at which molecules return to the surface being proportional to their speeds, which in turn vary as $T^{\frac{1}{2}}$. The rate of evaporation, however, is determined by the fraction of molecules in the liquid which have enough energy to escape from the attractions of their neighbours. This fraction is approximately proportional to $e^{-\lambda/RT}$, where λ is the latent heat.

Thus $$c_1' e^{-\lambda/RT} = c_2 p$$

approximately, or $$p = \text{const.} e^{-\lambda/RT},$$

whence $$\frac{d \ln p}{dT} = \frac{\lambda}{RT^2},$$

the formula which has already been derived (p. 74).

Next we may discuss melting. When the molecules of a liquid orient themselves into regular geometrical arrays, solidification occurs, and the potential energy (which decreases in liquefaction of vapour) runs down still farther. To escape from solid into liquid, molecules require extra energy (derived from their neighbours by the hazards of collision), while to pass from liquid to solid, molecules need to assume special orientations. The two corresponding rates of transfer are separate and independent functions of temperature, and could be plotted as curves. Where the lines cut equilibrium is possible. This is at the melting-point. The rates are separately and independently influenced by pressure, whence the functional relation between melting-point and pressure.

The balance liquid $\leftrightarrows$ solid is affected by pressure, and that of liquid $\leftrightharpoons$ vapour by temperature. By choosing the correct temperature a vapour pressure can be defined which will be equal to the equilibrium pressure for the system solid $\leftrightharpoons$ liquid at that same temperature. This defines the position of the triple point.

The entropy of a solid is less than that of a liquid, a fact which is reflected in the evolution of heat when the liquid solidifies. It is an expression of the ordered state of the solid, in which the regular orientation of the molecules contrasts with the more random configuration of the liquid.

Certain aspects of the phenomena of condensation and crystallization which the thermodynamic discussion ignores are accounted for by kinetic principles. In the last paragraphs the balancing of the rates of transfer from phase to phase was assumed to occur at plane interfaces which do not modify their character as the change of state proceeds. This assumption is perfectly correct when the equilibrium involves large amounts of established phases. Such quantities, however, are not present during the first formation of an entirely new phase; and the conditions become very different, with the result that special phenomena make their appearance.

If droplets of liquid are to form in the midst of vapour, or minute crystals in the midst of liquid, they must grow from nuclei which, in the first instance, have to be produced by the chance encounters of molecules with appropriate velocities and orientations. The incipient nuclei are subject to two opposing influences. In virtue of the attractive forces, they tend to grow, and in virtue of the thermal motion they tend to disperse. These tendencies exist at all temperatures, but above the point of condensation or crystallization they come into balance while the agglomerates are still few, minute, and transitory. The existence of the nuclei amounts to no more than an increased probability of finding small groups of molecules closer together than they would be in the absence of attractive forces, and is a direct consequence of the Boltzmann principle.

As the temperature falls the chance of a considerable gathering together of molecules increases. Growth of the clusters brings about two effects. First, the tendency to redisperse becomes greater with the number of independently moving molecules in the aggregate. But, secondly, the tendency to capture yet more molecules increases also, since the decrease in potential energy accompanying capture is, up to a point, greater the more complete and ordered the central nucleus has become. Thus, on growth of a nucleus, one factor favours and another antagonizes still further enlargement. At high temperatures the first factor predominates, at lower temperatures the second. There may be a definite size below which a nucleus will redisperse and above which it will grow. The lower the temperature the smaller is the critical size.

The temperature at which spontaneous condensation or crystallization will occur is ill defined. It depends upon the probability that somewhere in the system there is formed by a series of chance encounters a nucleus which exceeds the critical size and is large enough to grow continuously rather than to redisperse. Crystallization or condensation thus becomes more likely the lower the temperature, the longer the time allowed for the consummation of chance events, and the greater the volume in which the chances are awaited. Large quantities of liquid are in fact much less given to supercooling than small ones, and if a considerable number of small tubes of liquid are sealed up and left at a constant temperature somewhat below their melting-point, one can observe a functional relation between the time of waiting and the number which have crystallized at the end of it.

If there is present in the system a foreign body capable of attracting the molecules which are to condense or crystallize, it may constitute a sort of base on which a veneer of the new phase is deposited. Small numbers of molecules then imitate a nucleus which could only have been formed from much larger numbers in the pure substance. Hence the efficacy of dust particles and the like in promoting phase changes.

Principles similar to those which provide the interpretation of phase relations for single substances apply to the equilibria existing in systems of several. They can be illustrated conveniently by reference to the melting of a pair of substances, A and B, for the two cases respectively where these are immiscible and completely miscible in the solid phase, though miscible in all proportions in the liquid. The pressure may be taken as constant.

Pure solid A and pure liquid can coexist at one temperature only, that, namely, at which the rate of melting at the interface equals the rate of crystallization. These opposing rates are functions of temperature and are equal only for a single value of the latter. If a liquid phase of A is diluted by the addition of B, the rate of passage of molecules of A from the solid to the liquid will not be sensibly affected, but the rate of passage from the diluted liquid to the solid will usually be lowered. Thus passage from solid to liquid will come to predominate and melting will occur. In other words, the melting-point of A will be lowered by the presence of B. If the temperature is reduced, both the opposing rates fall, but the rate of passage from solid to liquid falls more than the reverse rate (since it depends upon the probability that molecules should acquire enough energy to escape from the ordered array of the solid state). Thus a new balance can be established at a lower temperature. The greater the reduction of the equilibrium temperature, the greater is the proportion of B required to produce it. As this proportion mounts there comes a point where there is enough in the liquid to hold in check the loss of B molecules from a crystal of solid B itself. Two solid phases now occur in the equilibrium state, and no further adjustments are possible while they persist. The temperature remains set at the eutectic point: A and B can crystallize out together in the proportion which leaves the composition of the liquid constant. A degree of freedom has disappeared. If more B is added to the liquid it occasions too great a rate of deposition on the solid and equilibrium cannot be maintained unless the temperature is raised. If this is done, solid A has to disappear, there being too little A in the liquid to balance its solution.

When, in contrast, A and B are miscible in the solid state, both may exist in each phase over a complete range of temperature. Suppose the rate of passage from solid to liquid at a given temperature balances the reverse rate for each of the two substances. Now let the temperature fall so that the balance is disturbed. Suppose A now crystallizes faster than it dissolves. The liquid becomes depleted of A and the solid enriched until a new steady state is reached in respect of A. This would in general lead to an A/B ratio in the liquid quite incompatible with the existence of pure B, or with any arbitrarily chosen proportion of B in the solid. But, since the concentrations of B both in liquid and solid are now independently adjustable, a new equilibrium for B also is attainable. In brief, any maladjustment of the relative rates in either direction is rectifiable both for A and for B by changes in their proportions in liquid and in solid, while any general disturbance of rate of melting relative to rate of crystallization is rectifiable by change of temperature.

Quantitative discussion of melting-point lowering: illustration of some general principles

It is now desirable to consider quantitatively the influence of a substance B on the melting-point of another substance A. The essential methods are illustrated, and the result assumes a conveniently simple form if attention is restricted to the special case where B is present in small proportion in a large amount of A, B thus constituting a 'solute' and A a 'solvent'. A and B are further assumed to be miscible in the liquid state but not in the solid state.

The principle of the calculation will first be stated. At the equilibrium temperature we have

$$\Delta G = G_2 - G_1 = 0.$$

For pure A, as has been seen, G_2 and G_1 are unaffected by the extent of the transformation, since affluence of fresh molecules of a uniform species to solid or liquid makes no difference to the properties of either. When the solid consists of pure A and the liquid is a mixture, G_1 retains its constant character but G_2 becomes a function of the proportions of A and B. In particular, it will be shown that the contribution to G from A drops as A becomes admixed with B. Addition of B thus lowers G_2 without changing G_1. At the melting-point of pure A, therefore, ΔG will no longer be zero and the equilibrium is disturbed. ΔG in fact becomes negative and there is a decrease in free energy accompanying melting, which thus tends to occur spontaneously. ΔG, however, depends upon temperature, and the departure from zero, which may be written $\delta(\Delta G)$, can be cancelled by a compensating change in ΔG brought about by a reduction in temperature. The change in temperature, δT, required to compensate the disturbance caused by the presence of B, measures the lowering of melting-point.

The detailed calculations fall into several parts, some of which involve results of general importance and validity.

1. We first require to know how the contribution to G from two components of a mixed phase are related. In general we may write

$$G = n_A \bar{G}_A + n_B \bar{G}_B,$$

where n_A and n_B are the respective numbers of gram molecules and $\bar{G}_A$ and $\bar{G}_B$ are the respective contributions per gram molecule.

$\bar{G}_A$ and $\bar{G}_B$ are themselves functions of the *proportions* of A and B, though not of their *absolute amounts*. As long as n_A and n_B are varied in such a way as to keep a standard *ratio*, $\bar{G}_A$ and $\bar{G}_B$ remain constant. Thus

$$dG = \bar{G}_A\,dn_A + \bar{G}_B\,dn_B.$$

On the other hand, as a quite general mathematical relation we have

$$dG = \bar{G}_A\,dn_A + \bar{G}_B\,dn_B + n_A\,d\bar{G}_A + n_B\,d\bar{G}_B.$$

Comparison of the last two equations reveals that

$$n_A\,d\bar{G}_A + n_B\,d\bar{G}_B = 0,$$

whence

$$d\bar{G}_A = -\frac{n_B}{n_A}\,d\bar{G}_B.$$

This relation is important, since for a dilute solution of B there is a simple expression for changes in $\bar{G}_B$, whereas for the purposes of the present calculation it is the corresponding changes in $\bar{G}_A$ which are required.

2. As has already been shown (p. 69),

$$G = G_0 + RT\ln a,$$

where a is the activity, and for a substance obeying the gas laws $a = c$ the concentration.

It will be shown in due course that for a dilute solution also a may be replaced by c. Moreover, apart from a constant which may be transferred to the G_0 term, $\ln c$ is, for a dilute solution, equal to to $\ln x$, where x is the *molecular fraction* of the solute:

$$x_B = n_B/(n_A + n_B) \sim n_B/n_A.$$

Thus for the solute B in the liquid

$$\bar{G}_B = \bar{G}_{0B} + RT\ln x_B.$$

From the result of the previous paragraph,

$$d\bar{G}_A = -\frac{n_B}{n_A}\,d\bar{G}_B = -x_B\,RT\,d\ln x_B.$$

Thus

$$\bar{G}_A = -RT\int x_B\,d\ln x_B = \text{constant} - RT x_B.$$

Thus the *change* in $\bar{G}_A$ as the molecular fraction of solute increases from zero to x_B is given by

$$\delta\bar{G}_A = -RT x_B.$$

The addition of B does not affect G_1 which refers to the solid. The change in ΔG is thus due to the change $\delta \bar{G}_A$ and for the present purpose the change in ΔG due to B may be written

$$\delta(\Delta G) = -RTx_B.$$

This must be compensated, if equilibrium is to be preserved, by a change of temperature.

3. It now remains to calculate the variation of ΔG with temperature. The result follows from the general equation

$$\Delta G - \Delta H = T\frac{\partial(\Delta G)}{\partial T}.$$

Division by T^2 and rearrangement gives

$$\frac{1}{T}\frac{\partial(\Delta G)}{\partial T} - \frac{\Delta G}{T^2} = -\frac{\Delta H}{T^2},$$

or

$$\frac{\partial}{\partial T}\left(\frac{\Delta G}{T}\right) = -\frac{\Delta H}{T^2}.$$

For *small* changes of temperature therefore

$$\delta\left(\frac{\Delta G}{T}\right) = -\frac{\Delta H}{T^2}\,\delta T,$$

or

$$\delta(\Delta G) = -\frac{\Delta H}{T}\,\delta T.$$

4. It now remains to choose δT so that the change in ΔG caused by the presence of B is compensated. Thus

$$-\frac{\Delta H}{T}\,\delta T - RTx_B = 0,$$

or

$$\delta T = -\frac{RT^2}{\Delta H}\,x_B = -\frac{RT^2}{\Delta H}\frac{n_B}{n_A}$$

approximately,

or

$$\delta T = -\frac{RT^2}{L_{\text{fusion}}}\frac{n_B}{n_A}.$$

This formula for the lowering of freezing-point provides the basis for the standard method of determining the molecular weights of dissolved substances.

Kinetic treatment of melting-point lowering

The following kinetic derivation of this same relation, although by no means rigid or complete, throws further light upon the nature of melting-point lowering.

When pure A is in equilibrium with its liquid, the rate of passage of molecules from solid to liquid, r_{12}, is equal to the rate of return, r_{21}. For the passage of a molecule from solid to liquid, energy must be supplied sufficient to detach it from the lattice. Let this be E_1 (calculated for convenience per *gram* molecule). The number of molecules which possess this energy at any moment is proportional to $e^{-E_1/RT}$. Thus

$$r_{12} = \rho_{12} e^{-E_1/RT},$$

where ρ_{12} to a first approximation is independent of temperature. The rate r_{21} may similarly be written

$$r_{21} = \rho_{21} e^{-E_2/RT},$$

where, however, E_2 is less than E_1 (and in fact approaches zero, since a molecule in the liquid does not necessarily require energy to attach itself to the solid). $E_1 - E_2$ is equal to $\Delta H = L_{\text{fusion}}$. Thus at the melting-point

$$\rho_{12} e^{-E_1/RT} = \rho_{21} e^{-E_2/RT},$$

so that

$$e^{-L_{\text{fusion}}/RT} = \rho_{21}/\rho_{12}.$$

Addition of B to the system does not affect ρ_{12}, since B does not enter the solid. On the other hand, the molecular fraction of A in the liquid is thereby reduced to $x_A = 1 - x_B$, and it is reasonable to suppose that ρ_{21} is reduced to $\rho'_{21} = \rho_{21}(1 - x_B)$. Thus in presence of B, r_{21} no longer balances r_{12}. If the temperature is lowered r_{12} falls more rapidly than r_{21} and equilibrium can be restored. At a temperature $T + \delta T$ we have

$$\rho'_{12} \sim \rho_{12},$$

$$\rho'_{21} = \rho_{21}(1 - x_B).$$

Therefore

$$e^{-L_{\text{fusion}}/R(T+\delta T)} = \frac{\rho_{21}}{\rho_{12}}(1 - x_B)$$

$$= e^{-L_{\text{fusion}}/RT}(1 - x_B),$$

$$1 - x_B = e^{-L_{\text{fusion}}/R\{1/(T+\delta T)-1/T\}}$$

$$= e^{-L_{\text{fusion}}(-\delta T)/RT^2}, \text{ approximately.}$$

$$\ln(1 - x_B) = \frac{\delta T \cdot L_{\text{fusion}}}{RT^2},$$

and if x_B is small we find, on expansion of the logarithm,

$$-x_B = \frac{\delta T \, L_{\text{fusion}}}{RT^2}.$$

Thus

$$\delta T = -\frac{RT^2}{L_{\text{fusion}}} x_B$$

$$= -\frac{RT^2}{L_{\text{fusion}}} \frac{n_B}{n_A}.$$

Lowering of vapour pressure and related phenomena

A case that has played an important part in the development of physico-chemical theory is that where a solvent A possesses a measurable vapour pressure. This is lowered by the presence of a solute B, as may be inferred from the following argument.

If a gram molecule of A is transferred by distillation from pure solvent to solution the work derivable (p. 61) is $RT \ln p_0/p$, where p is the vapour pressure of the solution and p_0 that of the solvent. Thus the free energy of the solution is seen to be less than that of the solvent by this amount. But by the argument given in a preceding section, the decrease in free energy of A caused by the presence of a small molecular fraction x_B of B is RTx_B. Thus we have

$$RT \ln p_0/p = RTx_B,$$

or

$$\ln p_0/p = x_B = n_B/n_A.$$

But $\ln p_0/p = \ln\{1+(p_0-p)/p\}$ and when $p_0-p = \delta p$ is small the logarithm may be expanded to the first term only, so that

$$\frac{\delta p}{p} = \frac{n_B}{n_A}.$$

What is called the *relative lowering of vapour pressure* is approximately equal, for sufficiently dilute solutions, to the ratio of the number of molecules of solute to the number of molecules of solvent.

This famous rule was first discovered experimentally by Raoult and provides another basis for the determination of the molecular weights of dissolved substances.

In some treatments of the thermodynamics of dilute solutions *Raoult's law* is taken as the fundamental datum, and the other relations concerning change in equilibrium temperatures are derived from it. For example, the elevation of boiling-point which is observed when a non-volatile solute is dissolved in a liquid may be calculated in terms of the relative vapour pressure lowering very simply.

At the boiling-point, the vapour pressure of the solvent attains a standard value equal to the fixed external pressure. The addition of a solute lowers the vapour pressure so that the system is no longer in equilibrium. The change is given by Raoult's law,

$$\delta p/p = n_B/n_A.$$

δp must be compensated by an increase in temperature if equilibrium is to be restored. Since $d\ln p/dT = L/RT^2$, it follows that for small changes

$$\frac{1}{p}\frac{\delta p}{\delta T} = \frac{L}{RT^2}, \quad \text{or} \quad \frac{\delta p}{p} = \frac{L\,\delta T}{RT^2}.$$

The two values of $\delta p/p$ must correspond, so that

$$\frac{L\,\delta T}{RT^2} = \frac{n_B}{n_A}, \quad \text{or} \quad \delta T = \frac{RT^2}{L}\frac{n_B}{n_A}.$$

Depression of freezing-point may be treated in an analogous fashion on the basis of the fact that at the equilibrium temperature of solid and liquid their respective vapour pressures are equal.

Osmotic pressure: dilute solutions and the gas laws

For perfect gases the influence of concentration on the free energy is simply expressed, as has been shown, in the form

$$G - G_0 = RT\ln c.$$

When the system is not a perfect gas the form is preserved by writing

$$G - G_0 = RT\ln a,$$

a being the activity. If from a theory of molecular interaction a can be expressed in terms of c, then all thermodynamic problems can be dealt with simultaneously.

For a binary system in which a molecular species A constitutes the solvent and a species B the solute, there is an important result, of which use has already been made in anticipation, namely that in the limit of sufficiently small concentrations $a_B = c_B$. This may be expressed by saying that in dilute enough solutions B *follows the perfect gas laws*. This statement is worthy of some detailed consideration. It implies that the solvent plays a role which is in certain respects analogous to the vacuum in which gas molecules move. In so far as deviations from the gas laws depend upon interactions of B molecules among themselves, an approach to conformity at infinite dilution is perfectly natural. What effect the mutual influences of solvent and solute, in general powerful, may have requires more careful examination. As we shall see, the analogy between a dilute solution and a gas is valid enough within certain well-defined limits, which, however, must not be transgressed.

The pathway to a clearer understanding of the matter lies through the study of the phenomenon called *osmosis*. Osmosis is sometimes dismissed as an obscure and secondary effect. It is, on the contrary, the most direct expression of the molecular and kinetic nature of solutions. This nature impels the molecules of a solute placed in contact with a solvent to diffuse until the concentration is everywhere the same. The spontaneous tendency towards the equalization of concentration can, according to a quite general principle (p. 60), be harnessed to yield work. If a solution of B in A is separated from pure A by a membrane which is permeable to A but not to B, then a pressure, known as the *osmotic pressure*, acts on the partition, whereby it is urged to move in such a direction that solvent flows in to dilute the solution. The motion of the partition may be opposed by a resistance against which work is done. If Π is the osmotic pressure and dV the volume of solvent which enters, the work done is $\Pi\,dV$.

Such semi-permeable membranes do in fact exist for solutions and solutes, just as they do for gases—warm palladium, for example, lets through hydrogen but no other gas. Their mode of action—and indeed their practical efficiency—is quite irrelevant. Given the specific permeability, the only matter which concerns us at the moment is the pressure which acts upon them.

Experiment shows that for dilute solutions the osmotic pressure of the solute is the same as the pressure which it would exert if it were present at the same concentration in the gas phase. Thus for a gram molecule $\Pi V = RT$. Free energy changes accompanying variation of concentration depend upon $\int \Pi\,dV$, and all the results for gases are transferable to dilute solutions.

While the empirical basis for the laws of dilute solutions is satisfactory as far as it goes, its theoretical interpretation is a matter of some subtlety, and has occasioned a good deal of difficulty in the past.

The problem is best approached by starting with a gaseous system and imagining the concentration to increase steadily until the liquid state is reached.

First let us consider a mixture of hydrogen and nitrogen. The total pressure is equal to the sum of the partial pressures of the two constituents. If the mixture were contained in a palladium vessel at a fairly high temperature, hydrogen inside and hydrogen outside would equalize their partial pressures and exert no resultant effect on the walls. The measured pressure in the vessel would now correspond to the partial pressure of the nitrogen. This would be analogous to the osmotic pressure of a solute. If the partial pressure of hydrogen inside the vessel were low, more hydrogen from outside could flow in to raise it, even though the measured partial pressure of nitrogen were high. (This, of course, is quite natural—though in the past some scepticism has been excited by the idea that solvent could flow into a container 'against the osmotic pressure' of a solute.) With an appropriate membrane the same considerations apply to a liquid mixture of B with A.

Suppose B starts as a gas unmixed with A. There are in each second a certain number of collisions between the molecules, which also make a certain number of impacts on the surface of the container. The calculation of the collision numbers has already been given. Reference to the derivation (p. 21) will show that the result would be in no way affected by the assumption that some foreign molecules, of A, were present. The cylindrical space swept out by the representative molecule of B becomes more and more bent as collisions with A increase, but apart from this these extra encounters of B with A have no relevance either to the collisions of B with B, or to the impacts of B on any surface which it may meet.

What does happen as the concentration of A becomes very high (and this can be shown very clearly with a mechanical model in which metal spheres of different kinds are agitated in a moving tray) is that pairs of B molecules tend to become hemmed in by the surrounding crowd of solvent molecules and caused to pommel one another repeatedly instead of wandering off to collide with new adversaries. But, as both theory and the mechanical model indicate, the total number of B, B collisions remains sensibly constant, although the ratio of repeated collisions to collisions of fresh pairs increases steadily as the state corresponding to that of a liquid is approached. (In the limit, if the concentration of A becomes so high that the system congeals to a glass, some of the B's would be kept locked in a perpetual clinch, but this would *not* be a state where B, B collisions could be said to have ceased.) In the same way the number of impacts of B on a membrane or partition is substantially unaffected by the presence of A, though, here again, in crowded systems the same B molecules make repeated hits, whereas in a gas the same number of hits would be made by a much more rapidly changing series of attackers.

As far, then, as the number of encounters goes, the mere impeding action of the solvent makes no major difference, and B behaves as though it were a gas.

Molecular interaction of A and B will also have little effect. Suppose A and B actually combined to give new molecules, such as BA or BA_4. The average kinetic energy of these would be precisely equal to that of B itself, and hence the kinetic pressure would be the same. If, then, at the two limits of complete independence and of definite chemical union the solute pressure is the same, it does not seem likely that any intermediate degree of attraction between A and B would alter it.

Thus we may reasonably conclude that deviations from the gas laws depend essentially upon solute-solute interactions, and that if the solute is at a low enough concentration it 'obeys the gas laws'.

The range of validity of this statement should now be evident. Osmotic pressure is equal to gas pressure, activity is equal to concentration, and the free energy is expressible by a formula which would apply to a gas. On the other hand, diffusion rates of solute through solvent bear no relation whatever to those of gas molecules through free space. In dynamic problems the strength and the weakness of the gas analogy become specially evident. As regards total numbers of encounters between solute molecules, the gas formula gives an adequate answer, provided that repeated impacts count as effective. If molecules reacted chemically at each collision, there would be no opportunity for repetitions, and the effective rate of reaction would come to depend upon a diffusion rate. For thermodynamic purposes, however, this limitation is not important, and the statement is essentially valid.

With uncharged molecules of ordinary size, activity and concentration converge at dilutions still within experimentally useful ranges. With charged ions, or with very long molecules, such as those of various polymerized substances, allowance for solute-solute interaction can hardly ever be disregarded, even in the most dilute solutions which it is convenient to employ.

FACTORS GOVERNING PHYSICAL
AND CHEMICAL EQUILIBRIUM

Further liquid-vapour relations

THE thermodynamic discussion of phase equilibria is based largely upon the principle that the free energy of unit mass of any given pure phase is constant at constant temperature. The corresponding assumption made in arguments resting directly upon the kinetic theory is that the rate of passage of molecules from one phase to another across an interface is independent of the absolute amounts of the phases present. For mixed phases the free energies and the rates depend upon concentrations, with corresponding modification of the equilibrium conditions.

When a new phase is in process of formation, it may be dispersed in droplets, or minute particles, so small that the free energy per unit mass is no longer independent of the state of mechanical division, and the phenomenon of delayed transformation—which is connected with this—may appear.

The transition from gas to liquid, in particular, follows a different course according as there is a continuous vapour-liquid interface present initially or not. If there is, then the liquid phase simply increases and the vapour phase decreases. If there is not, then droplets must form and grow safely past the limit of the region where redispersion is their likely fate.

In Fig. 9 the pressure-volume relations of a gas-liquid system are represented. A corresponds to a dilute unsaturated vapour. On compression at constant temperature the pressure and volume change more or less in accordance with Boyle's law and the curve AB is followed. Imagine the vapour to be tested at various points by being placed in contact with a continuous surface of its liquid. Up to B, the saturation point, it would take up liquid which would evaporate into it. At B there would be equilibrium, and if in presence of the liquid the pressure were infinitesimally raised, complete condensation would occur at constant pressure: the line BC would be followed to the point C. If pressure were raised further, the compression curve of the liquid, CD, would be traversed. The only variable quantity along BC would be the proportion of the two bulk phases, liquid and vapour.

Now suppose the compression from A to occur in complete absence of liquid. This time the point B possesses no special significance. From B to X the average aggregates of molecules in the vapour (formed in conformity with the Boltzmann principle) are too small to grow. Although beyond B the free energy per unit mass of vapour

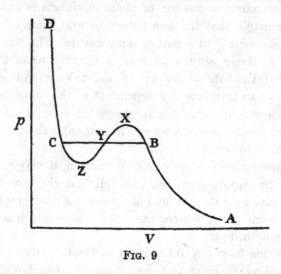

Fig. 9

has become greater than that of massive liquid, it has not become greater than that of the average spontaneously formed clusters. At X, however, these latter acquire the power to grow at the expense of vapour. As they do so they cause condensation. During the course of this process the state of affairs prevailing in the system is unstable with respect to time, but at any given moment there is an instantaneous relation between pressure and total volume such that both drop together, the vapour pressure falling as the vapour condenses to liquid and leaves the space unfilled with molecules. This phenomenon corresponds to the line XY, and though it represents a passage through a series of unstable states, it is none the less characterized by a perfectly definite pressure-volume relationship. The free energy per unit mass of vapour at a point along XY corresponds correctly to that of unit mass of liquid in its instantaneously realized state of dispersion, though this state is one which cannot persist in time. The droplets in fact coalesce to give liquid of lower free energy. If they could by some means be caused to retain their sizes, then the equilibrium represented by a point on XY would also be stable. When Y is reached we practically pass to the case previously considered and the rest of the condensation would follow the course YC.

Next we may consider the process from the other end, starting with liquid at D. The pressure is reduced along DC and the liquid dilates very slightly. At C under some conditions vapour can stream off from the liquid interface and evaporation occurs along CB. Under other conditions in a mass of heated liquid no vapour forms, since the minute bubbles which arise in the midst of the liquid redissolve. Their average size—and they constitute mere invisible holes in the normal texture of the liquid phase—is very small. It increases, however, as pressure drops along CZ. At Z spontaneous increase of these minute holes becomes possible. The pressure and the volume now increase as copious bubble formation occurs. The system passes through a series of states, once again unstable in respect of their permanence in time, but definite enough in that for each liquid–vapour ratio there is an average size of the growing bubble at which the free energy per unit mass of vapour equals that per unit mass of liquid, even though the distribution of sizes is continuously changing, and with it the equilibrium pressure. The curve ZY must be continuous with XY since obviously spontaneous evaporation and spontaneous condensation are reciprocal processes.

At any moment during the traversing of CZ on the one hand or of XB on the other, there is the possibility that a sufficiently large accidental fluctuation in the original phase may give rise to centres of the new one large enough to cause a rapid switch to ZY or to XY respectively, with consequent instability and rapid passage to those states of equilibrium corresponding to the presence of bulk phases. Such transitions correspond to the bumping of a superheated liquid on the one hand and, on the other, to the sudden relief of super-saturation in a vapour.

Van der Waals' equation and other equations of state. Critical phenomena

Somewhat similar relations find expression in equations, such as that of van der Waals, which, starting from the perfect gas laws, modify the expression $pV = RT$ by taking into account both the finite size of molecules and the existence of attractive forces between them, and seek to comprehend, at least in a rudimentary way, the whole fluid state of matter, liquid as well as gas.

Van der Waals' equation itself assumes the form

$$(p+a/V^2)(V-b) = RT.$$

The attractive forces between the molecules lower the momentum of the impacts which they make on the walls of a container, and hence reduce the pressure. To correct for this and obtain a value of p which would satisfy the equation of a perfect gas, van der Waals added the term a/V^2, where a is a constant. The correction is taken to be inversely proportional to V^2 since it is a function of the interaction of pairs of molecules and the numbers of close pairs will depend roughly on the square of the density. b is a correction for the finite size of the molecules, $(V-b)$ representing the free space in which movement can actually occur. The equation is approximate, and indeed of qualitative significance only, so that more elaborate arguments for the form of the correction terms are not worth entering into. Nevertheless, it gives an overall picture of important phenomena which is extremely valuable.

Rearrangement of the terms gives a cubic equation in V. For appropriate values of T, the cubic has three real roots. This means that for certain values of p, V has three values, as shown in Fig. 10, and indeed in Fig. 9, where the general form of the curve $DCZYXBA$ is just that given by van der Waals' equation in the region where it has three real roots. For large values of T there is only one root, and there is a definite transition temperature where the three roots become identical and above which two of them become imaginary. The family of curves corresponding to a series of increasing values of T is as shown in Fig. 10.

At temperatures above that where the three roots coalesce and the curve assumes the form 3, there ceases to be any region corresponding to XZ of Fig. 9, where unstable conditions prevail, and which can correspond to the growth of droplets or bubbles. Neither is there any possibility of a line such as BC short-circuiting the passage through unstable states when continuous phases are present. Pressure of vapour increases and volume decreases without any discontinuity, until the system is dense enough to be regarded as liquid. Conversely on reduction of pressure the liquid never generates bubbles of vapour, but thins down progressively till it can be called gas. There exists, in these circumstances, what is called *continuity of state*. The point above which the transformation of liquid to vapour becomes continuous in this way is called the *critical temperature*. Its existence is due to the fact that when thermal agitation is violent enough there is no pressure at which small liquid aggregates begin to grow spontaneously at the expense of vapour. They can grow gradually as the

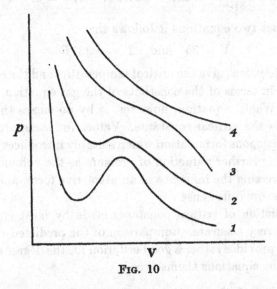

FIG. 10

pressure is raised and as the assembly passes through a series of states of stable equilibrium. At some not well-defined point we may choose to call the state liquid, but the single phase is in a sufficient condition of turmoil to be regarded equally well as highly compressed gas.

The mathematical derivation of the critical temperature in terms of the constants of the van der Waals' equation is simple. p is given as a function of V. Maxima and minima occur where $(\partial p/\partial V)_T = 0$. When there are three roots there are two such turning-points. The three roots coalesce at the critical point and here

$$\left(\frac{\partial^2 p}{\partial V^2}\right)_T = 0.$$

Differentiation with respect to V in the equation

$$(p+aV^{-2})(V-b) = RT$$

gives

$$\left(\frac{\partial p}{\partial V} - 2aV^{-3}\right)(V-b) + p + aV^{-2} = 0,$$

$$\frac{\partial p}{\partial V} = 2aV^{-3} - \frac{p+aV^{-2}}{V-b}$$

$$= 2aV^{-3} - RT(V-b)^{-2} = 0,$$

$$\frac{\partial^2 p}{\partial V^2} = -6aV^{-4} + 2RT(V-b)^{-3} = 0.$$

From the last two equations it follows that

$$V = 3b \quad \text{and} \quad T = 8a/27Rb.$$

These expressions give the critical temperature and the corresponding volume in terms of the constants of the gas equation.

Van der Waals' equation, however, is by no means the only one which yields the critical constants. Values for these are derivable from any analogous formulation which suitably introduces two terms, one favouring further reduction of pressure as the volume increases, that is, expressing the influence of an attractive force, and the other acting in the opposite sense.

The prediction of critical constants made by most equations of state is not very accurate. Comparison of the predicted values with experiment provides rather a good criterion for the degree of approximation of the equations themselves.

The direction of phase changes: range of existence of phases

Phase changes from solid to liquid or from liquid to vapour are accompanied by the absorption of heat. This is because there is more motion in the less constrained phase. There is also greater disorder, that is, higher entropy.

We write

$$\Delta G = G_2 - G_1,$$

$$\Delta H = H_2 - H_1,$$

where the subscript 2 refers to the phase of higher energy content.

Then

$$\frac{\partial(\Delta H)}{\partial T} = \frac{\partial H_2}{\partial T} - \frac{\partial H_1}{\partial T} = \Delta C_p,$$

where C_p is the specific heat measured at constant pressure. Therefore

$$\Delta H = \Delta H_0 + \int \Delta C_p \, dT,$$

where ΔH_0 is the latent heat at the absolute zero.

From the general relation (p. 67),

$$\Delta G - \Delta H = T \frac{\partial(\Delta G)}{\partial T}.$$

Rearrangement and division by T^2 gives

$$\frac{1}{T}\frac{\partial(\Delta G)}{\partial T} - \frac{\Delta G}{T^2} = -\frac{\Delta H}{T^2},$$

i.e.

$$\frac{\Delta G}{T} = -\int\left(\frac{\Delta H_0 + \int \Delta C_p\, dT}{T^2}\right)dT + J,$$

where J is an integration constant.

$$\Delta G = \Delta H_0 - T\int\left(\frac{\int \Delta C_p\, dT}{T^2}\right)dT + TJ.$$

When $T = 0$, $\qquad\qquad\qquad \Delta G = \Delta H_0.$

According to this, at the absolute zero the decrease in free energy will occur always in the direction in which the energy content decreases. That is to say, the molecular arrangements of the minimum potential energy prevail: all vapours will condense and all liquids will solidify.

What antagonizes this tendency and leads ultimately to the reversal of sign of ΔG is the influence of the entropy terms. In general liquids have more molecular freedom than solids, and vapours than liquids. As a result

$$T\int\frac{\left(\int \Delta C_p\, dT\right)}{T^2}dT$$

contains positive terms, and the influence of these in the expression for ΔG becomes more important as the temperature rises. Finally they reduce ΔG to zero. This means that the tendency to pass into a state of minimum potential energy is now opposed by the tendency to attain the more random conditions of the phase of higher energy (we shall return to this matter on p. 141).

The absolute position of the equilibrium point on the temperature scale depends upon the magnitude of the potential energy factor and upon the disparity in degree of randomness between the two phases.

Some interesting consequences follow from this conception. For example, the relative positions of melting- and boiling-points of a given compound vary very markedly with its chemical structure.

When the molecules are very far from spherical in form, rotations about the equilibrium positions are impossible in the solid, but become possible in the liquid. This means that large differences in entropy between liquid and solid exist. With molecules having something which approaches a spherical symmetry, rotations (or at any rate oscillatory angular displacements from the equilibrium orientation) can begin in the solid state without destruction of the ordered

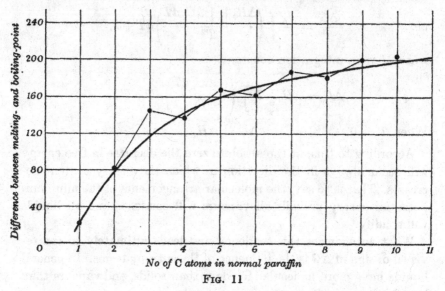

FIG. 11

crystalline lattice. Thus the solid can approach much more closely to the gas in entropy, and the melting process may be deferred until the temperature has risen quite close to the boiling-point. In some examples the range between melting- and boiling-point is extraordinarily small.

In Fig. 11 the difference between the melting-point and the boiling-point is plotted as a function of the number of carbon atoms for the series of the normal paraffins. The range expands steadily as the molecule becomes more elongated and its symmetry departs from that of methane.

The isomeric pentanes form an interesting series. With n-pentane the range is 168°, with the branched 2-methyl butane it is 187°, but with the completely symmetrical tetramethyl methane it falls abruptly to 30°, not very much more than for methane itself. Clearly the symmetrical molecule must possess in the solid state much of the freedom which other molecules only achieve by breaking loose from the crystal lattice.

The compound hexachlorethane has a melting-point and a boiling-point which practically coincide: in fact, at atmospheric pressure it sublimes before it melts. With benzene there is a liquid range of 74°, which rises to 202° in toluene. o- and m-xylene are also less symmetrical than benzene and the ranges are 171° and 193° respectively, but p-xylene achieves a certain measure of symmetry once more and the melting-point rises to within 125° of the boiling-point.

For the understanding of the range of existence of the solid and liquid states the entropy factor as affected by the degrees of freedom is thus of great importance. The position of the boiling-point itself is often to a predominant extent determined by the magnitude of the potential energy factor, as reflected in the latent heat, the other factors for gaseous and liquid states not being so different.

According to Trouton's rule the ratio of molecular latent heat to boiling-point is constant, and this statement is by no means very far from the truth. In so far as it holds it means that the expression

$$\ln p = \int \frac{\lambda}{RT^2} \, dT$$

can be integrated in the form

$$\ln p = C - \frac{\lambda}{RT},$$

with C varying little from liquid to liquid in comparison with λ. λ/T_b is in fact constant within about 10 per cent. for a quite wide range of liquids. (T_b = boiling-point at atmospheric pressure.)

Yet another aspect of this general question of the factors determining phase equilibria is illustrated by the partition of a solute between two solvents. In so far as the activity of a solute in a dilute solution is proportional to its concentration, the equilibrium distribution will be such that the ratio of concentrations in two immiscible solvents in contact is constant at constant temperature and pressure. The same principle will apply to the solution of a gas in a liquid. The concentration of dissolved gas is proportional to its partial pressure above the solution. These statements presuppose that the molecular complexity of the solute is the same in both phases, and that association or dissociation, ionic or otherwise, is excluded.

An interesting illustration of general thermodynamic and statistical principles is provided by the behaviour of very long molecules, such as those of a polymerized substance like rubber. In distribution experiments these substances show a marked tendency to go either wholly into one phase or wholly into the other. The distribution is, as always, determined by the free energy relationships. But a very long molecule, say of n segments, has an interaction energy with a solvent (or with molecules of its own kind) which is n times as great as that of a similar molecule of a single segment. By the equipartition law, however, it possesses only the normal allocation, $(3/2)kT$, of translational energy. Therefore its free energy relationships are determined almost entirely by the potential energy terms, and the concentration of molecules occurs where the potential energy is least. There is an almost complete displacement of the equilibrium in the one direction or in the other, as there would be even with small molecules in the neighbourhood of the absolute zero.

The discussion of these and other interesting matters demands intimate knowledge of the modes of motion of molecules and of the way in which these modes are affected by temperature. This is precisely what the simple conception of molecular chaos and of the kinetic theory cannot yield. As we have seen, these ideas, fruitful as they are, do not account for the variation of specific heats with temperature, nor indeed for the non-operation of certain degrees of freedom. Nor, moreover, do they yield any information about the magnitude of the constant J in the formula for the free energy change.

These limitations are removed by the introduction of the quantum theory. We shall have occasion to return to the whole problem after the quantum laws have been considered.

Chemical changes

The equilibrium in the solid state between such substances as the allotropic forms of tin or sulphur, or between the participants in any other chemical reaction, is governed by principles precisely similar to those which regulate the coexistence of phases such as solid and liquid.

The method of applying thermodynamics to the discussion of chemical equilibria in the gaseous state, or of equilibria in solutions, is also essentially the same. A small displacement of the equilibrium at constant temperature is envisaged, and, according as the pressure or the volume is fixed, dG or dF is equated to zero.

Consider the reaction

$$2H_2 + O_2 = 2H_2O.$$

For this transformation we have

$$\Delta U = 2U_{H_2O} - 2U_{H_2} - U_{O_2} = \sum nU,$$

$$\Delta H = \Delta U + \sum npV.$$

For constant volume and temperature

$$\Delta F = \Delta U - T\Delta S$$

$$= \Delta U - T \sum nS$$

$$= \Delta U - T \sum n(C_v \ln T - R \ln c + S_0).$$

Now let $2\delta x$ gram molecules of hydrogen react with δx of oxygen to give $2\delta x$ gram molecules of steam. The change in free energy is expressed by $dF = \delta x \Delta F$. If the concentrations of all the gases present correspond to an equilibrium state, then $\delta x \Delta F = 0$. Thus

$$\Delta U - T \sum n(C_v \ln T - R \ln c_{eq} + S_0) = 0.$$

In this equation the only terms which, for perfect gases, depend upon the concentrations at all are those contained in the set $\sum nR \ln c_{eq}$.

Therefore, since none of the other quantities could compensate a variation in this group as a whole, it must be constant. Thus, at constant temperature,

$$\sum n \ln c_{eq} = \text{constant}.$$

It may be written $\ln K_V$, where K_V is a function of the concentrations. In the specific example which was quoted above it assumes the form

$$K_V = c_{H_2O}^2 / c_{H_2}^2 c_{O_2}.$$

In general, if the concentrations do not correspond to equilibrium,

$$\Delta F = \Delta U - T \sum n(C_v \ln T - R \ln c + S_0).$$

But from the equilibrium condition itself

$$0 = \Delta U - T \sum n(C_v \ln T - R \ln c_{eq} + S_0),$$

whence by subtraction

$$\Delta F = RT[\sum n \ln c - \ln K_V].$$

If, for example, there were less steam and more hydrogen and oxygen than corresponded to equilibrium, $\sum n \ln c$ would be less than $\ln K_V$ (products being positive and reacting substances negative in the summation). ΔF would thus be negative and the reaction would go forward spontaneously.

The equation for ΔF is frequently known as the van 't Hoff reaction isotherm. (It must be noted that ΔF is the free energy change which *would* occur if molecular quantities reacted without change of their existing concentrations.)

The value of ΔF may be inserted in the general thermodynamic equation

$$\Delta F - \Delta U = T \frac{\partial(\Delta F)}{\partial T},$$

giving

$$RT \sum n \ln c - RT \ln K_V - \Delta U = T \left[R \sum n \ln c - R \ln K_V - RT \frac{d \ln K_V}{dT} \right],$$

whence

$$\frac{d \ln K_V}{dT} = \frac{\Delta U}{RT^2}.$$

In a precisely analogous manner, for a reaction at constant pressure, there is found the result

$$\frac{d \ln K_p}{dT} = \frac{\Delta H}{RT^2}.$$

These two equations illustrate a well-known principle of stable equilibrium. Substances formed endothermically become more stable as the temperature rises—a fact shown, for example, by the synthesis of nitric oxide from its elements in the high temperature arc. For such substances ΔU and ΔH are positive, that is, the energy or heat content of the products exceeds that of the starting materials. Thus the sign of $d \ln K/dT$ is positive, $\ln K$ (and hence K) increases with temperature, and, from the manner in which the equilibrium constant is written, this corresponds to a greater concentration of product.

The above equations (known as the two forms of the *reaction isochore*) may be integrated. For example

$$\ln K_V = \int_0^T \frac{\Delta U}{RT^2} dT + I.$$

Now

$$\frac{\partial(\Delta U)}{\partial T} = \frac{\partial U_{products}}{\partial T} - \frac{\partial U_{reactants}}{\partial T} = \sum n C_v = \Delta C_v,$$

so that

$$\Delta U = \Delta U_0 + \int \Delta C_v dT.$$

Thus

$$\ln K_V = \int_0^T \frac{\Delta U_0}{RT^2} dT + \int_0^T \frac{\left(\int \Delta C_v dT \right)}{RT^2} dT.$$

The corresponding equation for K_p involves ΔH_0 and $\int \Delta C_p dT$.

ΔU or ΔH is often large compared with ΔC_v or ΔC_p and then, especially since K varies rapidly in the relatively small range of temperatures where measurements are practicable, it is permissible to write approximately

$$\ln K_V = \text{constant} - \frac{\Delta U}{RT},$$

$$\ln K_p = \text{constant} - \frac{\Delta H}{RT},$$

$\ln K$ being a linear function of the reciprocal of the absolute temperature.

For reactions where ΔU or ΔH are small and ΔC_v or ΔC_p large this ceases to be admissible.

In the formation of steam from its elements ΔU is of the order 10^5 calories and ΔC_v under 10. Thus over a hundred-degree range, for example, the approximation would be a quite good one. In the dissociation of a weak acid in aqueous solution ΔH is of the order 10^3 and ΔC_p may be as much as 50 calories. Neglect of its influence here ceases to be even an approximately justifiable procedure and would lead to gross errors.

For these gaseous systems, as for condensed systems, the thermodynamic formulae provide complete information about the variation of equilibrium constant with temperature. They predict also the manner in which equilibrium is governed by concentration. They do not, however, provide information about the absolute values of the equilibrium constant. Knowledge of this depends upon the introduction of fresh conceptions.

The formulae which have been derived for gases apply also to dilute solutions in which the solute follows the gas laws in the sense which has been discussed previously. When these laws are not valid, the concentrations are formally replaced by activities and K is expressed in the form $\ln K = \sum n \ln a_{eq}$

and the free energy change becomes

$$\Delta G = RT[\sum n \ln a - \ln K].$$

One of the most important cases where activities and concentrations differ very widely is that of equilibria in solutions of ionized substances.

As has been shown, the condition that solutes should obey the gas laws is that the mutual interference of their molecules is negligible to a sufficient degree of approximation. Charged ions, acting upon one another according to the inverse square law of electrostatics, are subject to a mutual interference which only becomes negligible at dilutions far greater than those conveniently employed in ordinary chemical measurements. The calculation or experimental determination of the activity becomes, therefore, of great importance. Once its relation to the concentration has been discovered, it is applicable to all problems which depend upon the free energy, that is to say, to the formulation of chemical equilibria and the phase relationships of solutions.

The calculation of the activity itself is a problem quite independent of thermodynamics and rests upon an adequate theory of interionic forces, or, in general, of molecular interactions (see p. 278).

Achievements of the molecular and kinetic theory: conditions for further progress

From the foregoing pages it is already evident that much illumination about the nature of things does in fact proceed from the simple and easily intelligible assumption of a world of matter consisting of randomly moving particles, whose inherent tendency to dissipate themselves through all space is combated by attractive forces.

The essential character of thermal phenomena becomes clear, the conditions of coexistence of solids, liquids, and gases in systems of any number of chemical components are explained, the dependence of equilibria upon concentrations, upon pressure, and upon temperature is defined. The conceptions of entropy and free energy, of statistical equilibrium and energy distribution, provide quantitative laws which describe the perpetual conflict of order and chaos, and which prescribe in a large measure not only the shapes assumed by the material world but also the pattern of its possible changes.

The power, clarity, and beauty of these ideas are undeniable. Yet, great as is their range, it has its boundaries. It constitutes a brilliantly illuminated circle surrounded by obscurity. This darker region can, however, be penetrated and proves to contain many strange new things.

The first great limitation of what we might perhaps call the neo-Epicurean picture of matter is its lack of information about the character of the forces which hold the primordial motion in check.

They are made manifest by the energy changes which accompany their operation. For the theory these energies constitute the fundamental data, a circumstance which has fortunately proved to be no practical impediment, since in experimental science the direct measurement of them is precisely what has been most expedient.

Nevertheless, the nature of these all-important quantities must be inquired into and inferences made about the modes of interaction of the particles concerned.

But even when we accept the energies as known, there is still no way of understanding what determines the absolute position of an equilibrium, whether of solid and liquid, or of the participants in a chemical reaction. The influence of temperature and other variables can be precisely foretold, but all the knowledge is relative. The reason for this lies deep in the character of the theory.

Equilibria are determined by free energies which depend in turn upon entropies. Entropies are measures of the probability of given conditions, and probabilities are defined in terms of the number of modes in which assignments of molecules to states can be made.

The number of ways in which X molecules can be distributed among Y states depends upon the number Y. The extent of what constitutes an energy state has so far remained indefinite. If infinitesimally small differences in energy are detectable and significant, then the number of possible states is indefinitely large and the assignments are infinite. Thus an absolute entropy would be a meaningless quantity.

While changes in an indefinitely large quantity might themselves possess a relative significance and might possibly govern the behaviour of a single system, they could hardly predict the relations of one system with another. When the conceptual substratum is structureless and indefinite the search for a means of predicting the absolute position of equilibrium appears quite vain.

But if infinitesimally small differences of energy were not significant, and if the states were of finite range, then the number of assignments would become a definite number with an absolute meaning. The probability of one system in a given state could be compared with that of any other in another given state, and absolute prediction of equilibria would become a possibility. The substratum of the conceptual world would acquire a structure, and become, as it were, an atomic system rather than a continuum.

The imparting of this structure is precisely what the quantum theory does.

It has indeed already become obvious that the rules of Newtonian mechanics are insufficient for the full understanding of statistical phenomena. The division of molecular states into ranges of equal probability needed the assumption that equal ranges of momentum rather than of energy should be chosen (p. 31). There was no obvious reason for this. Even more striking is the complete failure of mechanics to account for the non-contribution of certain degrees of freedom to specific heats, and for the variation with temperature of the number of degrees of freedom which do so contribute.

It is the further investigation of this problem which leads most directly and simply into the realm of the quantum theory.

This theory, when established, indicates how the numbers of states may be defined, and it opens the way to a complete understanding of the absolute position of the equilibria between all forms of matter.

This fuller understanding, however, is purchased only by the sacrifice of the primitive and simple picture of a molecular chaos which rather resembles a swarm of bees and which is easy to visualize in a quite naïve fashion. It requires the introduction of abstract statements about a non-material substratum. The new picture loses in appeal to the senses what it gains in appeal to the intelligence.

In the course of the development of the quantum theory further sacrifices will be demanded from the naïve conception of particles. In particular it appears that for the correct calculation of entropies the identity of individual particles may even have to be disregarded. The intervening steps, however, will have so accustomed the inquirer to the shedding of his original ideas that he needs to feel no surprise—though possibly he experiences a slight regret. He does well, however, to remember that what emerges in the end is essentially a construction of the human mind by which various sets of facts are related in the most elegant and helpful way.

What is perhaps most important of all is to keep clear which parts of the construction are closely related to things of direct observation and experience, and how the hypothetical edifice expresses these relations.

PART II

CONTROL OF THE CHAOS BY THE QUANTUM LAWS

SYNOPSIS

As a result of a long series of intricate discoveries a solution is found to the problem of knowing where, in the absolute sense, the equilibrium lies between the different states of aggregation of matter and between the various configurations of atoms concerned in chemical transformations.

The first stage is the emergence of the quantum theory, according to which the energy of atomic or molecular systems varies discontinuously. The laws governing the various discrete series of possible energy states are discoverable from the study of such phenomena as specific heats and radiation. They evolve through different forms and are at length crystallized in rules whereby the energies are defined in terms of the permissible solutions of a semi-empirical differential equation, known as the wave equation.

The foundation of this equation is the discovery that on the scale of electronic or atomic phenomena particles obey dynamical laws which are neither precisely those followed by macroscopic masses nor yet those followed by light waves, but are of a special kind.

The new rules impart to the theory an abstract basis. Atoms and molecules can no longer be regarded as small-scale versions of ordinary objects. Furthermore, we have to conclude that there is no physical sense in treating different permutations of individual atoms or molecules within a given energy state as even theoretically distinguishable systems.

Given these apparent sacrifices of the primitive simplicity, energy states become, in compensation, like so many exactly defined boxes, the allocation of molecules to which can be treated by the laws of probability. Absolute equilibria are now seen to be governed by the interplay of two major factors: on the one hand, the tendency of atoms and molecules to assume a condition of minimal potential energy, and on the other hand, their tendency to fill impartially all energetically equivalent states.

With the equilibrium of solid and vapour, for example, the potential energy factor favours condensation of all the molecules to solid. But in the vapour the range of possible energy levels is much greater and molecules populate states according, as it were, to the housing conditions, quantum levels representing in effect accommodation. Expressed in another way, the atoms and molecules escape from the restraints imposed by the forces acting upon them in so far as they achieve fuller self-realization in conditions where more modes of motion and more quantum levels offer them opportunity. The quantum theory having given exact formulations of the accommodation ranges, these statements can be translated into precise terms which lead to a quantitative treatment of all types of equilibrium.

At this stage the nature of the forces and of the interaction energies still remains unknown.

VI

THE QUANTUM RULES

Specific heats

THE most direct route to an understanding of the rules governing
the energy content of substances is by the study of specific heats.

From the formula for the pressure of a perfect gas, $p = \frac{1}{3}nm\bar{u}^2$,
and the relation $pV = NkT$, it follows that the kinetic energy in
the three translational degrees of freedom is $\frac{3}{2}kT$. The allocation for
each is thus $\frac{1}{2}kT$. The equipartition law provides that where the
energy is shared between s square terms, the average amount in a
molecule is $\frac{1}{2}skT$. The molecular heat, C_v, is therefore

$$N\frac{\partial}{\partial T}\left(\tfrac{1}{2}skT\right) = \tfrac{1}{2}sR.$$

The assumptions about s which have to be made to account for
observed values of C_v are at first sight not unplausible. For the inert
gases, $s = 3$, so that they apparently contain translational energy
only. For the stable diatomic molecules such as oxygen, nitrogen,
and hydrogen, $s = 5$, while for less stable ones such as iodine, it is
closer to 7. It would seem that the extra two square terms appearing
with iodine represent vibrational energy (kinetic energy $\frac{1}{2}m\dot{x}^2$ and
potential energy $\frac{1}{2}ax^2$, where x is the extension from the equilibrium
position, m the effective mass, and a an elastic constant).

The two square terms, other than those for the translational energy,
which occur with the diatomic gases are evidently connected with
rotation. They are two rather than three, since one of the three axes
of reference is that joining the two atoms, and about this particular
axis the molecule will possess just the same kind of inertia as if it
were monatomic. Given that monatomic substances do not in fact
show rotation, there is no reason why diatomic substances should
show it about the axis in question.

The three vibrational degrees of freedom reasonably attributable
to a monatomic solid should account for a constant specific heat of 6,
in fair accord with the law of Dulong and Petit.

In one sense these interpretations of the specific heats of simple
substances are very successful. But deeper reflection shows that
something fundamental is missing. There is no reason in Newtonian

mechanics why all the possible degrees of freedom should not be operative in all cases, three translational and three rotational degrees with monatomic gases, and with diatomic gases these six together with a vibrational degree of freedom. Some extraneous factor must

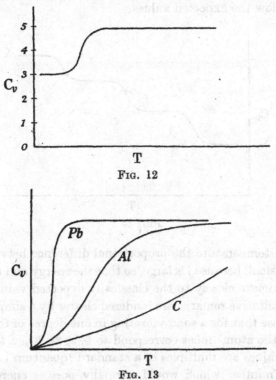

FIG. 12

FIG. 13

dictate whether or not a mechanically possible motion does in fact contribute to the energy.

This factor cannot be connected with permanent structural capacities or disabilities of molecules, since the operation or non-operation of degrees of freedom depends upon the temperature. At low temperatures the specific heat of hydrogen falls from 5 to 3, and that of metals drops to zero. Fig. 12 illustrates the behaviour of hydrogen, Fig. 13, that of some typical solids, and Fig. 14 shows the ideal course of the complete curve for a diatomic molecule.

These phenomena are fully interpreted by the quantum theory. This theory declares that the possible energy states of a molecule, or of a mechanical system in general, form not a continuous but a discrete series.

Suppose the energy of three successive states is 0, ϵ_1, and ϵ_2. At a certain low temperature, δT, the atoms or molecules, which according to the classical theory should possess a quota of less than ϵ_1, possess in fact nothing. Thus the energy content and the specific heat fall below the expected values.

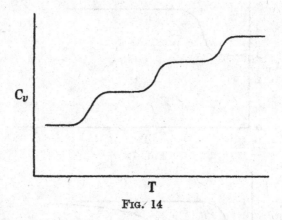

$$C_v$$

$$T$$

Fig. 14

At a high temperature the proportional difference between ϵ_j and ϵ_{j+1} is quite small because j is large, so that the energy and the specific heat approximate closely to the classically expected values.

These qualitative remarks are rendered clearer by a simple calculation. Suppose that for a solid vibrating in one degree of freedom the energies of the atoms must correspond to 0, ϵ, 2ϵ, 3ϵ,..., that is, the successive values are multiples of a standard 'quantum'. Out of N atoms, the number which would normally possess energy greater than $j\epsilon$ is given by the Maxwell–Boltzmann law to be $Ne^{-j\epsilon/kT}$ (see p. 133). The number with energy between 0 and ϵ is the difference between those with energy greater than 0 and those with energy greater than ϵ, that is to $N-Ne^{-\epsilon/kT}$. These, according to the rules of the quantum theory, contribute nothing. The number with energy greater than ϵ but not greater than 2ϵ is similarly $Ne^{-\epsilon/kT}-Ne^{-2\epsilon/kT}$ and these contribute ϵ only. The sum total of the contributions is thus

$$0(N-Ne^{-\epsilon/kT})+\epsilon(Ne^{-\epsilon/kT}-Ne^{-2\epsilon/kT})+2\epsilon(Ne^{-2\epsilon/kT}-Ne^{-3\epsilon/kT})+...$$

$$= N\epsilon(e^{-\epsilon/kT}+e^{-2\epsilon/kT}+e^{-3\epsilon/kT}+...)$$

$$= \frac{N\epsilon e^{-\epsilon/kT}}{1-e^{-\epsilon/kT}},$$

the series in the brackets being a geometrical progression with the
first term $e^{-\epsilon/kT}$ and the common ratio $e^{-\epsilon/kT}$ also. The total energy
is in fact contributed by three degrees of freedom, so that

$$E = \frac{3N\epsilon e^{-\epsilon/kT}}{1-e^{-\epsilon/kT}} = \frac{3N\epsilon}{e^{\epsilon/kT}-1}.$$

When $T = 0$, $E = 0$ and the energy vanishes. The behaviour at
high temperatures may be seen by expanding the exponential term:

$$E = \frac{3N\epsilon}{1+(\epsilon/kT)+...-1}.$$

When T is large enough this tends to the value

$$E_{\text{limit}} = \frac{3N\epsilon}{\epsilon/kT} = 3NkT = 3RT.$$

If ϵ is small enough, the same result holds for all temperatures, and
in the limit when ϵ vanishes, that is, when the series of possible energy
states is continuous, the energy is $3RT$ over the whole range and the
specific heat is constant with the value $3R$.

In general the value of C_v is given by the relation

$$C_v = \frac{\partial E}{\partial T} = \frac{3N\epsilon e^{\epsilon/kT}\epsilon/kT^2}{(e^{\epsilon/kT}-1)^2}.$$

This is the Einstein specific heat formula.

The course predicted by the equation for C_v as a function of temperature is similar to that shown in Fig. 13. The limiting values are
found, as before, by expansion.

$$C_v = \frac{3N\epsilon\{1+(\epsilon/kT)+...\}(\epsilon/kT^2)}{\{1+(\epsilon/kT)+...-1\}^2}.$$

When $\epsilon \to 0$ or $T \to \infty$ this becomes

$$C_v = \frac{3N\epsilon(\epsilon/kT^2)}{(\epsilon/kT)^2} = 3Nk = 3R.$$

When T approaches zero C_v becomes

$$\frac{3N\epsilon^2 e^{\epsilon/kT}}{kT^2(e^{\epsilon/kT})^2} = \frac{3N\epsilon^2}{kT^2\{1+(\epsilon/kT)+(\epsilon^2/2!k^2T^2)+...\}}$$

$$= \frac{3N\epsilon^2}{kT^2+\epsilon T+(\epsilon^2/2!k)+\text{terms in } 1/T, 1/T^2...}$$

$$= \frac{3N\epsilon^2}{0+0+(\epsilon^2/2!k)+\infty+\infty+...} = 0.$$

It will be observed that the course of the specific heat-temperature curve is determined by the value of ϵ/T. If $\epsilon = 0$, the specific heat remains $3R$ down to $T = 0$, and the greater the value of ϵ, the higher is the temperature at which C_v first begins to approach its limit. Inspection of Fig. 13 suggests therefore that the values of ϵ vary considerably from one solid substance to another.

The history of the stages by which the rules governing the magnitude of the energy quanta and the spacing of the possible states were discovered is a slightly tangled one. We may begin the elucidation of these rules by stating in a general way that the results regarding the specific heats of solids justify two assumptions: (a) that the successive increments of energy accompanying the passage from one possible state to the next are equal, and (b) that these increments are proportional to the frequency of the vibrational motion with which the energy is associated.

Measuring the energy of a given vibrational degree of freedom from $T = 0$, and taking no account of any which may in one way or another remain stored in atoms or molecules at the absolute zero, we write

$$E - E_0 = n\epsilon = nh\nu,$$

where n is an *integer*.

C_v may be determined as a function of temperature for various solids and may be represented approximately by the Einstein formula from which the size of the quantum may be calculated. This is found to be proportional to the vibration frequency of the solid, which can be estimated in various approximate ways.

A qualitative illustration of this proportionality is easily provided. The frequency of a simple harmonic motion is given by the formula

$$\nu = \frac{1}{2\pi} \sqrt{\left(\frac{restoring\ force\ per\ unit\ displacement}{mass}\right)}.$$

We may compare the three substances whose specific heats are represented in Fig. 13. Diamond is excessively hard, infusible, and involatile, lead soft and fusible, while aluminium is intermediate in character. Thus the frequencies, in so far as they depend upon the strength of the forces holding the atoms in the crystal, will tend to be in the order

$$\nu_C > \nu_{Al} > \nu_{Pb}.$$

Precisely the same order is indicated by the relative masses. If then $\epsilon = h\nu$, we should have

$$\epsilon_C > \epsilon_{Al} > \epsilon_{Pb},$$

and thus, from what has been said, we should expect lead to retain its specific heat of about 6 to much lower temperatures than carbon, and aluminium to lie between the other two. This is precisely what is found, and constitutes a striking verification of the rule relating quantum size to frequency.

The quantitative side of the matter is less definite as far as solid crystals are concerned, because the vibrations of the solid are in reality very complex and can only be described in rough approximation by a single frequency. In fact a complicated spectrum of frequencies must be invoked to do justice to the finer details of behaviour. Nevertheless, the operation of the first of the quantum rules is clearly shown by what has been described.

Radiation and the quantum laws

As it happened, the law that the vibrational energies increase by equal multiples of $h\nu$ had already been much more accurately, though perhaps less simply vindicated by Planck's study of radiation problems. We shall find it expedient to defer detailed discussion of radiation, but enough will be said here to indicate its place in the evolution of the quantum theory.

Matter absorbs and emits radiation, good absorbers being also good emitters, as shown by the fact that a blackened piece of metal glows more brightly when heated to a high temperature than a corresponding piece which has been polished. A cavity surrounded by matter at a given temperature reaches an equilibrium state and becomes filled with radiation of all wave-lengths (as can be shown by spectral analysis of what emerges from an opening in a furnace). The energy per unit volume and the distribution of energy among the various wave-lengths prove to be functions of temperature alone, and spectroscopic examination of the so-called black body radiation escaping through a small opening in an enclosure where thermal equilibrium prevails, reveals the character of this function, which is of great significance. The intensity passes through a maximum at a certain wave-length, and this maximum itself not only becomes higher but is displaced towards shorter wave-lengths as the temperature rises.

The theoretical treatment of the distribution problem is based upon principles analogous to those which determine the partition of molecules among energy states. That average condition is supposed to be realized in nature which can be achieved in the largest number of ways: where many possibilities exist, they contribute largely, where few exist they contribute little.

Radiation contained in an enclosure must satisfy certain relations analogous to those governing, for example, the modes of vibration of the air in an organ pipe, and the wave-lengths of the admissible components are governed by geometrical boundary conditions.

There are many more numerical possibilities for the accommodation of short waves than of long ones, so that the number of admissible frequencies in a given interval $d\nu$ increases rapidly with ν itself. If, then, energy distribution were governed solely by available modes, the intensity should increase continuously with frequency. There should be no maximum, and the actual existence of one reveals the operation of a second factor which discourages the location of energy in the shorter wave-lengths.

If now the quantum law is introduced, specifying that a vibration of frequency ν contains energy in quanta $h\nu$, then for high frequencies the total available energy can provide relatively few quanta only. If these were allocated, the number of ways would be limited. When the frequency is high, therefore, there are too few quanta, while when it is low there are too few states to which they can be allotted. The maximum variety of assignments is possible when an intermediate number of moderate-sized quanta are shared among the modes, that is to say, at a wave-length neither too long nor too short.

As the temperature rises the energy supply increases and more quanta of greater size become available for distribution. The maximum is displaced to higher frequencies. The law of this displacement is

$$\lambda_{max} T = \text{constant.}$$

It can be shown to define *precisely* the form of the dependence $\epsilon = h\nu$, and this same form also leads to the correct form of curve for the relation of intensity and wave-length at any given temperature. These results will be shown in greater detail later on when Planck's law is considered in the light of still further developments which make possible a derivation far more satisfactory than the original one.

The problem of rotations

The form of the quantum law which makes increments of vibrational energy proportional to frequency deals adequately with the requirements of the specific heat problem. The absence of vibrations in molecules such as O_2 at ordinary temperatures is explained by the tightness of binding of the atoms and the consequent high frequency, which corresponds to a quantum too great for appreciable occurrence. The specific heat-temperature relations of solids are accounted for as has been seen, the difficulties of detail which arise being connected simply with the determination of the true frequency-spectrum.

The question next arises how rotational energy states are to be specified. The proportionality of energy and frequency is here meaningless, since a rotation possesses no natural frequency, but merely one depending upon the energy, and which vanishes as this energy falls to zero.

The obvious course is to throw the formulation of the rule for vibrations into some form which does not involve the frequency explicitly, and to seek to generalize this rule in the modified version. A provisional solution of the problem on these lines was in fact soon found, in a way which came rather naturally to those versed in applied mathematics.

It had long been known that the laws of dynamics assume their simplest and most elegant form when the so-called Hamiltonian coordinates are employed as the fundamental variables. These are position coordinates, usually written q_1, q_2,..., on the one hand, and momentum coordinates, usually written p_1, p_2,..., on the other.

The law $\epsilon_n = nh\nu$ for a simple harmonic motion can be expressed in Hamiltonian coordinates in the form

$$\int_0^T p \, dq = nh,$$

where n is an integer, and $T = 1/\nu$, is the periodic time. This is the Sommerfeld–Wilson equation.

The identification of the two versions is easily made. A particle executing a simple harmonic motion follows the equation

$$x = a \sin 2\pi\nu t,$$

where x is the displacement at time t and a is a constant. In Hamiltonian coordinates

$$p = m\dot{x} = 2\pi\nu ma \cos 2\pi\nu t,$$
$$dq = dx = 2\pi\nu a \cos 2\pi\nu t \, dt,$$
$$\int_0^T p \, dq = 4\pi^2\nu^2 ma^2 \int_0^{1/\nu} \cos^2 2\pi\nu t \, dt = 2\pi^2\nu ma^2.$$

If then we write $\qquad 2\pi^2\nu ma^2 = nh,$

it is the same thing as to put

$$2\pi^2\nu^2 ma^2 = nh\nu,$$

and the expression on the left is none other than the energy. (For, total energy = K.E.+P.E. = constant: when P.E. = 0, K.E. = maximum: therefore, total energy = maximum value of K.E. = maximum value of $\frac{1}{2}m\dot{x}^2 = \frac{1}{2}m(2\pi\nu a)^2 = 2\pi^2\nu^2 ma^2$.)

The formula $\int_0^T p \, dq = nh$ can be applied immediately to a rotation.

Here, $\qquad\qquad p =$ angular momentum

and $\qquad\qquad dq = d\theta,$

where θ is an angular coordinate. For a complete rotation

$$\int_0^T p \, dq = \int_0^{2\pi} \text{ang. mom. } d\theta = nh,$$

whence angular momentum $= nh/2\pi$.

If the moment of inertia is I and the angular velocity ω,

$$I\omega = nh/2\pi,$$

whence the energy is given by

$$E = \frac{1}{2}I\omega^2 = n^2h^2/8\pi^2 I.$$

Here, it is to be noted, what increases by equal steps is not the *energy* but the *angular momentum*. This formula does in fact give satisfactory results in the discussion of rotational specific heats. But it does not yet tell the whole story.

The angular momentum rule found its most accurate and striking application in Bohr's interpretation of the hydrogen spectrum, where an analogous postulate was made about the permissible states of an electron circulating around a nucleus, and it may be regarded apart from a trivial correction as established.

Translational energy

The question of translational energy remains. To a superficial view this might appear less urgent than that of rotational and vibrational energy. There is no obvious need to account for any significant departure of the specific heats of monatomic gases from the classical values. Yet a discrete series of translational energies must be postulated if we are to retain the fundamental statistical principle.

If the translational distribution were infinitely fine-grained, and the rotational and vibrational ones were coarse-grained, the infinite increase in entropy which would accompany the passage of energy into the translational form would ensure that this process prevailed to the entire exclusion of the reverse change. Any prospect of defining absolute entropies and of calculating where a chemical equilibrium lies would vanish again. And, anyhow, it seems unlikely that one kind of motion, not differing on close analysis much from the others, should be exempt from what appears to be so fundamental a law of nature.

In the formulation of quantum rules the characteristic of the motion which always enters explicitly is its *periodicity*. The only sense in which a particle executing translational motion can be said to have a period is in relation to its impacts on the sides of a containing vessel. If the particle moves parallel to the x-axis, in a cubical box of side l_x, it repeats its motion each time it completes a path of length $2l_x$. We might try using this fact to determine a periodic time insertion in the Sommerfeld–Wilson equation.

$$\int_0^T p\, dq = \int_0^T mv_x\, dx = \int_0^{2l_x} mv_x\, dx = nh,$$

$$mv_x 2l_x = nh,$$

$$n\left(\frac{1}{2}\frac{h}{mv_x}\right) = l_x.$$

This gives a form of quantization for translational energy in one degree of freedom, which does in fact agree with that later formulated on the basis of a more general theory.

Wave mechanics

The more general theory of quantum states developed from certain surprising discoveries in physics. These could be summed up in the statement that light, first regarded by Newton as corpuscular and then for a long time believed to consist of some kind of wave motion, possesses a character which is both particulate and undulatory at the same time.

Absorption and emission of light, according to one view, occur in quanta of magnitude $h\nu$. Einstein regarded the quanta themselves as having some of the dynamic properties of particles. For such particles the term *photons* was introduced. By the theory of relativity, the mass and the energy of a particle are connected by the equation:

$$E = mc^2, \text{ where } c \text{ is the velocity of light (see p. 230).}$$

For a photon $\qquad\qquad mc^2 = h\nu$.

Thus $\qquad\qquad\qquad mc = h\nu/c$.

Since photons move with velocity c themselves, mc represents their momentum.

The photon theory achieves many brilliant results.

1. Calculation of the momentum reversal occurring when photons impinge upon a surface gives the pressure exerted by radiation, just as impact of gas molecules accounts for gas pressure. The results are in complete accord with experiment.

2. Consideration of the statistics of photons yields Planck's law of energy distribution in the simplest possible way (p. 155).

3. What is called the *Compton effect* is accounted for. When photons are scattered by matter their momentum is changed, presumably in accordance with the ordinary laws of impact. Since the momentum is $h\nu/c$, a calculable change of frequency is observed in the scattered radiation.

4. An otherwise very difficultly interpretable character of the *photo-electric effect* is explained. When ultra-violet light falls on a metal surface, electrons are emitted. The kinetic energy of these photo-electrons increases with the frequency of the light but is independent of the intensity (though the latter determines the *rate* of emission). The energy of the photon thus appears as though it were concentrated in packets which increase in size with the frequency, and which become more numerous, but no bigger, as the light intensity increases.

But the idea of photons does not dispense with the need for the wave theory of light, which is categorically demanded by the phenomenon of interference. Therefore, a great abnegation of naïve realism is imposed, and it has to be accepted that light behaves in accordance with rules which are unlike those describing the behaviour of bullets on the one hand or of the waves of the ocean on the other. Spatial distribution of light intensities (in interference, diffraction, and so on) follows undulatory laws: intimate interaction of light with matter seems to be governed by the photon properties which are something like those of bullets—though not by any means exactly like.

In the sense of the general thesis that the unknown is to be explained in terms of the known, it appears that the kinds of known things by which the unknown are to be interpreted have advanced a considerable stage in sophistication.

With the blurring of the distinction between waves and particles, the status of what have hitherto been accepted indisputably as particles becomes open to question once more. In a theoretical study L. de Broglie examined the conditions under which singularities in interfering trains of waves might be propagated according to the laws of moving mass points. His considerations led to the view that there could be important correspondences if the wave-length of the hypothetical waves and the momentum of the hypothetical particle were related by the equation

$$\lambda = h/mv.$$

This led rapidly to the discovery that beams of electrons are in fact subject to interference and that they behave in respect of this phenomenon as though an electron possesses, not indeed a constant wavelength, but one related to its momentum in precise accordance with the above equation.

If this result is extended to particles of macroscopic mass, the predicted wave-length is so small that the divergence from rectilinear propagation of a stream of such masses is quite negligible.

A mass moving with one degree of freedom in an enclosure of length l_x must, to be in a steady state, possess such a momentum that an integral number of half wave-lengths fit into l_x. Thus

$$n\lambda/2 = l_x \quad \text{or} \quad nh/2mv_x = l_x,$$

which is just the relation for the quantization of translational motion inferred from the tentative application of the Sommerfeld–Wilson equation.

Beams of electrons having many of the properties of minute masses (p. 164), but being capable of diffraction in a manner only describable by the equations of wave motion, the assumption of a general duality of behaviour is evidently worth exploring. Since a rule for the specification of possible translational states can be guessed from that which defines vibrational and rotational states, and since the same rule follows also from the ascription of a wave-length h/mv to the moving mass, it is fairly evident that all the consequences of the Sommerfeld–Wilson equation should be derivable from some form of wave theory.

Such a theory is embodied in the wave equation of Schrödinger. The propagation of a wave in three dimensions is represented by the expression

$$\frac{\partial^2\psi}{\partial x^2}+\frac{\partial^2\psi}{\partial y^2}+\frac{\partial^2\psi}{\partial z^2}+\frac{4\pi^2\psi}{\lambda^2}=0,$$

where ψ is the quantity which varies periodically in space and time.†

If for the wave-length is substituted the value h/mv, then this equation describes the motion of free particles of momentum mv, subject to interference after the manner of electrons.

The meaning of ψ will be considered more closely in a later section. At the present juncture it suffices to say that it is of the form

$$\psi = \psi_0\, e^{2\pi i \nu t},$$

† The following considerations show how this equation represents a wave.

$\psi = \psi_0 \sin 2\pi(x/\lambda - \nu t)$ is periodic in x when t is constant (instantaneous picture of a wave) and periodic in t when x is constant (each point vibrates). If the eye is kept fixed on a point such that $x/\lambda - \nu t = 0$, then ψ remains constant: thus the eye is following a disturbance which travels at a rate given by $x/t = \lambda\nu = v$. This is the characteristic of wave propagation along the x-axis.

Partial differentiation of the first equation gives

$$\frac{\partial^2\psi}{\partial x^2} = -\frac{4\pi^2}{\lambda^2}\,\psi.$$

For the three-dimensional problem the wave may be represented by the expression

$$\psi = \psi_0 \sin 2\pi\left\{\frac{(x^2+y^2+z^2)^{\frac{1}{2}}}{\lambda} - \nu t\right\}$$

$$= \psi_0 \sin R.$$

Then

$$\frac{\partial^2\psi}{\partial x^2} = -\frac{4\pi^2}{\lambda^2}\,\psi_0(\sin R)x^2(x^2+y^2+z^2)^{-1}+\frac{2\pi}{\lambda}\,\psi_0(\cos R)[(x^2+y^2+z^2)^{-\frac{1}{2}}-x^2(x^2+y^2+z^2)^{-\frac{3}{2}}]$$

$$\frac{\partial^2\psi}{\partial x^2}+\frac{\partial^2\psi}{\partial y^2}+\frac{\partial^2\psi}{\partial z^2} = -\frac{4\pi^2}{\lambda^2}\,\psi_0(\sin R)[(x^2+y^2+z^2)(x^2+y^2+z^2)^{-1}]+$$

$$+\frac{2\pi}{\lambda}\,\psi_0(\cos R)[(x^2+y^2+z^2)^{-\frac{1}{2}}-(x^2+y^2+z^2)^{-\frac{1}{2}}]$$

$$= -\frac{4\pi^2}{\lambda^2}\,\psi_0\sin R = -\frac{4\pi^2}{\lambda^2}\,\psi.$$

where ψ_0 is a function of the spatial coordinates only, the periodic variation with time being represented in the conventional way by the term $e^{2\pi i \nu t}$.† ν is the frequency, and i has the usual meaning of $\sqrt{(-1)}$. If one writes

$$\bar{\psi} = \psi_0 e^{-2\pi i \nu t}, \quad \text{then} \quad \psi\bar{\psi} = \psi_0^2.$$

ψ_0^2 is the square of a wave amplitude which in any wave phenomenon expresses the intensity. Thus ψ_0^2 is the density of distribution of particles in an interference experiment, or the probability of finding a given particle in the region specified by the spatial coordinates of ψ. For the present we only need to bear in mind that if $\psi = 0$ there are no particles.

If masses are contained in an enclosure, then solutions of the equation are only possible for certain integral relations between the wave-length and the linear dimensions of the container itself. For example, if the motion is confined to the x-axis, then

$$\frac{\partial^2 \psi}{\partial x^2} = -\frac{4\pi^2}{\lambda^2} \psi, \quad \text{whence} \quad \psi = A \sin \frac{2\pi x}{\lambda},$$

where A is a constant. Since there is no particle outside the container, ψ must be zero when $x = 0$ and when $x = l_x$. The value of $\sin(2\pi x/\lambda)$ is zero whenever $2x/\lambda$ is integral, that is when $2l_x/\lambda = n$ or $n\lambda/2 = l_x$: thus the permitted values of the wave-length and of the momentum are defined.

The relations are not quite so easily understandable for vibrating and rotating systems, but Schrödinger made the remarkable discovery that the wave equation is applicable quite generally if handled according to the following prescription.

From the total energy E of the particle under study is subtracted the potential energy U which it may possess in virtue of its presence in any field of force. The balance $E-U$ is the kinetic energy, which is $\frac{1}{2}mv^2$. Thus

$$m^2v^2 = 2m(E-U),$$

and since

$$1/\lambda^2 = m^2v^2/h^2 = 2m(E-U)/h^2,$$

---

† The appropriateness of the form $\psi = \psi_0 e^{2\pi i \nu t}$ to represent the time variation is seen by differentiation.

$$\frac{\partial^2 \psi}{\partial t^2} = -4\pi^2\nu^2\psi,$$

or

$$\frac{\partial^2 \psi}{\partial t^2} + a\psi = 0.$$

This is the simplest representation by a differential equation of a quantity vibrating in time.

it follows from the general equation that

$$\frac{\partial^2 \psi}{\partial x^2}+\frac{\partial^2 \psi}{\partial y^2}+\frac{\partial^2 \psi}{\partial z^2}+\frac{8\pi^2 m(E-U)}{h^2}\,\psi = 0.$$

This semi-empirical equation has become famous in virtue of the many remarkable properties which it possesses. Only certain of these concern us at the moment. The first set is connected with the possibilities of solution. The differential equation only possesses finite, single-valued solutions for certain quite definite values of the energy, E. These values, known as characteristic or proper values (German, *Eigenwerte*), specify the possible quantum states of the system.

The quantization of translational energy has already been considered. For vibrational systems the equation is found to yield physically admissible solutions only for values of E defined by the relation $E = (n+\frac{1}{2})h\nu$. The successive energy levels differ by $h\nu$ as required. The lowest value occurs when the integer n is zero, so that $E_0 = \frac{1}{2}h\nu$. Schrödinger's equation, unlike the quantum rule which it has superseded, predicts the existence of a so-called *zero-point energy*. The assumption that there is such a thing is in fact required for the explanation of certain phenomena, so that in this respect the new equation possesses an important advantage.

For a rotating system with a moment of inertia I, the permitted energy states are given by the relation

$$E = \frac{n(n+1)h^2}{8\pi^2 I}.$$

The previous rule was expressed by the equation

$$I\omega = \frac{nh}{2\pi}, \quad \text{so that} \quad E = \frac{1}{2}I\omega^2 = \frac{n^2 h^2}{8\pi^2 I}.$$

The new one replaces n^2 by $n(n+1)$. We may write

$$E = \frac{n(n+1)h^2}{8\pi^2 I} = \frac{(n+\frac{1}{2})^2 h^2}{8\pi^2 I} + \text{constant}.$$

Thus the succession of states according to the Schrödinger rules is governed by the values of $(n+\frac{1}{2})^2$ instead of n^2. In the interpretation of certain spectroscopic phenomena this also proves to be an essential emendation.

Before proceeding we may illustrate the manner in which the specification of the quantum states occurs.

Succession of vibrational states

For a simple harmonic motion the acceleration is related to the displacement from the equilibrium position by the equation

$$\ddot{x} = -\mu x,$$

so that
$$U = m \int_0^x \mu x \, dx = m\mu x^2/2.$$

Also
$$x = A \sin \sqrt{\mu} t = A \sin 2\pi\nu t,$$

so that
$$\mu = 4\pi^2\nu^2 \quad \text{and} \quad U = 2\pi^2\nu^2 m x^2.$$

The Schrödinger equation thus becomes

$$\frac{\partial^2\psi}{\partial x^2} + \frac{8\pi^2 m}{h^2}(E - 2\pi^2\nu^2 m x^2)\psi = 0. \tag{1}$$

This may be thrown into the form

$$\frac{\partial^2\psi}{\partial X^2} + (a - X^2)\psi = 0 \tag{2}$$

by the substitutions

$$X^2 = \frac{4\pi^2 m\nu}{h} x^2 \quad \text{and} \quad a = \frac{2E}{h\nu}.$$

The equation (2) has finite, single-valued solutions only for values of a which are of the form $(2n+1)$, n being a positive integer. Thus, for acceptable solutions,

$$2E/h\nu = (2n+1),$$

or
$$E = (2n+1)\frac{h\nu}{2} = (n+\tfrac{1}{2})h\nu.$$

The method of proving these statements belongs to the standard theory of differential equations and will be quoted as an example.

In (2) let
$$\psi = e^{-\frac{1}{2}X^2} v,$$

where v is an appropriate function. Then

$$\frac{\partial\psi}{\partial X} = v'e^{-\frac{1}{2}X^2} - ve^{-\frac{1}{2}X^2}X,$$

$$\frac{\partial^2\psi}{\partial X^2} = v''e^{-\frac{1}{2}X^2} - v'e^{-\frac{1}{2}X^2}X - v'e^{-\frac{1}{2}X^2}X - v(e^{-\frac{1}{2}X^2} - X^2e^{-\frac{1}{2}X^2})$$

$$= e^{-\frac{1}{2}X^2}(v'' - 2v'X - v + X^2v),$$

where v' and v'' are the differential coefficients of v with respect to X. Substitution in (2) and division by $e^{-\frac{1}{2}X^2}$ gives

$$(v''-2v'X-v+X^2v)+(a-X^2)v = 0,$$

or $\qquad\qquad v''-2v'X+(a-1)v = 0.$ \hfill (3)

Now, further, let v be represented by the general power series

$$v = \sum b_n X^n. \qquad\qquad (4)$$

Substitution gives for the coefficient of X^n in (3) the values

$$(n+2)(n+1)b_{n+2}-2nb_n+(a-1)b_n.$$

(It will be seen that in v'', for example, the term in X^n has been derived by differentiation twice of the term in X^{n+2}.) For (3) to be general the coefficient of X^n, and that of any other power of X in it, must equal zero, so that we have

$$\frac{b_{n+2}}{b_n} = -\frac{(a-1-2n)}{(n+2)(n+1)}.$$

If v is to constitute a finite series, the coefficient of some power of X in (4) must become zero, so that

$$a-1-2n = 0,$$

or $\qquad\qquad 2n+1 = a,$

which establishes the required result.

Rotational states

The general case of the rotator free to move in three dimensions is more complicated, but is treated according to similar principles. The Schrödinger equation is first expressed in spherical polar coordinates, r, θ, and ϕ. For the rotation of a rigid body about its centre of gravity, r is constant and is included in a term representing the moment of inertia, I. The conditions for physically admissible solutions lead to the result

$$E = \frac{h^2}{8\pi^2 I} n(n+1),$$

where n is a positive integer.

A new factor enters here. The function ψ itself begins to assume considerable direct importance. For each admissible value of E there proves to be not a single value of ψ, but a *set of $2n+1$ values*. This is expressed by saying that the *statistical weight* of the nth energy level is $2n+1$, and, in accordance with the general principle, it is supposed that the population of this level will be correspondingly dense.

The factor $(2n+1)$ comes, in this present mode of calculation, from the conditions for solution of the differential equation. In the more primitive form of the quantum theory it may be arrived at by considerations which, if less general and indeed in a certain sense less precise, are easier to relate to a more naïve picture of molecular events. The body is imagined to rotate on its axis, and the axis is conceived to possess not an indefinite number but a finite number of possible *spatial orientations*. n is, in fact, supposed to represent a vector the projection of which on a given axis must also be a whole number. This projection may, according to the angle between it and the vector, possess any value from $+n$ to $-n$, including zero (when the vector and the line of projection are perpendicular). Thus there are $2n+1$ possible values altogether.

The orientation does not normally affect the energy, so that the nth rotational level is, as it is called, a $(2n+1)$-fold *degenerate* one.

Degenerate states: statistical weight

The dependence of the whole statistical theory upon the idea that natural phenomena are largely determined by the possible ways of filling molecular states has already been abundantly illustrated. It is evident, therefore, that if several states of equal energy exist, they really should be regarded as multiple. Their availability does depend upon their number and not upon the accident that they are associated with energies which are quantitatively the same. The case of the orientations of the molecular axis was specially simple to visualize, since one can see clearly that if there are many possible orientations, then the chance that molecules possess motions which avail themselves of this freedom is correspondingly greater.

The example is important, since it shows also that the number of solutions of the ψ equation, which correspond to a given value of the permitted energy value E, is the expression for the statistical weight in the wave-mechanical formulation.

Conclusion

The need for defining the range of individual energy states arose as a logical necessity in the consideration of statistical problems. The discoveries which have been made in the study of specific heats, radiation, and various other matters have shown how the range can be defined.

The formulation of the rules has gone through various phases, the most comprehensive statement being that based upon the wave equation. This expresses the possible energy levels for any kind of system, and provides information about their multiplicity. The equation is not itself based upon any explicit theory of the nature of things, except in so far as it contains a general implication of the wave-particle duality in systems of minute enough dimensions.

The duality referred to, while cutting us off completely from the possibility of describing the invisible in terms of the visible, has the great simplicity that translational motion becomes subject in a not wholly unexpected way to the quantum rules. All kinds of molecular states fall, as a result, into discrete series, and the calculation of absolute probabilities acquires a meaning. There is thus a prospect of answering the fundamental question as to what determines the forms, physical and chemical, into which atoms and molecules eventually settle down.

VII

THE ABSOLUTE POSITION OF EQUILIBRIA

Further statistical considerations

THE great principle which governs the domain of equilibria and which so largely determines the nature of things is that states of equivalent energy are occupied in proportion to the accommodation they afford. Before, however, chemical and physical equilibria can be calculated from the fundamental energy data there is still an important problem to be solved about the assessment of this accommodation.

According to the principle which has already been invoked, the number of ways in which N molecules can be assigned to a series of energy levels, so that the occupation numbers are N_1, N_2,..., respectively is given by

$$W = \frac{N!}{N_1! \, N_2! \dots}.$$

The entropy, which we shall now write more specifically as $S_{\text{Boltzmann}}$, or S_B is given by

$$S_B = k \ln W.$$

Replacement of the factorials in W by Stirling's approximation, $\ln N! = N \ln N - N$, gives

$$S_B = k(N \ln N - \sum N_1 \ln N_1).$$

The condition that W shall be a maximum subject to the two conditions

$$\sum N_1 = N \quad \text{and} \quad \sum N_1 \epsilon_1 = E$$

gives

$$N_1 = \frac{N e^{-\epsilon_1/kT}}{\sum e^{-\epsilon_1/kT}} = \frac{N e^{-\epsilon_1/kT}}{f} \quad \text{(see p. 30).}$$

(see p. 30).

Substitution of this value for N_1 leads to the expression

$$S_B = kN \ln f + kNT \, d\ln f/dT.$$

In making the substitution it is to be noted that since

$$f = \sum e^{-\epsilon_1/kT}, \qquad \frac{df}{dT} = \sum \frac{\epsilon_1 e^{-\epsilon_1/kT}}{kT^2},$$

so that

$$\sum \epsilon_1 e^{-\epsilon_1/kT} = kT^2 \, df/dT.$$

Thus

$$E = \sum N_1 \epsilon_1 = \sum \frac{\epsilon_1 N e^{-\epsilon_1/kT}}{f} = kNT^2 \frac{d\ln f}{dT}.$$

The entropy, total energy, and free energy are thus given by

$$S_B = kN \ln f + kNT d\ln f/dT,$$

$$E = NkT^2 d\ln f/dT,$$

$$F_B = E - TS_B = -kNT \ln f.$$

The sequence of quantum states being known, f can be calculated in absolute terms. S_B, E, and F_B are thus known.

Although the resulting formulae are correct enough in so far as they are applied to the calculation of *changes* of entropy or free energy, for example, in consideration of shifts of chemical equilibrium with temperature, they still prove to be wrong in absolute magnitude.

The error lies not in the calculation of f, nor in the identification of S with $k \ln W$, but in the way in which W itself is computed.

What is called the Boltzmann statistics must be replaced by a new form which is naturally enough called quantum statistics. This new method of computation leaves the form of S unchanged, but makes a considerable difference to the constant term in the entropy formula, that is, to the absolute value of the entropy. It will be considered in more detail in the following section.

Before proceeding to investigate the need for a change in the definition of the probability, we shall find it convenient first to formulate some values of f, the partition function, in various simple cases.

The total partition function is represented by

$$f = \sum e^{-\epsilon/kT},$$

the sum being taken over all possible states.

In a given state the vibrational, rotational, and translational energies can often be regarded in a sufficient degree of approximation as independent, so that

$$\epsilon = \epsilon_V + \epsilon_R + \epsilon_T.$$

Moreover, each kind itself is represented by a whole series of states, and partial partition functions may be defined as follows:

$$f_V = \sum e^{-\epsilon_V/kT}, \quad f_R = \sum e^{-\epsilon_R/kT}, \quad f_T = \sum e^{-\epsilon_T/kT}.$$

The total partition function, f, is simply the product of the factors

f_V, f_R, and f_T, as may be seen by inspection of the following formula, in which, for brevity, $1/kT$ is written β.

$$\begin{aligned}
f_V f_R f_T &= (e^{-\beta \epsilon_{V_1}} + e^{-\beta \epsilon_{V_2}} + \ldots)(e^{-\beta \epsilon_{R_1}} + e^{-\beta \epsilon_{R_2}} + \ldots)(e^{-\beta \epsilon_{T_1}} + e^{-\beta \epsilon_{T_2}} + \ldots) \\
&= e^{-\beta \epsilon_{V_1}} e^{-\beta \epsilon_{R_1}} e^{-\beta \epsilon_{T_1}} + e^{-\beta \epsilon_{V_2}} e^{-\beta \epsilon_{R_1}} e^{-\beta \epsilon_{T_1}} + \ldots \\
&= e^{-\beta(\epsilon_{V_1} + \epsilon_{R_1} + \epsilon_{T_1})} + e^{-\beta(\epsilon_{V_2} + \epsilon_{R_1} + \epsilon_{T_1})} + \ldots.
\end{aligned}$$

The quantities multiplying β in the final series represent every combination of every possible value of the energies in the partial series. They therefore include every possible state. The final sum is therefore, by definition, f.

By an analogous argument the translational partition function itself may be split into three factors, each representing the contribution of a single degree of freedom. Thus

$$f_{T_{xyz}} = f_{T_x} f_{T_y} f_{T_z}.$$

For the series of energy levels characteristic of a single vibrational degree of freedom $\epsilon = (n + \frac{1}{2})h\nu$. The absolute zero may be taken as the energy reference point, so that the successive energies on the new scale are $nh\nu$ and thus

$$f_V = \sum_0^\infty e^{-nh\nu/kT},$$

a geometrical series of which the sum is

$$f_V = \frac{1}{1 - e^{-h\nu/kT}}.$$

The number of molecules in the nth and higher states, that is, the number with energy greater than ϵ_n is given by

$$\begin{aligned}
\frac{Ne^{-nh\nu/kT}}{f_V} + \frac{Ne^{-(n+1)h\nu/kT}}{f_V} + \ldots &= \frac{Ne^{-nh\nu/kT}}{f_V}(1 + e^{-h\nu/kT} + \ldots) \\
&= \frac{Ne^{-nh\nu/kT}}{f_V} f_V \\
&= Ne^{-\epsilon_n/kT}.
\end{aligned}$$

The correct treatment of rotations is somewhat difficult and gives rise to series which have no simply expressible sum. For a rigid rotator which is a solid of revolution of moment of inertia I and which, for reasons which will emerge later, must not consist of two identical atoms, the following simple treatment is possible.

The successive energy levels are represented by the formula

$$\epsilon_n = n(n+1)h^2/8\pi^2 I.$$

For the nth level there are $(2n+1)$ possibilities corresponding to $(2n+1)$ orientations of the axis. Thus

$$f_R = \sum (2n+1)e^{-n(n+1)h^2/8\pi^2 IkT}.$$

If the levels are fairly close and n is considerable, this sum may be expressed approximately by an integral

$$\int_0^\infty (2n+1)e^{-An(n+1)}dn \quad \text{where} \quad A = h^2/8\pi^2 IkT.$$

Let $n^2+n = x$, then the integral becomes

$$\int_0^\infty e^{-Ax}\,dx = 1/A.$$

Thus for this simple case

$$f_R = 8\pi^2 IkT/h^2.$$

Translational states are defined by the equation

$$nh/2mv_x = l_x \quad \text{so that} \quad \tfrac{1}{2}mv_x^2 = n^2h^2/8ml_x^2.$$

The unidimensional partition function is given by

$$f_{T_x} = \sum e^{-n^2h^2/8ml_x^2 kT}.$$

Since the energy steps are small the sum can be fairly well represented by an integral

$$\int_0^\infty e^{-B^2n^2}dn = \sqrt{\pi}/2B.$$

Thus

$$f_{T_x} = \frac{(2\pi mkT)^{\frac{1}{2}}l_x}{h}.$$

For the three degrees of freedom we have

$$f_{T_{xyz}} = \frac{(2\pi mkT)^{\frac{3}{2}}}{h^3} l_x l_y l_z = \frac{(2\pi mkT)^{\frac{3}{2}}V}{h^3},$$

where V is the volume.

The translational partition function unavoidably involves the volume, since only by relating the motion of the particle to the dimensions of the container in which it moves can any formulation of the quantum states be reached. This necessary connexion exists both in the older quantum theory and in the wave mechanical theory which replaced it.

For future reference it will be convenient here to tabulate the partition functions so far calculated.

Vibrational energy in one degree of freedom

$$f_V = 1/(1 - e^{-h\nu/kT}).$$

Rotational energy, rigid rotator, solid of revolution

$$f_R = 8\pi^2 IkT/h^2.$$

(This is a two-dimensional case.)

Translational energy in three degrees of freedom

$$f_{T_{xyz}} = \frac{(2\pi mkT)^{\frac{3}{2}}V}{h^3}.$$

The use of partition functions of the type just derived, or in appropriate cases more complex ones, in the calculation of absolute entropies, leads to incorrect results. The fault lies, as has been stated, with the mode of definition of the probability. The question of a modified definition and a reconsideration of what constitutes the number of assignments to states must now be considered.

Quantum statistics

The treatment of particles by the principles of wave mechanics results in a complete blurring of their identity as individuals. The calculation of statistical probability is profoundly modified thereby and a reformulation of Boltzmann's rules for calculating entropy becomes necessary.

So long as molecules were conceived as distinguishable units, the number of ways in which a given distribution over a series of energy states may be realized could be given by the expression

$$\frac{N!}{N_1! \, N_2! \dots},$$

where N_1 is the number in the first state, N_2 the number in the second, and so on. When, however, identity is lost, the permutations of individuals within a given state becomes meaningless, and this formula becomes unsuitable as a measure of the probability. It may, however, be replaced by another which defines the possible numerical types of assignment of molecules to various states, no distinguishable characteristics being attributed to individuals.

The reason for this renunciation lies in the wave-like nature ascribed to the particle, which is defined as regards its quantum state by its relation to the whole of the vessel in which it is located. This applies to all particles which, whatever they may possess of definite localization in respect of phenomena such as molecular collisions, have lost it in respect of the application of the quantum rules. Therefore there is, at any rate for the purposes of this problem, no way of distinguishing them, and the operation of working out permutations of individuals in a sort of box loses its sense. There will be further opportunity of pondering on this principle as the subject develops.

In the new formulation different assignments are represented by the statements that there are N_1 molecules in state 1, N_2 in state 2, and so on, on the one hand, or, on the other hand, that there are N'_1 in state 1, N'_2 in state 2, and so on. But it makes no difference how for these two assignments, N_1, N_2,..., or N'_1, N'_2,..., are selected and which individuals constitute these groups.

The detailed calculations proceed along lines which differ in the earlier stages from those followed previously. In each possible state the energy of a molecule is determined by the sum of translational, rotational, and vibrational contributions. The translational quantization involves three-dimensional space coordinates as well as momentum coordinates and leads to the existence of very large numbers of closely spaced levels, many of the states being in fact of equal energy. Thus there will be in general g_j states of energy ϵ_j, and the N_j molecules which possess this energy will be shared among the g_j states. The problem is to know how many numerical types of allocation there are when N_j molecules are distributed among g_j states, permutations among the individuals being not only undetectable but assumed to be of no interest, and even meaningless.

The problem is that of sharing N_j objects among g_j boxes. To solve it we consider this procedure: place one box on the left, and then to the right of it in line place in any order the N_j objects and the remaining g_j-1 boxes. There are $(N_j+g_j-1)!$ orders possible. For any one arrangement, sweep up all the sets of objects and place each set in the box immediately to its left. This experiment may be done in any of the $(N_j+g_j-1)!$ ways. But although variations in the number of objects in a box are of interest, the orders of them among themselves, as of the boxes among themselves, are irrelevant. The total number of significant results of the experiment therefore is

$$(N_j+g_j-1)!/(N_j!)(g_j-1)!$$

Thus we have g_j states of energy ϵ_j with N_j molecules, g_k states of energy ϵ_k with N_k molecules, and so on. The combined probability of the complete distribution is given by

$$P = \frac{(N_1+g_1-1)!}{N_1!(g_1-1)!} \cdot \frac{(N_2+g_2-1)!}{N_2!(g_2-1)!} \cdots$$

Neglecting unity in comparison with $N_1, \ldots$ and applying Stirling's approximation, we obtain

$$\ln P = \sum_j [(N_j+g_j)\ln(N_j+g_j) - N_j \ln N_j - g_j \ln g_j].$$

The entropy may now be defined as

$$S = k \ln P.$$

For the state of greatest probability we now have the relations

$$\delta \ln P = 0, \qquad \sum \delta N_j = 0, \qquad \sum \epsilon_j \delta N_j = 0.$$

These are combined by the method explained on p. 29 and the rest of the calculation proceeds as before, yielding the result

$$N_j = \frac{g_j}{e^\alpha e^{\beta \epsilon_j} - 1}.$$

This differs from the Boltzmann expression only in the presence of the -1 in the denominator, and for low densities becomes indistinguishable from it.

We now proceed to a direct comparison of the two types of expression for the entropy itself.

That already derived is of the form

$$k \ln W = k(N \ln N - \sum N_1 \ln N_1).$$

This, however, is not based upon quite the same distribution as that we have just envisaged in calculating P. The number of molecules in each possible state is specified in the value of W, states which happen to be of equal energy being considered separately. In the expression for P all the molecules with energy ϵ_j are taken together. What must be compared with P is not the W previously calculated but W', a larger quantity. For W' we have in fact instead of

$$\frac{N!}{N_1! N_2! \ldots} \quad \text{the expression} \quad N! \frac{g_1^{N_1}}{N_1!} \frac{g_2^{N_2}}{N_2!} \cdots,$$

since each of the N_1 molecules with energy ϵ_1 has a choice of g_1 subcompartments, these extra options giving rise to $g_1^{N_1}$ fresh possibilities.

When N_j and g_j are large

$$\ln W' = N \ln N + \sum N_j \ln g_j - \sum N_j \ln N_j.$$

We may now compare $k \ln P$ with $k \ln W'$. In general, they are different in form and in magnitude. But in one important special case they approximate closely in form, while remaining of quite different absolute value. If the density is low, that is for gases, g_j is much greater than N_j, the number of translational states being enormous because of the fine-grained nature of the quantization. When $g_j \gg N_j$

$$\ln P = \sum (N_j + g_j) \ln g_j - \sum N_j \ln N_j - \sum g_j \ln g_j$$
$$= \sum N_j \ln g_j + \sum g_j \ln g_j - \sum N_j \ln N_j - \sum g_j \ln g_j$$
$$= \sum N_j \ln g_j - \sum N_j \ln N_j.$$

Thus
$$\ln W' = \ln P + N \ln N,$$
$$k \ln W' = k \ln P + k N \ln N,$$

or
$$S_{\text{Boltzmann}} = S_{\text{quantum}} + k \ln(N!).$$

This result is important. It shows that, for gases at least, the newly defined entropy possesses all the properties of the old except that its absolute magnitude is different.

When this absolute value is required, it may be obtained from the Boltzmann entropy by subtraction of $k \ln(N!)$. It is lower in so far as it is based upon neglect of distinctions which a non-quantum theory might have regarded as valid.

All previous conclusions about equilibria in gases and in dilute solutions retain their applicability, but the convention now adopted will affect assessments of the absolute entropy. The distribution law itself is not appreciably affected. When the density is such that g can no longer be regarded as large compared with N further complications enter which will not be dealt with at this stage.

Absolute calculation of equilibria

With expressions for partition functions appropriate to various special cases, the calculation of absolute values of equilibrium constants becomes possible. The most important step in creating this possibility was, of course, the defining of what constitutes a molecular state. For some purposes the more correct idea of absolute probability is also essential.

Four examples of major importance will cover in principle a large part of the physico-chemical nature of things. They are:

1. The integration constant of the vapour-pressure equation.
2. The transition temperature for a reaction taking place in a condensed phase: for example, an allotropic change, or the melting of a solid phase.
3. The equilibrium constant of a chemical equilibrium in the gas. phase.
4. The equilibrium of radiation and matter.

The first shows in the simplest way how a balance is established between order and chaos when molecules endowed with a primeval tendency to roam all space are impelled by attractive forces to agglomerate. The second and third illustrate the subtle interplay of factors which governs the transformation from one possible atomic pattern to another; and the fourth defines the sharing of energy between atomic and molecular systems and what was once called the ether, but later came to be regarded, for some purposes, as space containing photons.

As a first step we proceed to the calculation of the absolute entropy of a perfect gas.

Absolute entropy of a monatomic gas

$$S_{\text{Boltzmann}} = kN \ln f + kNT \, d\ln f/dT, \tag{1}$$

where
$$f = \frac{(2\pi mkT)^{\frac{3}{2}} V}{h^3}, \tag{2}$$

since there is no energy to be considered except that of the translational motion.
$$S_{\text{quantum}} = S_{\text{Boltzmann}} - k \ln(N!) \tag{3}$$

$$pV = kNT. \tag{4}$$

Insertion of (1), (2), and (4) into (3) gives

$$S_{\text{quantum}} = \tfrac{5}{2}kN \ln T - kN \ln p + kN \ln \frac{(2\pi m)^{\frac{3}{2}} k^{\frac{5}{2}} e^{\frac{5}{2}}}{h^3}.$$

All the quantities in this expression are known, so that the entropy is calculable in absolute measure.

Absolute calculation of vapour pressures

What is perhaps the first and simplest task to attempt in the light of knowledge of the absolute entropy is the calculation of the equilibrium between the gaseous and the solid phase for a monatomic substance. This equilibrium expresses the first departure of matter from the state of random dispersion and the first emergence of the forms of things we meet in the ordinary world.

For the equilibrium of solid and vapour,

$$G_1 = G_2 \quad \text{(p. 71)},$$

where G_1 is the free energy for a gram molecule of the solid and G_2 that for one of vapour.

$$G_1 - H_1 = T\frac{\partial G_1}{\partial T},$$

whence
$$G_1 = -T\int_0^T \frac{H_1}{T^2}\,dT + TJ \quad \text{(see p. 101).}$$

Since
$$\frac{\partial G}{\partial T} = -S,$$

$$\frac{\partial G_1}{\partial T} = -S_1 = J \quad \text{when } T = 0$$

(see also later, p. 145).

With a condensed phase, it is by no means unreasonable to assume that at the absolute zero $S_1 = 0$. With the disappearance of molecular identities there will be only one way in which particles can be assigned to their lowest state, unless, for some special reason connected with geometrical or other molecular characteristics, the ground level can be regarded as multiple.

With this reservation, to which further reference will be made, J may be set equal to zero.

Then
$$G_1 = -T\int_0^T \frac{H_1}{T^2}\,dT.$$

For the vapour
$$G_2 = H_2 - TS_2,$$

where
$$S_2 = \tfrac{5}{2}kN\ln T - kN\ln p + kN\ln\frac{(2\pi m)^{\frac{3}{2}}k^{\frac{5}{2}}e^{\frac{5}{2}}}{h^3} \quad \text{(p. 139).}$$

When $G_1 = G_2$, $p = \pi$, the vapour pressure. Thus

$$-T\int_0^T \frac{H_1}{T^2}\,dT = H_2 - \tfrac{5}{2}kNT\ln T + kNT\ln\pi - kNT\ln\frac{(2\pi m)^{\frac{3}{2}}k^{\frac{5}{2}}e^{\frac{5}{2}}}{h^3}.$$

But
$$H_2 = U_2 + p_2 V_2 = \tfrac{3}{2}kNT + kNT = \tfrac{5}{2}kNT,$$

and therefore

$$\ln \pi = -\int_0^T \frac{H_1}{kNT^2}\,dT - \tfrac{5}{2} + \tfrac{5}{2}\ln T + \ln \frac{(2\pi m)^{\frac{3}{2}}k^{\frac{5}{2}}e^{\frac{5}{2}}}{h^3}$$

$$= -\int_0^T \frac{H_1\,dT}{kNT^2} + \tfrac{5}{2}\ln T + \ln \frac{(2\pi m)^{\frac{3}{2}}k^{\frac{5}{2}}}{h^3}, \quad \text{since} \quad \ln e^{\frac{5}{2}} = \tfrac{5}{2},$$

or if we write

$$\ln \pi = \phi(T) + i,$$

$$i = \ln \frac{(2\pi m)^{\frac{3}{2}}k^{\frac{5}{2}}}{h^3}.$$

This gives the absolute value of the integration constant of the vapour-pressure equation

$$\ln \pi = \int \frac{\lambda}{RT^2}\,dT + i.$$

For monatomic substances, then, the vapour pressure is determined by λ and by i, where the latter appears from this discussion to be a function only of the atomic weight and of universal constants.

Calculated and observed values for various monatomic vapours are in general agreement and sometimes correspond very closely.

A certain reservation must be made about the possible multiplicity of the lowest states of the molecules or atoms. If this is taken into account, the formula becomes

$$i = \ln \frac{(2\pi m)^{\frac{3}{2}}k^{\frac{5}{2}}}{h^3}\frac{g_0}{g_c},$$

where g_0 is the so-called statistical weight of the ground level in the gas and g_c that of the ground level in the crystal.

An analogous discussion for diatomic molecules leads to the formula

$$i = \ln \frac{(2\pi m)^{\frac{3}{2}}k^{\frac{7}{2}}}{h^5}\frac{8\pi^2 I}{\sigma}\frac{g_0}{g_c},$$

where I is the moment of inertia of the molecule and σ is the symmetry number, a geometrical characteristic which determines what orientations count as separate states.

Factors determining the balance of solid and vapour

It is now possible to form a general picture of the factors which determine the tendency of a given substance to exist as solid or to escape as vapour. The equation for the vapour pressure is

$$\frac{d\ln\pi}{dT} = \frac{\lambda}{RT^2} = \frac{\lambda_0 + \int \Delta C_p\, dT}{RT^2},$$

i.e.
$$\ln\pi = -\frac{\lambda_0}{RT} + \int \frac{\left(\int \Delta C_p\, dT\right)}{RT^2}\, dT + i.$$

λ_0 measures the potential energy of attraction. The greater this attraction, the lower the vapour pressure. Neither thermodynamics nor statistics vouchsafes any information about the magnitude of λ_0.

ΔC_p may be written in the form $\alpha + \beta T + ...$, and its contribution to the second term on the right of the above equation depends upon the signs of α and β. As long as ΔC_p is positive λ increases, since $d\lambda/dT = \Delta C_p$. The energy of the vapour is then increasing more rapidly than that of the solid, and thus still more energy runs down when condensation occurs. But other influences are at work: consider, for example, a negative value of β. A negative β means that C_p of the vapour increases less rapidly than that of the solid. Increase of C_p means that fresh degrees of freedom come into operation (or that existing ones come more fully into operation). If this happens less rapidly in the vapour than in the solid then, from the point of view of entropy, the solid is gaining in disorder relatively to the vapour. Additional possibilities for existence in the solid state are thus opening up, and the further rise in vapour pressure with increasing temperature is antagonized. Some of the facilities previously offered by vapour only are now offered by existence in the solid state.

An increase of energy of the solid relatively to the vapour has thus two opposing effects. In so far as it represents an increasing facility of escape from the attractive forces it favours evaporation: in so far as it leads to a more disordered state in the solid it favours condensation, or at any rate lessens the tendency to evaporate.

The influence of the constant term i is connected purely with the entropy. If i becomes larger, the vapour pressure becomes greater also. Scrutiny of the formulae reveals several points of interest. As the mass of the molecules or atoms increases, the vapour-pressure constant increases also. This is because a greater mass leads to a more fine-grained quantization of the translational energy in the gas phase. There are thus relatively more possibilities of existence as vapour for the heavier than for the lighter particles. The effect of this factor itself is to favour the existence of heavier molecules in the gaseous state. The fact that the vapour pressure itself usually *decreases* with increasing mass is due to the influence of λ, acting in strong opposition to that of i.

A large ratio g_c/g_0 lowers i, and thus operates in the direction of diminishing the vapour pressure. If there are multiple ground states in the crystal which are not possible in the gas, then the solid phase is thereby favoured. But the question of these relative statistical weights is not very well understood.

For diatomic molecules and polyatomic molecules generally, the moment of inertia appears in the integration constant of the vapour-pressure equation, because the existence of rotations in the gas, while there are usually none in the solid, favours the evaporation of molecules by offering more possibilities of distribution in the gas phase. The greater the moment of inertia, the smaller the rotational quantum and the more numerous the levels. Hence the increase of i with I revealed in the formula.

This whole subject illustrates in a very interesting way how what might metaphorically be called the tendency of matter to achieve self-realization in the dynamical sense opposes the discipline of the attractive forces and determines the partition between different states of aggregation. Quite a vivid mechanical picture can be formed of this conflict, though it must be remembered that the principles underlying the dynamical behaviour are in fact highly abstract.

Transition temperatures

The equilibrium between two solid phases, whether the same in chemical composition or not, is governed by the condition

$$\Delta G = 0.$$

As shown on p. 101,

$$\frac{\Delta G}{T} = - \int \frac{\Delta H}{T^2}\, dT + J.$$

ΔH may be expressed as a power series in the following manner:

$$\Delta H = \Delta H_0 + aT + bT^2 + ...,$$

$$\frac{\partial(\Delta H)}{\partial T} = \Delta C_p = a + 2bT + ...$$

Since at the absolute zero the specific heats of solid phases vanish, a must be zero. Thus

$$\Delta G = -T \int \frac{\Delta H_0 + bT^2 + \cdots}{T^2}\, dT + TJ$$

$$= \Delta H_0 - bT^2 + \cdots + TJ,$$

$$\frac{\partial(\Delta G)}{\partial T} = -2bT + \cdots + J.$$

When $T = 0$, $\qquad\qquad \dfrac{\partial(\Delta G)}{\partial T} = J.$

But $\qquad \dfrac{\partial G}{\partial T} = -S$ and $\dfrac{\partial(\Delta G)}{\partial T} = -\Delta S.$

If, therefore, the change of entropy vanishes at the absolute zero, the integration constant J is zero.

In many solid substances the entropy probably does vanish when $T = 0$. Permutations of individual molecules, as has been seen, do not count in the world of quantum mechanics, and at the absolute zero all molecules occupy the lowest possible levels, which they can do in one way only. If the entropy is proportional to the logarithm of unity it of course vanishes. The lowest energy levels may possibly be multiple for one reason or another, and if they are so to different extents for the initial and final phases of a given transformation, J may in fact not be accurately zero. Nevertheless, in many cases it will be.

The original arguments of Nernst on this subject were based upon the view that ΔG and ΔH are equal not only at the absolute zero itself (as they must be from the general thermodynamic equation) but also in the immediate vicinity of this point, so that $\partial(\Delta G)/\partial T$ and $\partial(\Delta H)/\partial T$ vanish when $T = 0$. From these conditions it follows that J will be zero. Nernst put forward the hypothesis about the temperature coefficients as a fundamental thermodynamic principle (sometimes spoken of as the *third law of thermodynamics*). While there are reservations about the complete validity of this view, it is certainly one which is always nearly true, and may often be accurately so.

In so far as the Nernst postulate may be accepted,

$$\Delta G = -T \int \frac{\Delta H}{T^2}\, dT.$$

If ΔH is known from experimental observations and is represented as a function of temperature, then ΔG may also be so expressed with the aid of the same data.

For example, if experimentally determined heat quantities are given by

$$\Delta H = \Delta H_0 + bT^2 + ...,$$

the free energies will be given by

$$\Delta G = \Delta H_0 - bT^2 + ...$$

When G is plotted as a function of T, the point at which it becomes zero can be read off from the graph. This point is the transition temperature at which one solid phase would be transformed into the other.

The nature of the factors which cause ΔG to drop from the value ΔH_0 at the absolute zero to nothing at the transition temperature has already been partially discussed (p. 140). As is now evident, the omission of any reference in the previous account to the influence of the term TJ was immaterial. Its influence is unimportant since it usually vanishes. The sign of b is important. Suppose in the transition $I \rightarrow II$, ΔH_0 is positive. When $T = 0$, $\Delta G = \Delta H_0$ and I is stable. If b is positive, ΔG gradually falls to zero as the temperature rises, and II now becomes stable in spite of its higher potential energy.

Now

$$\frac{\partial(\Delta H)}{\partial T} = 2bT + ...$$
$$= C_{p_{II}} - C_{p_I}$$
$$= \Delta C_p$$

and

$$2b + ... = \frac{\partial(\Delta C_p)}{\partial T}$$
$$= \frac{\partial}{\partial T}(C_{p_{II}} - C_{p_I}).$$

If b is positive it means that the specific heat of the second phase *increases* with temperature more rapidly than that of the first. In other words, the second phase develops new degrees of freedom more rapidly, and offers relatively more accommodation in the way of molecular energy states. This, essentially, is why it eventually becomes the preferred form.

Chemical equilibrium in gases

We may now turn to the further consideration of gaseous equilibria. Thermodynamic principles lead to the equations

$$d \ln K_V / dT = \Delta U / RT^2$$

and $$d \ln K_p / dT = \Delta H / RT^2.$$

We may take the latter as typical and integrate it thus:

$$\ln K_p = \int \frac{\Delta H}{RT^2} \, dT + I.$$

Knowledge of the absolute magnitude of K_p depends upon the evaluation of I.

The first important step in this direction depends upon a theorem due to Nernst which shows that

$$I = \sum ni,$$

where the i's are the integration constants of the respective vapour-pressure equations for the various substances participating in the equilibrium,

$$\ln \pi = \int \frac{\lambda}{RT^2} \, dT + i.$$

The proof is simple and depends upon the fact that $J = 0$ for a transformation in the solid state.

As shown on p. 105 for a gaseous reaction,

$$\Delta F = RT(\sum n \ln c - \ln K_V),$$

and similarly $$\Delta G = RT(\sum n \ln p - \ln K_p).$$

If the corresponding chemical change occurred between substances in the solid state we should have

$$\Delta G_{\text{solid}} = -T \int \frac{\Delta H_{\text{solid}}}{T^2} \, dT + \text{zero}.$$

Now suppose all the solids participating in the reaction to be in equilibrium with their respective vapours. There is no change in free energy when evaporation occurs at the saturated vapour pressure. Consequently the value of

$$\Delta G_{\text{sat vap}} = \Delta G_{\text{solid}},$$

but $$\Delta G_{\text{sat vap}} = RT \sum n \ln \pi - RT \ln K_p$$

and $$\ln \pi = \int \frac{\lambda}{RT^2} \, dT + i.$$

Therefore

$$\Delta G_{\text{sat vap}} = RT \sum n \int \frac{\lambda}{RT^2} dT + RT \sum ni - RT \ln K_p = \Delta G_{\text{solid}},$$

and thus

$$RT \sum n \int \frac{\lambda}{RT^2} dT + RT \sum ni - RT \int \frac{\Delta H_{\text{gas}}}{RT^2} dT - RTI$$
$$= -T \int \frac{\Delta H_{\text{solid}}}{T^2} dT + 0.$$

The integrals contain no terms independent of T except the integration constants which have been written explicitly. Moreover, the above equation must be identically true for all values of T, and one may therefore equate coefficients of like powers of T on the two sides. The coefficients of the first powers are seen to be the various integration constants themselves, so that

$$R \sum ni - RI = 0$$

or
$$I = \sum ni.$$

Thus
$$\ln K_p = \int \frac{\Delta H}{RT^2} dT + \sum ni.$$

Similarly
$$\ln K_V = \int \frac{\Delta U}{RT^2} dT + \sum ni_0.$$

i_0 differs from i by a constant, the logarithm of the saturated vapour *concentration* differing from that of the pressure in a way calculable from elementary theory.

The two constants i and i_0 are, in principle, calculable in terms of the absolute entropies, so that these two formulae define the equilibrium constants themselves in terms of known quantities.

Gaseous equilibrium constants from partition functions

Before the significance of the various factors expressed in the equations for $\ln K$ is discussed further, an alternative method of deriving the equilibrium constant will be considered. In this it is related to the partition functions of the various molecules taking part in the reaction. It will suffice to deal with K_V.

In simple examples the problem is accessible to frontal attack. The following derivation applies in the first instance to an equilibrium between free atoms and diatomic molecules of a single species formed from them, but the conditions for generalizing the proof are fairly evident.

Let there be in all M atoms of one species and N of another, and let the numbers which exist in the free state be X and Y respectively. Let there be Z molecules formed by the union of one atom of each kind. Then we have the balance-sheet

$$X + Z = M,$$
$$Y + Z = N.$$

There is only one way in which all the atoms could be free and only one in which they could all be combined, but there are many ways in which some can be free and some combined. The total number of permutations by which the balance-sheet specified above can be realized is given by the expression

$$W_0 = \frac{M! \, N!}{X! \, Y! \, Z!}.$$

This follows from the fact that the total number of permutations of the two kinds of atom is $M! \, N!$. For each of these, the first X of one kind and the first Y of the other kind could be detailed for the free state and the residual Z of each sent to form molecules. But the order within the groups X, Y, and Z is immaterial.

The atoms of the two species and the molecules are now assignable to their several energy states, and this process leads to additional combinations

$$\frac{X!}{X_1! \, X_2! \, \ldots}, \qquad \frac{Y!}{Y_1! \, Y_2! \, \ldots}, \qquad \frac{Z!}{Z_1! \, Z_2! \, \ldots},$$

where X_1 is the number of atoms of the first kind in the first of their possible energy levels, and so on.

The probability of the whole assignment—atoms to the free or combined conditions, atoms and molecules to their various energy states—is given by

$$W = W_0 W_x W_y W_z = \frac{M! \, N!}{X_1! \, X_2! \ldots Y_1! \, Y_2! \ldots Z_1! \, Z_2! \ldots}.$$

Stirling's approximation gives

$$\ln W = M \ln M - M + N \ln N - N - \sum X_1 \ln X_1 + \sum X_1 -$$
$$- \sum Y_1 \ln Y_1 + \sum Y_1 - \sum Z_1 \ln Z_1 + \sum Z_1. \quad (1)$$

We also have

$$\sum X_1 + \sum Z_1 = M, \quad (2a)$$
$$\sum Y_1 + \sum Z_1 = N, \quad (2b)$$

for the constancy of the total numbers of atoms, and for the constancy of the total energy we have

$$\sum \epsilon_1 X_1 + \sum \eta_1 Y_1 + \sum u_1 Z_1 = E, \tag{3}$$

where ϵ, η, and u are the energies of the corresponding atoms and molecules in the levels denoted by the subscripts.

The condition that W shall be a maximum subject to $(2a)$, $(2b)$, and (3) is

$$-\delta \ln W = \sum (1+\ln X_1)\delta X_1 + \sum (1+\ln Y_1)\delta Y_1 +$$
$$+ \sum (1+\ln Z_1)\delta Z_1 - \sum \delta X_1 - \sum \delta Y_1 - \sum \delta Z_1 = 0,$$
$$\sum \delta X_1 + \sum \delta Z_1 = 0,$$
$$\sum \delta Y_1 + \sum \delta Z_1 = 0,$$
$$\sum \epsilon_1 \delta X_1 + \sum \eta_1 \delta Y_1 + \sum u_1 \delta Z_1 = 0.$$

The last four equations are, according to the method of undetermined multipliers illustrated already on p. 29, multiplied separately by unity, α_x, α_y, and β, and are then added and rearranged. The result is

$$\sum (\ln X_1 + \alpha_x + \beta\epsilon_1)\delta X_1 + \sum (\ln Y_1 + \alpha_y + \beta\eta_1)\delta Y_1 +$$
$$+ \sum (\ln Z_1 + \alpha_x + \alpha_y + \beta u_1)\delta Z_1 = 0.$$

By an argument similar to that used before, the separate coefficients of δX_1, δY_1, and δZ_1 are therefore equal to zero. Thus

$$\ln X_1 = -\alpha_x - \beta\epsilon_1,$$
$$\ln Y_1 = -\alpha_y - \beta\eta_1,$$
$$\ln Z_1 = -\alpha_x - \alpha_y - \beta u_1,$$

whence

$$X_1 = e^{-\alpha_x}e^{-\beta\epsilon_1},$$
$$Y_1 = e^{-\alpha_y}e^{-\beta\eta_1},$$
$$Z_1 = e^{-(\alpha_x+\alpha_y)}e^{-\beta u_1}.$$

It follows that

$$X = \sum X_1 = e^{-\alpha_x} \sum e^{-\beta\epsilon_1} = e^{-\alpha_x}f'_x,$$
$$Y = \sum Y_1 = e^{-\alpha_y} \sum e^{-\beta\eta_1} = e^{-\alpha_y}f'_y,$$
$$Z = \sum Z_1 = e^{-(\alpha_x+\alpha_y)} \sum e^{-\beta u_1} = e^{-(\alpha_x+\alpha_y)}f'_z,$$

and thus

$$\frac{Z}{XY} = \frac{f'_z}{f'_x f'_y} \frac{e^{-(\alpha_x+\alpha_y)}}{e^{-\alpha_x}e^{-\alpha_y}} = \frac{f'_z}{f'_x f'_y}.$$

As before, $\beta = 1/kT$.

The functions f_x', f_y', and f_z' are of the form of ordinary partition functions except in so far as they require adjustment in respect of the energy zeros.

When a molecule is formed from its constituent atoms there is a release of energy, q, of which equation (3) took no account. That expression rests therefore on the convention that the levels belonging to the u series are measured from a zero which is uniformly q higher than the normal molecular level. If we wish to drop this tacit convention and to use normal partition functions, we must increase the values u_1, u_2,... by q. Thus

$$f_z = \sum e^{-(u_1+q)/kT} = f_z' e^{-q/kT}.$$

Therefore $$f_z' = f_z e^{q/kT}.$$

But q is the energy released and is thus $-\Delta U_0/N$, where ΔU_0 is the conventional increase of energy per gram molecule. Thus

$$f_z' = f_z e^{-\Delta U_0/RT}.$$

f_x' and f_y' may be replaced by f_x and f_y since there is no adjustment of the reference levels to be undertaken. We therefore have

$$\frac{Z}{XY} = \frac{f_z}{f_x f_y} e^{-\Delta U_0/RT}.$$

Reference to the formulae on p. 135 reminds us that all the partition functions will contain V, the volume, once for each species (from the translational quantization). We may conveniently extract it from f by writing $f = f_0 V$. Transposition of the last formula then gives

$$\frac{(Z/V)}{(X/V)(Y/V)} = \frac{f_{0z}}{f_{0x}f_{0y}} e^{-\Delta U_0/RT}.$$

The last equation may be written

$$K_N = \Pi(f_0)e^{-\Delta U_0/RT},$$

where $\Pi(f_0)$ represents products and quotients of partition functions (with V omitted) built up on precisely the model of the equilibrium constant. For example, for the equilibrium $2H_2+O_2 \rightleftharpoons 2H_2O$ the value of $\Pi(f_0)$ would be

$$f_{0(H_2O)}^2/f_{0(H_2)}^2 \cdot f_{0(O_2)}.$$

K_N denotes an equilibrium constant in which concentrations are expressed not in the usual way as numbers of gram molecules per unit volume but as numbers of molecules per unit volume.

An inspection of the steps of the above derivation reveals what changes will be introduced by a modification in the form of the chemical equation. If the reaction were of a type where two atoms of the first species and one of the second entered the molecule, we should have for the balance-sheet of the total numbers of atoms

$$X + 2Z = M,$$
$$Y + Z = N.$$

Equation ($2a$), which is differentiated and multiplied by α_x, would then contain $2\sum Z_1$ and give rise to a term $2\alpha_x$ in the coefficient of $\sum \delta Z_1$. The cancellation of the α terms then requires the presence of X^2 and of f_x^2 in the final expression.

It will be noted that the above calculations, like those given on p. 28, amount to the derivation of a distribution law. The result is therefore unaffected by the question of the constant defining the absolute entropy.

More formal consideration of the foregoing theorem

As shown on p. 132, the Boltzmann free energy is related to the partition function by the equation

$$F = U - TS = -RT\ln f.$$

The quantum-mechanical requirement gives for the absolute entropy a value less by $k\ln(N!) = k(N\ln N - N)$ so that $kT(N\ln N - N)$ must be added to F.

Since, moreover, chemical changes are to be taken into account, and since these involve releases or uptakes of energy in forms which do not constitute part of the ordinary thermal energy—being in fact changes in the internal electronic energy, a term U_0 will be added to F. U_0 represents the non-thermal energy which changes only when the chemical nature of the molecule is altered.

Thus $$F = U_0 - RT\ln f + RT\ln N - RT.$$

f, the complete partition function, contains a factor V, the volume occupied by a gram molecule, and may be written $f_0 V$.

Thus $$F = U_0 - RT\ln f_0 + RT\ln(N/V) - RT.$$

In a chemical change, with the usual meaning of $\sum n$,

$$\Delta F = \sum n(U_0 - RT\ln f_0 + RT\ln c - RT + RT\ln N),$$

since $1/V = c$, the concentration of the gram molecule occupying the volume V.

If the gases are in equilibrium and if a small chemical transformation takes place such that dx times each of the quantities represented in the chemical equation reacts, then

$$(\Delta F)\, dx = 0,$$

$$(\Delta F)\, dx = \sum n(U_0 - RT\ln f_0 + RT\ln c - RT + RT\ln N)\, dx +$$

$$+ \frac{d}{dx}(\sum nRT\ln c)\, dx.$$

The last term must be added, since it is in fact impossible to conduct the small virtual displacement without slight changes in c. But

$$\frac{d}{dx}(\sum nRT\ln c)\, dx = RT \sum n\left(\frac{1}{c}\frac{dc}{dx}\right) dx = RT \sum n\, dx$$

since $dc/c = dx$, dx being the fraction of each gram molecule which reacts, and therefore the fractional change in concentration. Therefore

$$(\Delta F)\, dx = \sum n(U_0 - RT\ln f_0 + RT\ln c + RT\ln N)\, dx = 0,$$

$$\Delta F = \sum nU_0 - \sum nRT\ln f_0 + \sum nRT\ln(N/V) = 0,$$

whence
$$\sum n\ln\frac{N}{V} = \sum n\ln f_0 - \frac{\Delta U_0}{RT},$$

since $\sum nU_0 = \Delta U_0$, the heat of reaction at $T = 0$.

$\sum n\ln(N/V) = \ln K_N$, where K_N is the equilibrium constant with concentrations expressed as number of molecules in unit volume. (V is the volume of one gram molecule, while N, Avogadro's number, is the number of molecules in one gram molecule.)

This formula is that arrived at in the last section, $\ln \Pi(f_0)$ being identical with $\sum n\ln f_0$.

General discussion of the factors determining the position of a gaseous equilibrium

In the light of the formulae derived in the preceding sections we may once more make a general survey of the competing factors which determine the chemical make-up of things.

We begin by collecting together the relevant equations: they are

$$\ln K_V = \int \frac{\Delta U}{RT^2}\, dT + \sum ni_0, \qquad (1)$$

$$\ln K_N = \sum n\ln f_0 - \frac{\Delta U_0}{RT}, \qquad (2a)$$

$$K_N = \Pi(f_0)e^{-\Delta U_0/RT}. \qquad (2b)$$

K is written, as usual, with the reaction products in the numerator. Thus in (1) a large value of ni_0 favours completeness of reaction. The influence of this term might be roughly but vividly expressed by saying that those molecules which have the greatest inherent tendency to escape from the solid state into the vapour have also the greatest tendency to be formed in chemical transformations in the gas phase. The reason is that such molecules have open to them the widest range of states and that populations in general increase with the number of available states.

The operation of the same principle is revealed in a slightly different way in ($2a$). Other things being equal, K increases in proportion as the partition functions of products exceed those of the initial substances, and they may do this in so far as the selection of available states offered to them is wider.

(1) and (2) both express the powerful influence exerted by $\sum n$ itself. Values of i_0 and values of f_0 vary, of course, considerably from one substance to another, but not so much that the sign of $\sum ni_0$ or of $\sum n\ln f_0$ will not usually be positive or negative according as there are more molecules of product or more molecules of the initial substances in the chemical equation. In other words, products formed with an increase in the number of molecules are favoured, while those formed with a decrease in this number are not. That is to say, that there is a factor rendering decomposition reactions essentially more probable than synthetic reactions. The interpretation of this is obvious. The latter require the fortuitous encounter of a larger number of particles to unite to a smaller number, a process of comparative rarity compared with the spontaneous break-up of a more complex structure into chaotically dispersed fragments.

As T approaches infinity, the factor $e^{-\Delta U/RT}$ tends towards unity and the two influences which have just been considered, namely the relative numbers of available states open to reactants and products on the one hand, and on the other, the relative probabilities of congregation or dispersal, dominate the situation entirely. Conversely, as T drops, the relative effect of these factors declines until at very low temperatures the influence of ΔU_0 becomes paramount.

ΔU_0 represents the amount by which the potential energy of the structures formed in the reaction exceeds that of the original ones. K increases as $(-\Delta U_0)$ becomes numerically greater, that is, as the running down of potential energy accompanying the reaction grows larger. In a static world everything would tend to be in a state of minimum potential energy and this condition is more nearly realized as the absolute zero is approached, the formative tendencies of the attractive forces being here less and less combated by the disruptive action of molecular motions and the blind urge to explore all possible states of existence.

It is interesting to observe the form in which the potential energy factor is expressed in $(2\,b)$. Consider the simple example of a reaction $A \rightleftharpoons B$. The equilibrium constant $K = [B]/[A]$ contains a term $e^{-\Delta U_0/RT}$. The chance that a molecule of B is in its jth energy state is normally $e^{-\epsilon_j/kT}$ or $e^{-E_j/RT}$, counting on a gram-molecular basis. If, however, B is formed from A in a reaction, it receives, as it were, a bonus of energy E_0 and only needs $E_j - E_0$ to attain its jth state, the chance of which now becomes $e^{-(E_j - E_0)/RT}$. Thus every single state of B increases in probability by $e^{+E_0/RT}$ and, since $E_0 = -\Delta U_0$, by $e^{-\Delta U_0/RT}$.

There is still another matter illustrating the interplay of energy and entropy factors which the equations (1) and (2) reveal. (1) may be written

$$\ln K = -\frac{\Delta U_0}{RT} + \int \frac{\left(\int \sum n C_v \, dT \right)}{RT^2} \, dT + \sum n i_0 \qquad (3)$$

since

$$\frac{\partial (\Delta U)}{\partial T} = \sum n C_v$$

so that

$$\Delta U = \Delta U_0 + \int \sum n C_v \, dT.$$

Thus

$$\frac{d \ln K_V}{dT} = \frac{\Delta U_0}{RT^2} + \frac{\int \sum n C_v \, dT}{RT^2}.$$

Differentiation of $(2\,a)$ gives (since the temperature variations of K_V and K_N are the same)

$$\frac{d \ln K_N}{dT} = \frac{\Delta U_0}{RT^2} + \sum n \frac{d \ln f_0}{dT}.$$

From the result on p. 132 the final terms of the last two equations are seen to be consistent. The thermal energy E is $RT^2 d \ln f/dT$, so that $\sum n \, d \ln f_0/dT = \sum n E/RT^2$. $\sum n E$ is the amount by which ΔU will differ from ΔU_0, and so is $\int \sum n C_v \, dT$.

Equation (3) brings one factor more clearly to light. If the specific heats of products increase with temperature more rapidly than those of the reactants, the formation of products is favoured (cf. p. 142).

an average over cavity fields,

$$U_{a\to i}(\vec{u}) = \mathcal{Z}^{-1} \sum_{\{\vec{h}_{j\to a}\}_{j\in J}} \left[\sum_{\vec{s}\in T(\{\vec{h}_{j\to a}\})} \delta_{\vec{u},\vec{u}_{a\to i}(\vec{s})} \right] \prod_{j\in J} H_{j\to a}(\vec{h}_{j\to a}) . \qquad (8)$$

The pre-factor $\mathcal{Z}^{-1}$ is a normalization constant. Note that each cavity field configuration $\{\vec{h}_{j\to a}\}$ contributes $|T(\{\vec{h}_{j\to a}\})|$ terms. As a result, contradictory messages never contribute to eq. (8).

The BP eqs. (7) and (8) are equivalent to the so-called sum-product (or belief network, or Bayesian network) equations [188, 431]. One can try to solve them by iteration, starting from randomly chosen beliefs and then updating $U_{a\to i}$ sequentially on randomly chosen (a,i) edges. In some cases the process converges to a unique solution, independently of the updating scheme. When the belief propagation equations converge, one can use the resulting beliefs to estimate the histogram of local fields, using:

$$H_j(\vec{h}) \simeq \sum_{\{\vec{u}_{b\to j}\}_{b\in A(j)}} \delta_{\vec{h},\vec{h}_{j\to a}} \prod_{b\in A(j)} U_{b\to j}(\vec{u}_{b\to j}) , \qquad (9)$$

and this histogram can be used for decimation.

3.4   AN EXAMPLE OF BELIEF PROPAGATION: 3-COL

For the sake of clarity, let us work out BP on a simple example of the 3-COL problem ($q = 3$), for which the part of the factor graph is shown in figure 4. Since function nodes are connected to two variable nodes only (constraints represent edges in the original graph), there is only one variable node j above function node a. For a given configuration of incoming warnings $\{\vec{u}_{b\to j}\}$, we can make a table of allowed values s_j, and for each of them compute the outgoing warning $\vec{u}_{a\to i}(s_j)$. The only possible warnings are $(1,0,0)$, $(0,1,0)$, $(0,0,1)$, since a function node can only forbid one color: the value of the other variable connected to the function node.

- Suppose that $\vec{u}_{b_1\to j} = (1,0,0)$, $\vec{u}_{b_2\to j} = (0,1,0)$, and $\vec{u}_{b_3\to j} = (0,0,1)$. Then $h_{j\to a} = (1,1,1)$ and we find a contradictory message. No satisfiable configuration exists for s_j. According to the procedure given above, this configuration does not contribute to $U_{a\to i}$.
- Suppose that $\vec{u}_{b_1\to j} = \vec{u}_{b_2\to j} = (1,0,0)$, and $\vec{u}_{b_3\to j} = (0,1,0)$. Then $h_{j\to a} = (1,1,0)$, and the only possible coloring assignment for j is $s_j = 3$. For this configuration, we have only one possible outgoing warning: $\vec{u}_{a\to i} = (0,0,1)$.
- Suppose that $\vec{u}_{b_1\to j} = \vec{u}_{b_2\to j} = \vec{u}_{b_3\to j} = (1,0,0)$. Then $h_{j\to a} = (1,0,0)$, and there are two possible colors for s_j, namely the values 2 and 3. For the first one we have $u_{a\to i} = (0,1,0)$, and for the second one $u_{a\to i} = (0,0,1)$. Both contribute with equal weight to $U_{a\to i}$.

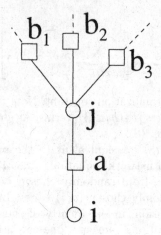

FIGURE 4 An example of the coloring problem. This part of the factor graph is the one necessary to compute the messages (warning and belief) passed from the function node a to the variable node i.

- All other configurations are simple color permutations of the three cases mentioned above, and are handled analogously.

From eqs. (7) and (8), we can easily deduce the equation giving the probability distribution $U_{a \to i}$ in terms of all distributions $\{U_{b_l \to j}; l = 1, 2, 3\}$. Parametrizing $U_{a \to i}$ according to the three possible messages as

$$U_{a \to i}(\vec{u}) = \eta^1_{a \to i} \delta_{\vec{u},(1,0,0)} + \eta^2_{a \to i} \delta_{\vec{u},(0,1,0)} + \eta^3_{a \to i} \delta_{\vec{u},(0,0,1)} , \qquad (10)$$

we find

$$\eta^p_{a \to i} = \frac{\prod_{l=1}^3 (1 - \eta^p_{b_l \to j})}{\sum_{r=1}^3 \prod_{l=1}^3 (1 - \eta^r_{b_l \to j})} . \qquad (11)$$

This expression can be understood easily: $\eta^p_{a \to i}$ equals the probability that color p is forbidden for node i, which means that node j has already taken this color, $s_j = p$. Now, node j can take color p if and only if it is not forbidden by any incoming warning: the numerator in eq. (11) simply calculates the probability that none of the incoming messages forbids color p, and the denominator guarantees normalization. Note that configurations in which all variables b_l take the same color r are counted twice, namely in the expressions for both values of $p \neq r$. According to the discussion above, this is correct because we have two new configurations for s_j, and two corresponding messages $\vec{u}_{a \to i}$ can be sent.

Finally, note that due to the symmetry among colors, a trivial solution to the BP equations is $\eta^p_{a \to i} = 1/q$ for all edges $(a, i) \in E$ and all colors p. However, in a recursive coloring algorithm some variable nodes would be assigned a color at the outset. This would explicitly break the symmetry.

4 SURVEY PROPAGATION

4.1 CLUSTERING

Unfortunately, the belief propagation dynamics are known not to converge for the random version of many combinatorial problems (including again 3-SAT and q-COL) in the region of the parameters near the SAT/UNSAT threshold. Recently, as discussed in earlier chapters, it has become possible using tools from statistical physics to gain an understanding of what happens in the solution space around the threshold [392, 395, 397]. Well below the threshold, where the number m/n of constraints per variable is relatively small, a generic problem has exponentially many solutions, which tend to form one giant cluster: for any two solutions, it is possible to find a connecting path via other solutions that requires short steps only (each pair of consecutive assignments in the path is close together in Hamming distance). Close to the critical threshold, however, the solution space breaks up into many smaller clusters. Solutions in separate clusters are generally far apart. In addition, the cost function $C[\vec{x}]$ has exponentially many local minima, separated from each other by large cost "barriers." These local cost minima are exponentially more numerous than the solution clusters. As seen explicitly in the example of error correcting codes in chapter 3, their metastability causes them to act as traps for local search algorithms.

According to the statistical physics analysis, which considers the infinite size limit $n \to \infty$, there exist exponentially many widely separated clusters of solutions. Within a given cluster of solutions, we may identify two types of variables: those that are frozen in one single assignment for all configurations belonging to the cluster, and those—unfrozen—that fluctuate from solution to solution inside the cluster. Note also that the variables that are frozen within one solution cluster may change their value when we go to another cluster, where they may even be unfrozen. While in general the distinction above can only provide an approximate description of clusters, it appears from numerical experiments that in many hard random CSPs, such as k-SAT or q-COL, this type of approximation is already rather accurate.

4.2 THE JOKER ASSIGNMENT

Survey propagation (SP) turns out [79, 427] to be able to deal with this clustering phenomenon for large (finite) sizes n. Although the original derivation uses subtle statistical physics ideas, one can also develop it more directly in algorithmic terms. The key intuition is that we no longer work with individual solutions $\vec{s} \in S_C$, but rather with entire clusters of solutions. The variables that are frozen within a cluster retain one single value $s_j \in \{1, ..., q\}$ in our description. Other variables may take several values within the cluster. For handling these, we introduce an additional joker value denoted by "$\star$" so that within each cluster, $s_j \in \{1, ..., q, *\}$. We can then generalize the constraint to this enlarged space and work out the corresponding belief propagation equations.

An even finer description, useful for general CSPs, would use multiple jokers to describe the set of allowed values for the variable, so that $s_j \in \mathcal{P}$ where $\mathcal{P}$ is the ensemble built from all subsets of $\{1, ..., q\}$. Within each cluster one could then assign exactly one of these generalized values to each variable. However, we will not develop this "derivation" in further detail, since in any event it does not give any rigorous construction. Rather, we will directly write the SP equations themselves, in terms of the original variables $\vec{x} \in \{1, ..., q\}^n$, and then analyze them.

4.3 GENERALIZED MESSAGES

We first need to define the generalizations of the warnings, local fields, and cavity fields used in survey propagation. In order to simplify the presentation, we shall often omit the "generalized" qualifier, and use the same notation for generalized warning as we used for warnings in the BP section. The reader should bear in mind that in the context of SP, all these messages are taken to be "generalized" messages.

For a given CSP, we define the warning:

Definition 4.1. *For a given edge (a, i) of the factor graph, with $a \in A$ and $i \in I(a)$, let S be a set of possible values for the variables $(x_j)_{j \in J}$ "above" a. $(J = I(a) \setminus i.)$ The* **warning** *is the q-dimensional vector $\vec{u}_{a \to i}(\vec{x}) \in \{0, 1\}^q$ with components:*

$$u^p_{a \to i}(S) = \min_{(x_j) \in S} C_a \left[(x_j)_{j \in J} \middle| x_i \leftarrow p \right] , \quad p = 1, ..., q .$$

This generalized warning is also known in the literature as *cavity bias* [395]. Note that the set of possible warnings is enlarged in SP: for the example of 3-COL, the *null message* $(0, 0, 0)$ is added to $(0, 0, 1)$, $(0, 1, 0)$ and $(1, 0, 0)$. As in section 3.4, the non-null messages are sent if the node "above" a function node is assigned a fixed color in the solution cluster. Correspondingly, the null message is sent if this vertex is not fixed to a single color, that is, if it is has the joker value.

Based on these warnings, we define local and cavity fields according to definitions 3.2 and 3.3, with the single configuration argument $\vec{x}$ again replaced by a set S of configurations. Using the J and B_j notation from section 3.3,

$$h^p_i(S) = \max_{a \in A(i)} u^p_{a \to i}(S) \text{ and}$$

$$h^p_{j \to a}(S) = \max_{b \in B_j} u^p_{b \to j}(S) . \tag{12}$$

4.4 HISTOGRAMS

Histograms of warnings and fields are now defined over clusters rather than over individual solutions. Letting S_C^α represent solution cluster α, and n_{cl} the total number of clusters, the histogram of local fields is given by

$$H_i(\vec{h}) = \frac{1}{n_{cl}} \sum_{\alpha=1}^{n_{cl}} \delta_{\vec{h},\vec{h}_i(S_C^\alpha)} \cdot \tag{13}$$

The histogram of the generalized warning on an edge (b,j) is called the *survey*, denoted by $Q_{b \to j}(\vec{u})$. It is defined in terms of the clusters of solutions for the cavity graph $G^{(j)}$. Calling $S_C^{\alpha,(j)}$ the corresponding solution cluster, and $n_{cl}^{(j)}$ the corresponding number of clusters, one defines:

$$Q_{b \to j}(\vec{u}) = \frac{1}{n_{cl}^{(j)}} \sum_{\alpha=1}^{n_{cl}^{(j)}} \delta_{\vec{u},\vec{u}_{b \to j}(S_C^{\alpha,(j)})} \cdot \tag{14}$$

4.5 SURVEY PROPAGATION EQUATIONS

Based on these definitions, one can easily infer the generalized recurrence equations for the (approximate) probabilities $Q_{a \to i}(\vec{u})$ that implement the solutions in this enlarged configuration space. These SP equations lead to a small, yet fundamental modification of the BP equations. The basic assumption is again that incoming warnings are independent. In this case, however, contradictory messages have to be explicitly forbidden. Keeping figure 3 in mind, we use the incoming set of surveys $\{Q_{b \to j}\}$ with $b \in B_j$ and $j \in J$ to calculate the cavity field distributions exactly as in eq. (7):

$$H_{j \to a}(\vec{h}) = \sum_{\{\vec{u}_{b \to j}\}_{b \in B_j}} \delta_{\vec{h},\vec{h}_{j \to a}} \prod_{b \in B_j} Q_{b \to j}(\vec{u}_{b \to j}) \cdot \tag{15}$$

Recall that these fields lead to contradictions if and only if $\vec{h}_{j \to a} = (1,1,\ldots,1)$ for at least one j. Therefore, we introduce the ensemble of all *non-contradictory* cavity field configurations,

$$\mathcal{M}_{a \to i} = \left\{ \{\vec{h}_{j \to a}\}_{j \in J} \mid \forall j \in J : \vec{h}_{j \to a} \in \{0,1\}^q, \ \vec{h}_{j \to a} \neq (1,\ldots,1) \right\} \cdot \tag{16}$$

Then, for an element of $\mathcal{M}_{a \to i}$—one specific set of $\{\vec{h}_{j \to a}\}_{j \in J}$—we again define

$$T(\{\vec{h}_{j \to a}\}) = \{(s_j)_{j \in J} \mid \forall j \in J : h_{j \to a}^{s_j} = 0\} , \tag{17}$$

the set of allowed configurations for the variable nodes above function node a. Now the difference with respect to BP arises: since all elements of $T(\{\vec{h}_{j \to a}\})$

give rise to a single outgoing warning, they all belong to the same cluster. The new warning is thus computed on the set of allowed configurations, and is given by $\vec{u}_{a \to i}(T(\{\vec{h}_{j \to a}\}))$. Its distribution follows:

$$Q_{a \to i}(\vec{u}) = \tilde{Z}^{-1} \sum_{\{\vec{h}_{j \to a}\} \in \mathcal{M}_{a \to i}} \delta_{\vec{u}, \vec{u}_{a \to i}(T(\{\vec{h}_{j \to a}\}))} \prod_{j \in J} H_{j \to a}(\vec{h}_{j \to a}) \ . \tag{18}$$

Equations (15) and (18) are the SP equations. Note that eq. (18) produces a dramatic change in the iteration of the probabilities compared with the BP eq. (8): every allowed cavity field configuration contributes only one term to the sum. Note also that contradictory messages have to be excluded "explicitly" by summing only over $\mathcal{M}_{a \to i}$. In BP, for each configuration of input messages one takes the full collection of possible outputs, thereby introducing a bifurcation mechanism that can easily become unstable. In SP, on the contrary, the presence of multiple outputs is collapsed into a null message (which in the example of graph coloring does not even exist in the belief propagation formalism). A variable receiving a message with at least two zero components will be "unfrozen" in the corresponding cluster.

The eqs. (15)–(18) provide a closed set of equations for the surveys. Practically, this recurrence defines a map

$$\Lambda : \left\{ Q_{a \to i}^{\text{old}} \right\}_{a \in A(i)}^{i \in I} \mapsto \left\{ Q_{a \to i}^{\text{new}} \right\}_{a \in A(i)}^{i \in I} \tag{19}$$

and we look for a fixed point of this map, obtained numerically by starting with some (random) initial $\left\{ Q_{a \to i}^0 \right\}$ and applying Λ iteratively:

$$\left\{ Q_{a \to i}^{SP} \right\} = \lim_{N \to \infty} \overbrace{\Lambda \circ \cdots \circ \Lambda}^{N \text{ times}} \left\{ Q_{a \to i}^0 \right\} . \tag{20}$$

Such a fixed point will be called a *self-consistent* set of surveys.

4.6   AN EXAMPLE OF SURVEY PROPAGATION: 3-COL

For the 3-COL example, because of the additional null message, the warning distribution now reads [78]

$$Q_{a \to i}(\vec{u}) = \eta_{a \to i}^{\star} \delta_{\vec{u},(0,0,0)} + \eta_{a \to i}^1 \delta_{\vec{u},(1,0,0)} + \eta_{a \to i}^2 \delta_{\vec{u},(0,1,0)} + \eta_{a \to i}^3 \delta_{\vec{u},(0,0,1)} \ , \tag{21}$$

and the SP equations corresponding to figure 4 are given by

$$\eta_{a \to i}^p = \frac{\prod_{l=1}^3 (1 - \eta_{b_l \to j}^p) - \sum_{r \neq p} \prod_{l=1}^3 (\eta_{b_l \to j}^{\star} + \eta_{b_l \to j}^r) + \prod_{l=1}^3 \eta_{b_l \to j}^{\star}}{\sum_{r=1}^3 \prod_{l=1}^3 (1 - \eta_{b_l \to j}^r) - \sum_{r=1}^3 \prod_{l=1}^3 (\eta_{b_l \to j}^{\star} + \eta_{b_l \to j}^r) + \prod_{l=1}^3 \eta_{b_l \to j}^{\star}} , \tag{22}$$

for $p \in \{1, 2, 3\}$. Then $\eta_{a \to 1}^{\star}$ can be computed by normalization:

$$\eta_{a \to i}^{\star} = 1 - \left(\eta_{a \to i}^{1} + \eta_{a \to i}^{2} + \eta_{a \to i}^{3} \right) . \tag{23}$$

The interpretation of this equation is again straightforward. We explain it for color 1: now $\eta_{a \to i}^{1}$ is given by the probability that s_j is forced to take value 1, that is, by the probability that the cavity field equals $\vec{h}_{j \to a} = (0, 1, 1)$, conditioned on non-contradictory cavity fields. The numerator calculates the unconditioned probability. The first term includes all cases where $h_{j \to a}^{1} = 0$: (0,0,0), (0,0,1), (0,1,0), (0,1,1). The second term then excludes those cases where $h_{j \to a}^{1} = h_{j \to a}^{r} = 0$, summed over $r \neq 1$: (0,0,0), (0,0,1) for $r = 2$; (0,0,0), (0,1,0) for $r = 3$. Finally, the third term includes (0,0,0) again since it was double-counted in the second term. The denominator then provides for conditioning on non-contradictory fields, by giving the probability that $\vec{h}_{j \to a} \neq (1, 1, 1)$. The counting of possible cases follows a similar inclusion-exclusion principle as for the numerator.

Note that as with belief propagation, the symmetry among colors leads to a trivial solution: $\eta_{a \to i}^{\star} = 1$ for all edges (a, i) of the factor graph, that is, only null messages are sent. Clearly, this is not the correct solution in the clustered regime, and the color symmetry is not valid at the level of solution clusters. In fact, it is the appearance of a nontrivial solution for the $\eta_{a \to i}^{p}$ that marks the onset of clustering.

4.7 AN EXAMPLE OF SURVEY PROPAGATION: K-SAT

In the case of SAT, $q = 2$: possible $\vec{u}$ warnings are $(0,0)$, $(1,0)$, $(0,1)$, and $(1,1)$. As any clause can be satisfied by any given variable (choosing the variable's value according to whether or not the corresponding literal is negated), the $(1, 1)$ message will never appear. Moreover, for a given edge (a, i), the sign of the literal at i will completely determine whether $(1, 0)$ or $(0, 1)$ can appear on $\vec{u}_{a \to i}$. So we can parametrize distributions $Q_{a \to i}$ with only one real number $\eta_{a \to i}$, namely the probability of the nontrivial $\vec{u}_{a \to i}$ message, that is, a message other than $(0, 0)$. The probability of $(0, 0)$ is then $1 - \eta_{a \to i}$. The corresponding equations have been written and implemented in Braunstein et al. [79] and at the Survey Propagation web site [536]. In the case of 3-SAT, they read

$$\eta_{a \to i} = \prod_{j \in J} \left[\frac{\Pi_{j \to a}^{u}}{\Pi_{j \to a}^{u} + \Pi_{j \to a}^{s} + \Pi_{j \to a}^{0}} \right] , \tag{24}$$

where

$$\Pi_{j \to a}^{u} = \left[1 - \prod_{b \in A_a^u(j)} (1 - \eta_{b \to j}) \right] \prod_{b \in A_a^s(j)} (1 - \eta_{b \to j}) ,$$

$$\Pi^s_{j \to a} = \left[1 - \prod_{b \in A^s_a(j)} (1 - \eta_{b \to j}) \right] \prod_{b \in A^u_a(j)} (1 - \eta_{b \to j}) \, ,$$

$$\Pi^0_{j \to a} = \prod_{b \in B_j} (1 - \eta_{b \to j}) \, . \tag{25}$$

$A^u_a(j), A^s_a(j)$ are the two sets into which $A(j)$ decomposes $(A(j) = A^u_a(j) \cup A^s_a(j))$ where the indices s and u refer to the neighbors b for which the literals (b, j) and (a, j) agree and disagree, respectively. This separation corresponds to the distinction of which neighbors cause the clause a to be satisfied or unsatisfied by variable j.

For example, the product $\prod_{b \in A^s_a(j)} (1 - \eta_{b \to j})$ gives the probability that no nontrivial message arrives at j from the function nodes $b \in A^s_a(j)$ (empty products are set to 1 by definition).

4.8 DECIMATION

Once convergence is reached in eq. (20) (we stop when $\max_{i \in I, a \in A(i)}$ $|Q^{old}_{a \to i} - Q^{new}_{a \to i}|$ becomes small enough), we can use the information computed so far to find a solution to the original problem [79, 395]. We can easily compute the (approximate) local field distributions $\{H_i\}_{i \in I}$ introduced in eq. (13) by considering all neighboring function nodes, and forbidding contradictory messages. Recall that in the cavity graph we delete the constraints containing variable i, whereas in H_i we have to restrict the sum to messages that can be extended to solutions of the complete problem. In the example of 3-COL, $H_i(\vec{h})$ is given by

$$H_i(\vec{h}) = \mathcal{Z}'^{-1} \sum_{\{\vec{u}_{a \to i}\}_{a \in A(i)}} \left(1 - \delta_{\vec{h},(1,1,1)} \right) \delta_{\vec{h},\vec{h}_i} \prod_{a \in A(i)} Q_{a \to i}(\vec{u}_{a \to i}) \, , \tag{26}$$

with $\vec{h}_i$ determined according to eq. (12).

Given the vector $\vec{h}_0$ with a 0 entry at component p and 1 at the other two components, the value $H_i(\vec{h}_0)$ gives the probability for a variable i to be frozen to a certain value p. A simple decimation procedure can then be implemented. Select the variable that is frozen with the highest probability, and fix it to its most frozen value. Then simplify the problem: certain constraints may already be satisfied independently of the values of other participating variables, and can be deleted from the problem instance. Other constraints might immediately fix single variables to one value (unit clause resolution). Reconverge the warning distributions on the smaller subproblem.

The decimation algorithm can lead to three types of behaviors:

1. The algorithm can solve the problem fixing all, or almost all variables (some variables may not need to be fixed, even if the problem is already solved).

2. The surveys converge at some stage to the trivial solution concentrated on null messages, $Q_{a \to i}(u) = \delta_{h,(0,0,0)}$ for all $(a,i) \in E$. In this case SP has nothing more to offer. Luckily, the resulting subproblems are generally under-constrained and then easy to solve by other means. Note that, for q-COL, the trivial solution always exists. In numerical experiments, we found that in the case of the existence of another solution, the latter was the correct one. Therefore, even if a trivial solution is found once, it is reasonable to restart the iteration of the SP equations. Only if no nontrivial solution can be found after several restarts does the subproblem need to be passed to a different solver.

3. The SP algorithm never converges, even if the initial problem was satisfiable.

On large random instances of 3-SAT [79, 395, 397, 429] and q-COL [78] in the hard SAT region, though not too close to the satisfiability threshold, numerical experiments show that the algorithm behaves as in case 2. The subproblems generated turn out to be very simple to solve by other conventional heuristics, such as WalkSAT [470] or unmodified belief propagation.

Case 3 generally occurs very close to the SAT/UNSAT transition. It is not yet clear whether this outcome appears due to the existence of finite loops in the original problem (which make the SP equations only approximate), due to the simple decimation heuristic that always fixes the most frozen variable, or due to problems that go beyond the validity of the SP equation itself.

5   WHAT'S NEXT

Among all the possible directions of research that may follow from the algorithm we have presented, we would like to highlight two in particular. The first is to formalize rigorously the notions suggested in section 4, establishing precise definitions for the clusters, and a corresponding derivation of the SP equations. The second, of great computational relevance, is to generalize SP. SP has been presented here in its purest form, but can be adapted to deal with correlations between warnings that arise from local problem structures such as small loops in the factor graph. Similar extensions have been considered for BP [533]. A further possible generalization would include diverse structures of the solution space. Notions of *replica symmetry breaking*, discussed in chapter 1, argue for considering clusters of solution clusters or even a hierarchical construction of clusters. Developing this might be a further step towards more fully applying theory and analysis from statistical physics to algorithmic methods.

ACKNOWLEDGMENTS

It is a pleasure to thank R. Mulet, A. Pagnani, and F. Ricci-Tersenghi for numerous discussions. M. Mézard and M. Weigt acknowledge the hospitality of the ICTP Trieste, where a portion of this work was done. The work has been supported in part through the EC "STIPCO" network, grant No. HPRN-CT-2002-00319.

CHAPTER 5

The Easiest Hard Problem:
Number Partitioning

Stephan Mertens

1 INTRODUCTION

The number partitioning problem (NPP) is defined easily: Given a list $a_1, a_2, \ldots, a_n$ of positive integers, find a partition, that is, a subset $\mathcal{A} \subset \{1, \ldots, n\}$, minimizing the discrepancy

$$E(\mathcal{A}) = \left| \sum_{i \in \mathcal{A}} a_i - \sum_{i \notin \mathcal{A}} a_i \right|. \tag{1}$$

A *perfect partition* is a partition with $E = 0$ for $\sum a_j$ even, or $E = 1$ for $\sum a_j$ odd.

Number partitioning is of considerable importance, both practically and theoretically. Its practical applications range from multiprocessor scheduling and the minimization of VLSI circuit size and delay [102, 504], to public key cryptography [387], to choosing up sides in a ball game [237]. Number partitioning is also one of Garey and Johnson's six basic NP-hard problems that lie at the heart of the theory of NP-completeness [191, 388], and is in fact the only one of these

Computational Complexity and Statistical Physics, edited by
Allon G. Percus, Gabriel Istrate, and Cristopher Moore, Oxford University Press.

problems that actually deals with numbers. Hence, it is often chosen as a base for NP-completeness proofs of other problems involving numbers, such as bin packing, multiprocessor scheduling [38], quadratic programming, and knapsack problems.

The computational complexity of the NPP depends on the type of input numbers $\{a_1, a_2, \ldots, a_n\}$. Consider the case where the values of a_j are positive integers bounded by a constant A. Then the discrepancy E can take on at most nA different values, so the size of the search space is $O(nA)$ instead of $O(2^n)$ and it is straightforward to devise an algorithm that explores this reduced space in time polynomial in nA [191]. Of course, such an algorithm does not prove $P = NP$: a concise encoding of an instance requires $O(n \log_2 A)$ bits, and A is not bounded by any power of $\log_2 A$. This feature of the NPP is called *pseudo-polynomiality*. The NP-hardness of the NPP becomes apparent when input numbers are of a size exponentially large in n or, after division by the maximal input number, of exponentially high precision.

To study typical properties of the NPP, the input numbers are often taken to be independent, identically distributed random variables. Under this probabilistic assumption, the minimal discrepancy E_0 is a stochastic quantity. For real-valued input numbers (infinite precision, see above), Karmarkar et al. [298] proved that the *median* value of E_0 is $O(\sqrt{n}2^{-n})$. Lueker [370] showed that the same scaling holds for the *mean* value of E_0. From numerical simulations [157] it is known that the standard deviation of E_0 is of the same order of magnitude as the mean: E_0 is *non-self-averaging*.

Another surprising feature of the NPP is the poor quality of heuristic algorithms [281, 448]. The differencing method, discussed below, is the best polynomial time heuristic known to date, and for real-valued a_j yields minimum discrepancies $O(n^{-\alpha \log n})$ with some positive constant α [532]. This is far above the true optimum, yet it is the best that one can get for large systems! The poor quality of polynomial time heuristics is a very peculiar feature that distinguishes the NPP from many other hard optimization problems such as the Euclidean traveling salesman problem [444], for which satisfactory approximation algorithms do exist.

The NP-hardness of the NPP tells us that for numbers a_j bounded by $A = 2^{\kappa n}$, the worst-case complexity of any exact algorithm is almost certainly exponential in n for all $\kappa > 0$. Numerical simulations show that the typical complexity on instances of the random ensemble is exponential only for $\kappa > \kappa_c > 0$. For $\kappa < \kappa_c$ it is polynomial. The critical value κ_c marks a transition point, where the random ensemble somehow changes its character. Below κ_c, typical instances seem to have a special property that can be exploited by an exhaustive algorithm. This abrupt change of an averaged quantity, as a parameter of a statistical ensemble is varied, is called a *phase transition* by analogy with the transitions observed in thermodynamic systems. Phase transitions in the average complexity have been observed in many NP-hard problems such as satisfiability [236, 319], or Hamiltonian circuit [91], and are discussed throughout this volume.

Their study forms the base of an emerging interdisciplinary field of research that encompasses the efforts of computer scientists, mathematicians and physicists [133].

The NPP illustrates the interdisciplinary character of the field. Fu [185], a physicist, first mapped partitioning to an infinite-range, antiferromagnetic spin glass, concluding (incorrectly) that this model did not have a phase transition. Gent and Walsh [194], computer scientists, demonstrated the existence of the phase transition using numerical simulations. They introduced the control parameter κ and estimated the transition point close to $\kappa_c = 0.96$. Mertens [389], a physicist, reconsidered Fu's spin glass analogy and derived a phase transition at $\kappa_c = 1 - (\log_2 n/2n) + O(1/n)$. Then Borgs, Chayes and Pittel [71], mathematicians, took over and established the phase transition and its characterization rigorously. The mathematical proofs for the phase transitions are another exceptional feature of the NPP. For other NP-hard problems such as satisfiability, much less is known from rigorous techniques and the sharpest results have been obtained by the powerful but non-rigorous techniques from statistical mechanics [395], as seen in the previous two chapters.

It is this combination of algorithmic hardness and analytical tractability that earns the NPP the description of *easiest hard problem*, a phrase coined by Brian Hayes [237]. In this chapter, we exploit the easiness of the NPP to provide an understanding of some of its remarkable properties.

2   ALGORITHMS AND COMPLEXITY

In view of the NP-hardness of the NPP, it is wise to abandon the idea of an exact solution and to ask instead for an approximate but fast heuristic algorithm. An obvious approach is to place the largest number in one of the two subsets, then continue to place the largest among the remaining numbers in the subset with the smaller total sum so far, until all numbers are assigned. The idea behind this *greedy heuristic* is to keep the discrepancy small with every decision. In the worst case, the two subsets could be perfectly balanced just before the last number is assigned: since numbers are assigned in decreasing order, this leads to the discrepancy scaling as $O(n^{-1})$ for real-valued a_j. That, of course, is extremely bad compared to the optimum discrepancy of $O(\sqrt{n}\,2^{-n})$. The time complexity of the greedy algorithm is given by the time complexity to sort n numbers, or $O(n \log n)$. Applied to the set $\{a_j\} = \{8, 7, 6, 5, 4\}$, the greedy heuristic misses the perfect solution and yields a partition $\{8, 5, 4\}\,\{7, 6\}$ with discrepancy 4.

The differencing method of Karmarkar and Karp [297], also called the KK heuristic, is another polynomial time approximation algorithm. The key idea of this algorithm is to reduce the size of the numbers. This is achieved by replacing the two largest numbers with the absolute value of their difference. This differencing operation is equivalent to committing the numbers to different subsets without actually fixing which subset each will go into. With each differencing

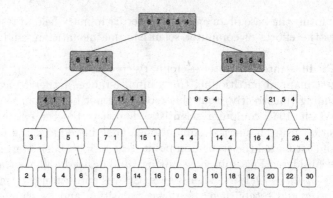

FIGURE 1 Search tree of the complete differencing algorithm. Left branch means "replace two largest numbers with their difference," right branch means "replace them with their sum." With appropriate pruning rules only the shaded nodes have to be visited to find the optimum solution.

operation the number of numbers decreases by one, and the final number is the discrepancy. Applied to $\{8, 7, 6, 5, 4\}$, the differencing method yields a discrepancy of 2 that results from the partition $\{8, 6\}$ $\{7, 5, 4\}$. Note that reconstructing the partition requires some additional bookkeeping that we did not mention in our brief description of the algorithm. Again the heuristic misses the perfect solution, but at least the outcome is better than the greedy result. Yakir [532] has proven that the differencing method applied to random real-valued $a_j \in [0, 1]$ produces mean discrepancies $n^{-\alpha \log n}$ with a constant $\alpha = 0.72$. Again this is much better than the greedy result, yet it is still far from the optimum. The time complexity of the differencing method is dominated by the complexity of selecting the two largest numbers. This is done most efficiently by sorting the initial list and keeping the order throughout all iterations, leaving us with a time complexity $O(n \log n)$.

Either one of these heuristics can be used as a base for an exact algorithm, analogous to the search tree methods analyzed in chapter 3. At each iteration, the greedy algorithm places a number in the subset with the smaller total sum so far. The only alternative is to place the number in the other subset. Exploring both alternatives means searching a binary tree that contains all 2^n possible partitions. In the KK heuristic, the corresponding alternative is to replace the two largest numbers by their sum rather than by their difference, equivalent to committing them to the same subet. Korf [341] calls the algorithms that explore both alternatives *complete greedy* and *complete differencing* algorithms. Figure 1 shows the search tree of the complete differencing method for our example $\{8, 7, 6, 5, 4\}$.

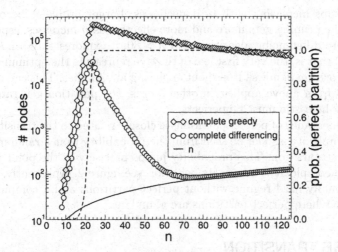

FIGURE 2   Number of nodes visited by the complete greedy and the complete differencing algorithms. Instances are sets (of cardinality n) of random 20-bit integers, each data point representing an average over 10^4 instances. The dashed line indicates the empirical probability that a given instance has a perfect solution.

Both complete algorithms have exponential time complexity in the worst case, but it is possible to prune parts of the search tree using simple rules. For the complete differencing method these rules are:

1. If fewer than 5 numbers are left, take the left branch (apply the differencing operation).
2. If the largest number in the set is larger than or equal to the sum of all the other numbers, stop branching: the best solution in this subtree is to place the largest number in one set, and all the other numbers in the other set.
3. If a perfect partition has been found, stop the process.

The first rule needs some thought, but it can in fact be proven that the KK heuristic always yields the optimum for $n \leq 4$. Similar pruning rules can be added to the complete greedy method. Figure 1 shows an example in which the rules chop off large parts of the search tree.

The question is how pruning affects the search in general and for large instances. Figure 2 shows the number of nodes visited by the complete greedy and the complete differencing algorithms while solving large instances of random 20-bit integers. For small values of n, the number of nodes grow exponentially with n, that is, the pruning shows only little effect on the performance. For systems beyond $n = 23$ the situation changes drastically: the number of nodes not only stops increasing with n, it decreases. Larger problems become easier to solve! The pruning gets more and more effective as n increases, especially for the complete differencing algorithm. For $n > 80$, it explores only n nodes of the search tree, that is, the very first leaf of the tree represents the optimum solution, and the algorithm "knows" it without exploring any further. This can only mean that rule 3 from above applies, in other words, the partition generated by the differencing heuristic must be perfect.

The appearance of perfect partitions is closely related to the transition in the average complexity, as can be seen from the probability that a random instance has a perfect partition. This probability jumps precisely at the point where the algorithmic complexity changes its behavior, see figure 2. Apparently, there is a computationally hard regime without perfect partitions and a computationally easy regime where perfect partitions are abundant.

3 PHASE TRANSITION

As we have seen in the preceding section, the average complexity of algorithms for the random NPP depends on the presence of perfect partitions. The probability of perfect solutions is a property of the ensemble of instances, and can be studied independently of algorithms. That is what we do in this section.

A partition $\mathcal{A}$ can be encoded by binary variables $s_j = \pm 1$: $s_j = +1$ if $j \in \mathcal{A}$, $s_j = -1$ otherwise. The cost function then reads $E = |D(s)|$ where

$$D = \sum_{j=1}^{n} a_j s_j \tag{2}$$

is the signed discrepancy. An alternative cost function is $H = D^2$ or

$$H = -\sum_{i,j} J_{ij} s_i s_j \qquad \text{with } J_{ij} = -a_i a_j. \tag{3}$$

H is the Hamiltonian of an infinite-range, antiferromagnetic spin glass, which has been studied by physicists [157, 185, 389] within the canonical framework of statistical mechanics. Here we follow another, very simple approach that has been used recently to analyze the multiprocessor scheduling problem [38].

The signed discrepancy D can be interpreted as the distance from the origin of a one-dimensional walk with steps to the left ($s_j = -1$) and to the right ($s_j = +1$), and with random stepsizes (a_j). The average number of walks that end at D reads

$$\Omega(D) = \sum_{\{s_j\}} \left\langle \delta \left(D - \sum_{j=1}^{n} a_j s_j \right) \right\rangle \tag{4}$$

where angular brackets denote averaging over the random numbers a. By the central limit theorem, for a fixed walk $\{s_j\}$ and large n, the sum $\sum_{j=1}^{n} a_j s_j$ is

Gaussian with mean

$$\langle D \rangle = \langle a \rangle \sum_j s_j \tag{5}$$

and variance

$$\langle D^2 \rangle - \langle D \rangle^2 = n(\langle a^2 \rangle - \langle a \rangle^2). \tag{6}$$

The sum over $\{s_j\}$ is basically an average over all trajectories of our random walk. For large n this average is dominated by trajectories with $\sum s_j = 0$, leading to $\langle D \rangle = 0$. Hence the probability density for ending the walk at distance D reads

$$p(D) = \frac{1}{\sqrt{2\pi n \langle a^2 \rangle}} \exp\left(-\frac{D^2}{2n \langle a^2 \rangle}\right). \tag{7}$$

Note that our walk involves only a sublattice of $\mathbb{Z}$ with lattice spacing 2: movements are confined to either the even or odd numbers, depending on whether $\sum a_j$ is even or odd. Hence the average number of walks ending at distance D, when D is of the same parity as $\sum a_j$, is given by

$$\Omega(D) = 2^n 2p(D) = \frac{2^{n+1}}{\sqrt{2\pi n \langle a^2 \rangle}} \exp\left(-\frac{D^2}{2n \langle a^2 \rangle}\right). \tag{8}$$

For the location of the phase transition we can concentrate on perfect partitions, that is, assume $D = 0$. If the a's are uniformly distributed κn-bit integers (for large n, without loss of generality $\sum a_j$ can be taken to be even),

$$\langle a^2 \rangle = \frac{1}{3} 2^{2\kappa n} \left(1 - O(2^{-\kappa n})\right) \tag{9}$$

and so

$$\log_2 \Omega(0) = n(\kappa_c - \kappa) \tag{10}$$

with

$$\kappa_c = 1 - \frac{\log_2 n}{2n} - \frac{1}{2n} \log_2\left(\frac{\pi}{6}\right). \tag{11}$$

This is our phase transition: according to eq. (10) we have an exponential number of perfect partitions for $\kappa < \kappa_c$, and no perfect partition for $\kappa > \kappa_c$. Our derivation is a bit sloppy, of course, but the result agrees with the rigorous theory of Borgs et al. [71].

From eq. (10), we expect the entropy $S = \log_2 \Omega(0)$ of perfect partitions for fixed but large n to be a linear function of κ. In fact this can already be observed for rather small problem sizes in figure 3. Linear extrapolation of the simulation data for $\log_2 \Omega(0)$ gives numerical values for the transition points $\kappa_c(n)$. Again the numerical data for small systems agree very well with the predictions of the asymptotic theory (fig. 4). The strong finite size corrections of order $\log n / n$ lead to the curvature of $\kappa_c(n)$ and they are responsible for the incorrect value $\kappa_c = 0.96$ that Gent and Walsh extrapolated from their simulations [194].

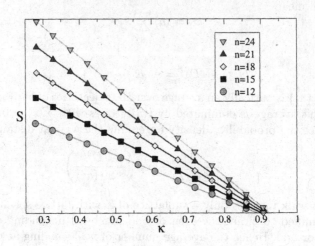

FIGURE 3 Entropy $S = \log_2 \Omega(0)$ of perfect partitions vs. κ. Theory (eq. (10)) compared to numerical enumerations (symbols).

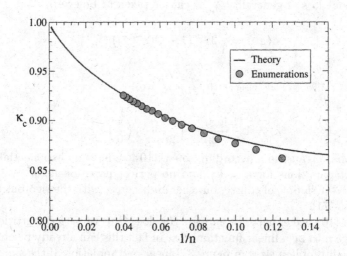

FIGURE 4 Numerical data for the transition points $\kappa_c(n)$ have been obtained by linear extrapolation of the data for $S = \log_2 \Omega(0)$ from figure 3. The solid line is eq. (11).

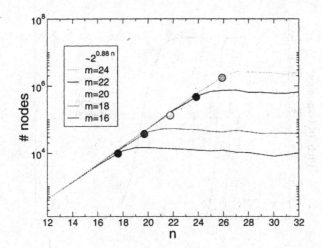

FIGURE 5 Partitioning m-bit numbers with the complete greedy algorithm: number of search nodes visited vs. n. The curves are averages over 10^4 random samples, and the symbols mark the values n_c given by eq. (12). The fitted curve $2^{0.88n}$ shows that pruning has almost no effect for $n < n_c$.

The phase transition at κ_c is a property of the instances. In contrast to the analysis of chapter 3, it is by no means clear how this transition affects the dynamical behavior of search algorithms. Note that even for $\kappa < \kappa_c$, the fraction of perfect partitions is exponentially small, and finding one of these is non-trivial.

In numerical experiments like the one shown in figure 2, the number $m = \kappa n$ of bits is usually fixed and n is varied. Then κ_c translates into a critical value $n_c = m/\kappa_c$ or

$$\frac{m}{n_c} = 1 - \frac{\log_2 n_c}{2n_c} - \frac{1}{2n_c} \log_2 \left(\frac{\pi}{6}\right). \tag{12}$$

For $m = 20$ this gives $n_c = 21.8$, in good agreement with the location of the hardest instances in figure 2. Figure 5 shows that the average time complexity of the complete greedy algorithm changes its dependence on n precisely at the values n_c given by eq. (12). It is well justified to classify the two regimes $\kappa < \kappa_c$ and $\kappa > \kappa_c$ as *easy* and *hard*.

4   EASY PHASE

The hallmark of the easy phase is the exponential number of perfect partitions, but the easy phase is not homogeneous: the number of perfect partitions increases with decreasing κ. This phenomenon might yield an interesting structure with

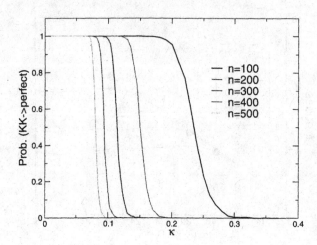

FIGURE 6 Probability of the Karmarkar-Karp differencing heuristic yielding a perfect partition.

regard to algorithms: the performance of an algorithm improves as one moves away from the phase boundary towards smaller values of κ. In fact, figure 2 indicates that complete differencing finds a perfect partition with its very first descent in the search tree if κ is small enough (n is large enough). Does this mean that the situation is reminiscent of satisfiability (see ch. 3) and the $\kappa < \kappa_c$ region disintegrates into two phases, one in which complete differencing hits a perfect solution on the first try and another one in which it needs to backtrack?

To test this hypothesis, we investigate the Karmarkar-Karp (KK) heuristic solution for the NPP. Recall that this solution is the first one generated by the complete differencing algorithm. Let D_{kk} be the discrepancy of the KK solution. Our hypothesis would then be: there is a value $0 \leq \kappa_{kk} \leq \kappa_c$ such that

$$\lim_{n \to \infty} \mathbf{Pr}(D_{kk} \leq 1) = \begin{cases} 1, & \kappa < \kappa_{kk}; \\ 0, & \kappa > \kappa_{kk}. \end{cases} \tag{13}$$

Figure 6 shows the result of a simulation of the KK algorithm. While there is a sharp transition at a value κ_{kk}, the value depends on n and seems to go to 0 as $n \to \infty$.

A simple argument explains why this happens. We know from the work of Yakir [532] that given real-valued input numbers $a_j \in [0, 1]$, the KK algorithm generates partitions with mean discrepancy $n^{-\alpha \log n}$ for some constant $\alpha > 0$. So if the numbers are integers with $m = \kappa n$ bits, we would expect that on average

$$D_{kk} = 2^{\kappa n} n^{-\alpha \log n}. \tag{14}$$

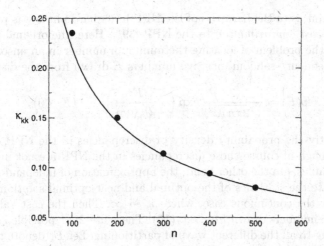

FIGURE 7   Threshold value below which Karmarkar-Karp differencing yields perfect solutions. Solid line denotes κ_{kk} from eq. (15), symbols show results of numerical simulations.

Therefore, $D_{kk} \leq 1$ as long as $\kappa \leq \kappa_{kk}(n)$ with

$$\kappa_{kk}(n) = \alpha \frac{(\log n)^2}{n \log 2}. \tag{15}$$

Figure 7 shows $\kappa_{kk}(n)$ compared with the results from simulations, where we have measured κ_{kk} as the value where the probability of generating a perfect partition is $1/2$. For α, we take the value 0.72 reported by Yakir for the average discrepancy of the KK solution.

Note that a similar argument suggests that even the greedy heuristic eventually yields perfect partitions for sufficiently small values of κ. At the value $m = 20$ used in figure 2, we expect the greedy heuristic to generate perfect partitions for $n > 839000$.

5 HARD PHASE

Figure 2 shows that in the easy phase, complete differencing outperforms complete greedy, and in view of the exponentially small fraction of perfect partitions, both algorithms outperform exhaustive search through all partitions. Figure 2 also indicates that complete greedy and complete differencing perform similarly to each other in the hard phase. In fact, in the hard phase neither is superior to blind random search, as we will see in this section.

A first hint as to the hardness of the NPP in its hard phase was provided by the *random cost approximation* to the NPP [391]. Here, the original problem is replaced by the problem of locating the minimum number in an unsorted list of 2^{n-1} "independent" random, positive numbers E drawn from the distribution

$$p(E) = \frac{2}{\sqrt{2\pi n \langle a^2 \rangle}} \exp\left(-\frac{E^2}{2n \langle a^2 \rangle}\right) \qquad (E \geq 0). \qquad (16)$$

This is exactly the probability density of discrepancies in the NPP, as seen in eq. (7), although of course those discrepancies in the NPP are not independent random variables. On the other hand, the approximation of independence allows us to calculate the statistics of the optimal and near-optimal solutions.

Consider the continuous case, where $\kappa \to \infty$. Then the cost values E are real, positive numbers drawn from eq. (16). There are 2^{n-1} possible cost values, corresponding to all the different ways of partitioning. Let E_k denote the $k+1$th lowest of these cost values, so that E_0 is the minimum (optimal) cost, followed by E_1 (lowest near-optimal) and so on. The probability density ρ_0 of E_0 can easily be calculated:

$$\rho_0(E_0) = 2^{n-1}p(E_0)\left(1 - \int_0^{E_0} p(E')dE'\right)^{2^{n-1}-1}. \qquad (17)$$

To get a finite right-hand side in the large n limit, E_0 must be small. Hence we may approximate

$$\rho_0(E_0) \simeq 2^{n-1}p(0)\left(1 - E_0 p(0)\right)^{2^{n-1}-1}$$
$$\simeq 2^{n-1}p(0)e^{-2^{n-1}p(0)E_0}.$$

This means that the rescaled minimum,

$$\varepsilon_0 = 2^{n-1}p(0)E_0 \qquad (18)$$

is an exponential random variable,

$$\rho_0(\varepsilon) = e^{-\varepsilon} \qquad (\varepsilon > 0). \qquad (19)$$

Along similar lines, one can show [187] that the density ρ_k of the $k+1$th lowest (kth near-optimal) rescaled number is

$$\rho_k(\varepsilon) = \frac{\varepsilon^k}{k!}e^{-\varepsilon} \qquad k = 1, 2, 3, \ldots. \qquad (20)$$

Figures 8 and 9 compare eqs. (19) and (20) with the probability density of the rescaled optimal and near-optimal discrepancies for the NPP in the $\kappa \to \infty$ limit of real-valued input numbers. The agreement is amazing, even for small values of n. In fact, eqs. (19) and (20) have been established as the asymptotic

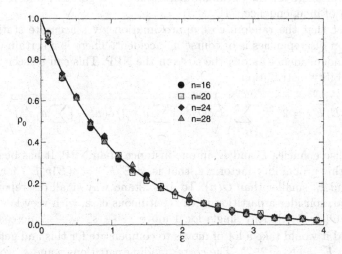

FIGURE 8 Probability density of the rescaled optimum discrepancy in the hard phase. Symbols: numerical simulations. Solid line: prediction by the random cost approximation.

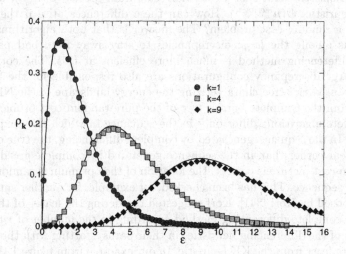

FIGURE 9 Probability densities of the kth best partition in the hard phase. Symbols: numerical simulations for $n = 24$. Solid lines: predictions by the random cost approximation.

probability measure for the optimum discrepancies—rigorously and without the assumption of independence [71].

The fact that the random cost approximation gives accurate statistics for the optimum discrepancies is of course no accident. There is a certain degree of statistical independence among the costs in the NPP. This can be seen from the joint probability distribution

$$p(E, E') = 2^{-2n} \sum_{\{s_j, s'_j\}} \left\langle \delta(E - |\sum_j a_j s_j|)\, \delta(E' - |\sum_j a_j s'_j|) \right\rangle \qquad (21)$$

of finding discrepancies E and E' in one instance of an NPP. It has been shown [390] that this probability factorizes, that is, $p(E, E') = p(E)p(E')$ for discrepancies E and E' smaller than $O(n)$. To understand why small discrepancies are uncorrelated, consider a partition in the continuous case, with very low discrepancy $E = O(\sqrt{n}\, 2^{-n})$. Any single local move $s_j \mapsto s'_j = -s_j$ increases E by $O(n^{-1})$, and it would take a lot of moves to compensate for this and get another discrepancy $E' = O(\sqrt{n}\, 2^{-n})$. The corresponding partitions s and s' would then have vanishingly small overlap, which leads to the factorization of $p(E, E')$.

The random cost problem is an algorithmic nightmare. No smart heuristic can be quicker than exhaustive search. This is the reason why there are no good heuristics in the hard phase of the NPP, and why complete algorithms cannot really take advantage of pruning rules. But there are differences in the quality of heuristic solutions: recall the result from greedy, $O(n^{-1})$, and from the KK heuristic, $O(n^{-\alpha \log n})$. How can these differences arise if the NPP is essentially a random cost problem? The answer is that both algorithms exploit correlations among the *large* discrepancies to stay away from bad partitions, and the differencing method is much more efficient at this. The correlations between large discrepancy configurations are also responsible for the fact that the complete barrier tree characterizing the energy landscape of the NPP looks different from the complete barrier tree of the pure random cost problem [483].

Complete algorithms differ only in the sequence in which they explore the partitions. In the sequence generated by complete differencing, the true optimum might appear earlier than in the sequence generated by complete greedy. But if the random cost picture is correct, the location of the optimum is random in any prescribed sequence. This has been checked, for example, for another smart algorithm proposed by Korf [341]. Korf suggested reordering the leaves of the search tree of the complete differencing method according to the number of *right turns* (violations of the differencing heuristic) in their paths, starting with those leaves that deviate least from the KK heuristic. In our example from figure 1 the leaves would be visited in the sequence $(2, 4, 4, 6, 0, 6, 8, 14, 8, 10, 12, 16, 18, 20, 22, 30)$, and in fact the perfect solution would appear earlier than in the order shown in figure 1. Numerical simulation, however, revealed that in the hard phase the position of the optimum in the sequence generated by this method is completely random—as predicted by the random cost problem [390].

Apparently, there is no way to overcome the random cost nature of the NPP in the hard phase. When the NPP is hard, it's very hard.

6 CONCLUSIONS

We have seen that random NPP has a phase transition in average complexity, and that this phase transition goes hand in hand with a transition in probability of perfect solutions. The control parameter κ of both transitions is the ratio m of the number of bits in the input variables to the number n of variables, and $\kappa_c = 1 - \log_2(n)/2n + O(n^{-1})$ is the critical value that separates the hard ($\kappa > \kappa_c$) from the easy ($\kappa < \kappa_c$) phase. Much more can be said about the phase transition, notably concerning the width of the transition window and the probability of perfect solutions inside that window. Another proven fact is the uniqueness of the solution in the hard phase. For all this (and much more), the reader is referred to the paper of Borgs, Chayes, and Pittel [71]. Their work answers most of the open questions on random NPP that are not related to algorithms. The major open problem is putting the random cost approximation on rigorous grounds and clarifying its relevance for algorithms. From a practical point of view it would be very nice to have a polynomial time algorithm that yields better results than the differencing method. After all, there is much room between $O(n^{-\alpha \log n})$ and $O(\sqrt{n}\, 2^{-n})$.

The NPP as shown here can be generalized and modified in various directions. An obvious generalization is to partition the numbers into $q \geq 2$ subsets. This is called the multiprocessor scheduling problem, and in physics parlance this corresponds to a Potts spin glass or to a walk with random stepsizes in $q - 1$ dimensions. The latter approach has been used to analyze an "easy-hard" phase transition in multiprocessor scheduling [38].

Another variant is the constrained NPP where the cardinality of the subsets is fixed. This is necessary for problems such as choosing up sides in a ball game [237], where both teams need to have the same number of players. The cardinality difference of the subsets is a control parameter that triggers another phase transition in computational complexity, giving rise to a two-dimensional phase diagram [72].

ACKNOWLEDGMENTS

Discussions with Heiko Bauke are gratefully acknowledged. Part of the numerical simulations have been performed on *Tina* [500] a 156-CPU Beowulf cluster built in-house at Otto-von-Guericke-Universität, Magdeburg.

CHAPTER 6

Ground States, Energy Landscape, and Low-Temperature Dynamics of $\pm J$ Spin Glasses

Sigismund Kobe
Jarek Krawczyk

1 INTRODUCTION

The previous three chapters have focused on the analysis of computational problems using methods from statistical physics. This chapter largely takes the reverse approach. We turn to a problem from the physics literature, the spin glass, and use the branch-and-bound method from combinatorial optimization to analyze its energy landscape. The spin glass model is a prototype that combines questions of computational complexity from the mathematical point of view and of glassy behavior from the physical one. In general, the problem of finding the ground state, or minimal energy configuration, of such model systems belongs to the class of NP-hard tasks.

The spin glass is defined using the language of the Ising model, the fundamental description of magnetism at the level of statistical mechanics. The Ising model contains a set of n spins, or binary variables s_i, each of which can take on the value *up* ($s_i = 1$) or *down* ($s_i = -1$). Finding the ground state means finding the spin variable values minimizing the Ising Hamiltonian energy (cost)

Computational Complexity and Statistical Physics, edited by
Allon G. Percus, Gabriel Istrate, and Cristopher Moore, Oxford University Press.

function, written in general as

$$E = - \sum_{i<j}^{n} J_{ij}\, s_i\, s_j \tag{1}$$

for given interaction strengths J_{ij}. This is a problem of nonlinear discrete optimization. When J_{ij} is positive, interactions are called *ferromagnetic*. In this case, there is a trivial solution: all spins are aligned, meaning they have identical signs. When J_{ij} is negative, interactions are called *antiferromagnetic*. Physically, the spins are often taken to lie on a lattice. For a square or cubic lattice with negative J_{ij} for neighboring pairs of spins and $J_{ij} = 0$ otherwise, the ground state is clearly the configuration where adjacent spins have opposite signs.

If all nonzero interaction strengths J_{ij} are equal, the system is said to be *ordered*. For ordered systems with antiferromagnetic interactions between nearest neighbors, but where neighbors of a given spin are also neighbors of each other, *frustration* prevents certain interactions from being "satisfied." An example is the triangular lattice: to quote the early work of Wannier, "antiferromagnetism does not fit into the triangular pattern" [517]. While the solution to the optimization problem is straightforward, one third of all interactions lead to conflicts that increase the ground-state energy. The structural sensitivity of antiferromagnetic order has been discussed by Sato and Kikuchi [453] for the face-centered cubic lattice. Other ordered systems have been considered by Liebmann [361].

The problem becomes more complex when *disorder* arises, and interaction strengths are not equal. Often, such systems can only be solved by numerical methods. The time complexity of an algorithm is defined by the growth of the solution time as a function of input size [230]. For many disordered spin models, it can be shown that finding ground states is NP-hard [31, 214], and so the time complexity likely grows faster than any polynomial. In order to address this, Kobe and Handrich [335] introduced a "misfit" parameter characterizing the degree of frustration, and used it to find exact ground states in a two-dimensional system of $n = 23$ hard disks (an amorphous Ising model) with distance-dependent antiferromagnetic interactions. Further exact results for two-dimensional ($n = 40$) and three-dimensional systems ($n = 30$) were obtained by Kobe [334] and Kobe and Hartwig [336] using the branch-and-bound method of combinatorial optimization [137, 354, 363].

Another concept for studying systems with disorder and competing interactions was introduced by Toulouse [502]. He analyzed the frustration effect in a two-dimensional lattice model with a random distribution of ferromagnetic and antiferromagnetic nearest-neighbor interactions J_{ij} of equal strength, known as the Edwards-Anderson $\pm J$ model. The system may be described by plaquettes representing elementary lattice regions, such as a unit cell on a square lattice. The quantity $\Phi = \prod_c J_{ij}$, taken over the contour c forming the perimeter of the plaquette, measures frustration: $\Phi = -1$ if the plaquette is frustrated, $\Phi = 1$ if it is not. The exact ground state is then associated directly with a matching [139] of frustrated plaquettes that minimizes the sum of lattice distances between matched pairs. The multiplicity of ground states, or degeneracy, comes from the total number of ways to create such a minimal matching. This approach only works for two-dimensional systems, but it is an efficient one, since minimal matching can be solved to optimality in polynomial time. In the years following Toulouse's work, the matching method of optimization was used widely [32, 52].

This chapter is organized as follows. In section 2, we introduce the branch-and-bound algorithm as a prototype for a numerical procedure of nonlinear discrete optimization. Then, in section 3, we describe the Edwards-Anderson $\pm J$ spin glass model and give an overview of numerical results for the ground-state energy and entropy. In section 4 the low-energy landscape of finite three-dimensional $\pm J$ spin glasses (consisting of clusters and valleys) is analyzed and visualized. The correlation with the real-space picture shows the existence of rigid spin domains in the ground state. We discuss dynamical consequences in section 5, focusing on the transition from one ground-state cluster to another by way of a saddle cluster. It can be shown that internal structure contributes to the slowing of relaxation processes. Finally, we point out the progress and challenges of complexity theory for a better microscopic understanding of glassy behavior.

2 BRANCH-AND-BOUND

The ground state of the Ising model with n spins $s_i = \pm 1$ is the spin configuration with energy

$$E_0 = \min_{s_i = \pm 1} \left(-\sum_{i<j}^{n} J_{ij}\, s_i\, s_j \right).$$

(2)

For interactions J_{ij} of arbitrary sign and magnitude, finding the exact ground state is an NP-hard problem. Since the number of states increases with 2^n, only for very small n can eq. (2) be solved by complete enumeration.

Complete algorithms for combinatorial optimization problems aim to reduce the numerical effort while still giving an exact solution. The general principle can be demonstrated for the branch-and-bound algorithm. The strategy of branch-and-bound is to exclude as many states with high energy values as possible, in an early stage of calculation [336]. Let us consider a small cluster with $n = 8$ spins and J_{ij} values of differing strengths. To simplify matters and without loss of generality, we take the case where there are only antiferromagnetic interactions, representing an amorphous antiferromagnetic cluster with dilution. The upper triangle of the interaction matrix $\mathbf{J} = (J_{ij})$ is given by

$$
\mathbf{J} = \begin{pmatrix}
0 & -5 & -2 & -5 & -6 & -1 & 0 & 0 \\
  & 0 & -10 & -4 & 0 & -2 & -1 & 0 \\
  &   & 0 & 0 & 0 & -3 & 0 & -1 \\
  &   &   & 0 & -3 & -5 & -7 & -4 \\
  &   &   &   & 0 & -4 & -5 & -8 \\
  &   &   &   &   & 0 & 0 & -1 \\
  &   &   &   &   &   & 0 & 0 \\
  &   &   &   &   &   &   & 0
\end{pmatrix}. \tag{3}
$$

In figure 1, a tree is constructed by successively fixing spin values. At the branching depth $l = 1$, spin number 1 is set to the positive direction $(+)$. At depth $l = 2$ the spin number 2 is fixed, and so on. At each branching node a configuration $(s_1, ..., s_l)$ and an energy value E_l is shown. For $l = 1$ the starting energy $E_1 = E_{\mathrm{id}}$ is chosen, where $E_{\mathrm{id}} = -\sum_{i<j} |J_{ij}|$ is a lower bound on the ground-state energy E_0 in eq. (2), representing the situation where all interactions are satisfied and no conflict is present. The other values E_l for $1 < l \leq n$ can be obtained by the following rule:

$$
E_l = E_{l-1} + 2 \sum_{k(\|l)}^{l-1} |J_{kl}|. \tag{4}
$$

In the example of eq. (3), where all interactions are antiferromagnetic, $k(\|l)$ denotes those k for which spin k has already been fixed in the same direction as the spin l. More generally, when both positive and negative J_{ij} values are present, the sum contains those contributions that arise due to conflicts of spin l with all spins fixed earlier. In figure 1 the values of the summation term in eq. (4) are given at the branching lines. From eq. (4) it follows that

$$
E_l \geq E_{l-1}. \tag{5}
$$

It is easy to recognize that all configurations of the system (modulo a global spin flip) and their associated energies can be found at the end of the fully branched tree, at $l = n$. The goal of the branch-and-bound strategy is to prune some branches. In order to do this, a heuristic is used to generate an approximate solution, that is, the greedyindexbranch-and-bound algorithm procedure of steepest descent shown in figure 1 where at each step the new contribution to the sum in eq. (4) is minimized. In our example, the energy of the resulting configuration is -47. This value is used as E_{bound}, and signals that branching can stop at any node where $E_l > E_{\mathrm{bound}}$. From eq. (5) it is certain that all branches pruned in this way can lead only to states with $E_n \geq E_{\mathrm{bound}}$, and so none of them can yield a solution to eq. (2). Therefore, in place of a complete enumeration of all states, the pruned tree in figure 1 can be used to search for

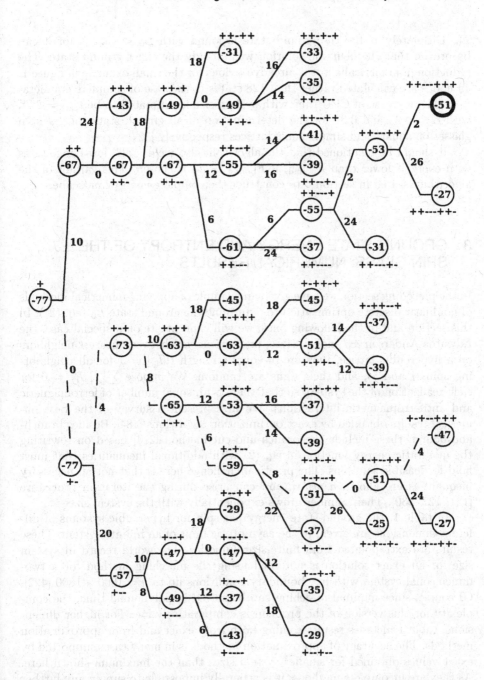

FIGURE 1   Branch-and-bound tree for a cluster with $n = 8$ spins given by the interaction matrix **J** in eq. (3); $E_{id} = -77$, $E_{bound} = -47$ obtained by steepest descent (dashed line). The exact ground state with the energy $E_0 = -51$ is marked in bold.

E_0. Ultimately, either the ground state is found with $E_0 < E_{bound}$, or it can be proven that the heuristic solution was already the exact ground state. The reduction of numerical effort is already obvious for the small example of figure 1: 49 nodes are calculated rather than 128 states in the case of complete enumeration. The increase of CPU time with system size is estimated to be $t_{calc} = 2^{\alpha n}$, with $\alpha = 0.23$ and 0.27 for the determination of all ground states of $\pm J$ spin glasses on square and simple cubic lattices respectively [327].

It should be mentioned that the algorithm also yields "all" low-lying states with energies lower than E_{bound}, if $E_{bound} > E_0$ is chosen. This variant of the algorithm is used in section 4 to construct the complete energy landscape.

3   GROUND-STATE ENERGY AND ENTROPY OF THE $\pm J$ SPIN GLASS: NUMERICAL RESULTS

In the preceding section we have given an example of applying numerical methods of nonlinear discrete optimization to determine the ground state E_0 (eq. (2)). In this section and in the following ones, we will concentrate on a special case, the Edwards-Anderson $\pm J$ model. Here, interactions are between nearest neighbors on a hypercubic lattice, they are of equal strength ($|J_{ij}| = J$ for all neighboring spins i and j), and their signs are random. We impose $\sum_{i<j}^{n} J_{ij} = 0$ for each realization of the system, so that there is an equal number of ferromagnetic and antiferromagnetic interactions. We first present a survey of the best numerical results obtained by exact optimization algorithms [484]. Besides branch-and-bound, these include the branch-and-cut method [214] based on rewriting the quadratic energy function in eq. (2) with additional inequalities that must hold for feasible solutions. The practical challenge here is that not all necessary inequalities are known *a priori*, and can arise during the iteration procedure [121, 122, 230]. Their number grows exponentially with the system size.

In table 1, the ground-state energy per spin for hypercubic systems of different dimensions are given, in the asymptotic limit of an infinite system. These results are extrapolated from finite-size numerics. The world record in system size for an exact solution is obtained using the matching method for a two-dimensional system with free boundary conditions up to $n = 1800 \times 1800$ [422]. Of course, since minimal matching can be solved in polynomial time, the complexity for this version of the problem is comparatively low. For higher dimensions, table 1 includes results coming both from exact and from approximation methods. The accuracy of approximation methods is in many cases supported by exact values obtained for smaller system sizes than the maximum shown here. As they are incomplete methods, it is generally impossible to supply any further evidence for their exactness [230]. However, the inclusion of such methods with a "high level of reliability" [262] provides the possibility of considering systems of larger size than would be otherwise available, and thus to extrapolate more convincingly to infinite lattices. Much less is known about the exact ground-state energy for other than hypercubic lattices.

TABLE 1 Ground-state energy per spin e_0 and entropy per spin s_0 of the hypercubic $\pm J$ spin glass in d dimensions. Results for infinite systems are extrapolated from system sizes up to $n = L_{max}^d$. Parenthetical numbers denote error bar in final digit(s).

d	method	L_{max}	e_0	s_0	Ref.
2	matching*	1800	−1.40193(2)		Palmer and Adler [422]
2	branch-and-cut	50	−1.4015(8)		De Simone et al. [122]
2	branch-and-bound	8	−1.40(6)	0.077(21)	Klotz [327]
2	genetic cluster appr.	40	−1.4015(3)		Hartmann [232]
2	genetic cluster appr.	40		0.078(5)	Hartmann [228]
2	transfer matrix×	11	−1.4024(12)	0.0701(5)	Cheung and McMillan [92]
2	expansion-fall-invasion-spring	10	−1.40169		Vogel et al. [513]
2	genetic	20	−1.401(1)		Gropengiesser [211]
2	flat histogram sampling	32	−1.4007(85)	0.0709(6)	Zhan et al. [537]
3	branch-and-bound	4	−1.778(14)	0.054(16)	Klotz [327]
3	extremal optimization	12	−1.7865(3)		Boettcher and Percus [61]
3	genetic	10	−1.787(3)		Gropengiesser [211]
3	genetic cluster appr.	8		0.051(3)	Hartmann [228]
3	genetic cluster appr.	14	−1.7876(3)		Hartmann and Rieger [230]
3	multicanonical sampling	12		0.04412(46)	Berg et al. [47]
4	genetic cluster appr.	7	−2.095(1)		Hartmann and Rieger [230]
4	genetic cluster appr.	6		0.027(5)	Hartmann [228]
4	extremal optimization	7	−2.093(1)		Boettcher and Percus [61]
5	extremal optimization+	4	−2.3511		Boettcher [57]

* free boundary conditions
× rectangular lattice ($L \times W$) with $L_{max} = 11$ (periodic boundary conditions) and $W_{max} = 10^4 \cdots 10^5$ (free boundary conditions)
+ without extrapolation

A shortcoming of presenting the ground-state energy per spin is that the value is not comparable across different dimensions, lattice types, etc. For that reason, a universal measure of frustration has been introduced by the *misfit* parameter

$$\mu_0 = \frac{1}{2}\left(1 + \frac{E_0}{\sum_{i<j}|J_{ij}|}\right), \qquad (6)$$

TABLE 2　Misfit parameter μ_0 of the $\pm J$ spin glass in d dimensions. Estimate for infinite system, extrapolated from numerics.

lattice	d	μ_0	E_0 from
honeycomb	2	0.09	Lebrecht and Vogel [356]
square	2	0.150	Table 1
triangular	2	0.22	Vogel et al. [511]
simple cubic	3	0.202	Table 1
hypercubic	4	0.24	Boettcher and Percus [61]
hypercubic	5	0.26	Boettcher [57]

representing the mean fraction of unsatisfied bonds in the ground state [337]. For the $\pm J$ spin glass, μ_0 values from numerical simulations are compiled in table 2. They may be compared with $\mu_0 = 1/3$ for the antiferromagnetic triangular or face-centered cubic lattice, and $\mu_0 = 1/2$ for fully frustrated hypercubic and face-centered cubic lattices in the limiting case of infinite dimensions [16, 120]. Moreover, it can be seen that the $\pm J$ spin glass is less frustrated on the honeycomb lattice, and more frustrated on the triangular lattice, than on the square one.

4　ENERGY LANDSCAPE

An advantage of the branch-and-bound algorithm is that it is very easy to implement a variant allowing the calculation of all near-optimal solutions. For these purposes, a certain $E_{\text{bound}} > E_0$ has to be chosen and fixed during the calculation. All states of the system with energy $E_i < E_{\text{bound}}$ can then be found. (Note that here, the subscript i denotes an excited state of the system, that is, with higher energy than the ground state, rather than an intermediate level of branch-and-bound as in section 2.) Through subsequent analysis of these states with respect to their neighborhood structure, the complete low-energy landscape in the high-dimensional configuration space can be obtained. The situation in this space is analogous to that of "fog in the mountains" in a real landscape: all areas below the upper limit of the fog are covered.

　　Let us first investigate the low-energy landscape of a three-dimensional system of size $n = 4 \times 4 \times 4$, with periodic boundary conditions. All $N = 1635796$ configurations up to the third excitation (fourth-lowest energy state) were calculated using branch-and-bound [231]. The configurations were then studied with regard to their one-spin neighborhood. Two configurations that differ in the orientation of only one spin are considered neighbors. Consequently, each of the N configurations can have at most n neighbors belonging to the set of N. The

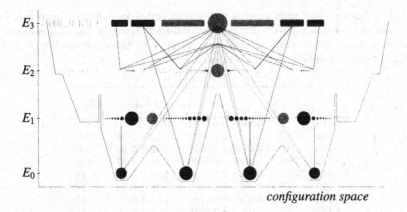

configuration space

FIGURE 2 Schematic picture of the exact low-energy landscape up to the third excitation for one system of size $n = 4 \times 4 \times 4$. Clusters are marked by circles of sizes proportional to the number of configurations in the cluster: the two ground-state clusters on the left, for instance, consist of 12 and 18 configurations. (Note that the scale is different for different energy levels, so the largest clusters in the first, second, and third excitations contain 819, 82,960, and 1,503,690 configurations, respectively.) Lines denote single spin-flip connections. All clusters connected to the same neighborhood structure are pooled in a box.

low-energy landscape is formed by all states of energy E_k, with $k \leq 3$. Due to the discreteness of the coupling constants J_{ij}, the energy values are degenerate. E_0 is the ground-state energy and $E_k = E_0 + 4k$ are the excitation energies.

An energy landscape is thus formed, consisting of clusters, valleys and saddles [328, 329, 330]. A set of configurations is called a *cluster* if a *chain* connecting them exists. The chain is built up by neighboring configurations with the same energy. The landscape is symmetric, due to eq. (2). Two clusters of different energies are "connected" whenever at least one configuration of the first cluster is a neighbor of one configuration of the second cluster. A schematic picture of this low-energy landscape is illustrated in figure 2. Finally, *valleys* can be associated with ground-state clusters. A valley consists of clusters that have connections to one single ground-state cluster. Different valleys are connected by *saddle* clusters, which mediate the transition over energy barriers.

Note that there is a broad distribution of realizations of the $\pm J$ systems. The ground-state energy of 8555 systems of the size $n = 4 \times 4 \times 4$ varies between $E_0 = -100$ and -128. The respective values for the mean ground-state entropy $\overline{s}_0$, the number of clusters $\overline{N}_{cl}$, and the number of ground states $\overline{N}_{gs}$, are given in table 3.

In table 4, corresponding values characterizing the structure of the first excitations of the same set of realizations are given. Here, the following average

TABLE 3   Characteristic properties of the ground states for $8,555$ realizations of systems of size $n = 4 \times 4 \times 4$ with different ground-state energies E_0. N_{sys} denotes the number of systems, $\overline{s}_0$ the mean ground-state entropy, $\overline{N}_{cl}$ the mean number of clusters, and $\overline{N}_{gs}$ the mean number of ground states.

E_0	N_{sys}	$\overline{s}_0$	$\overline{N}_{cl}$	$\overline{N}_{gs}$
-100	5	0.1153 ± 0.0091	10.20 ± 2.82	3848
-104	505	0.0974 ± 0.0231	6.19 ± 3.14	1088
-108	2769	0.0748 ± 0.0257	3.19 ± 1.99	326
-112	3541	0.0566 ± 0.0249	1.91 ± 1.17	114
-116	1358	0.0448 ± 0.0221	1.38 ± 0.74	45
-120	291	0.0371 ± 0.0210	1.18 ± 0.46	24
-124	52	0.0311 ± 0.0181	1.11 ± 0.32	14
-128	7	0.0259 ± 0.0235	1.33 ± 0.57	41
	8555	0.0623 ± 0.0285	2.47 ± 1.98	228

values are specified: $\overline{N}_1$ is the mean number of states in the first excitation: of these, $\overline{N}_s$ belong to saddle clusters and $\overline{N}_m$ are metastable states without direct connections to one of the ground states. $\overline{N}_{cl1}$ is the mean number of clusters: of these, $\overline{N}_{cls}$ are saddle clusters and $\overline{N}_{clm}$ are metastable clusters. It can be seen that systems with higher ground-state energies (i.e., higher frustration) also possess more complex energy landscapes with larger entropies and many clusters.

The relation between the energy landscape in configuration space and the spin structure in real space is demonstrated in figure 3. Here an example with $n = 6 \times 6 \times 6$ spins is shown. The ground states can be grouped into four clusters, similar to the situation in figure 2. Two clusters contain 5632 states and two clusters contain 1280 states. The degeneracy within the clusters is caused by the existence of *free spins* that feel no internal field and can thus be flipped without energy input. Let us first consider the two clusters that remain when one ignores the *mirror states* arising from a global spin flip. All spins in real space that are free in either of these clusters are marked by empty circles. The remaining spins are divided into two groups, marked by full circles and shaded triangles. In each of these groups the relative orientation of any given spin is fixed with respect to all others in the group. Due to this internal rigidity, the two groups are called *spin domains* [230]. When one includes the mirror states, there are four different orientations of the two spin domains, resulting in the four ground-state clusters in configuration space, see also Hed et al. [238]. Many of the free spins are situated physically between the spin domains. Thus, the low-energy excitations in figure 2 can be understood as a successive softening of the spin domains starting from the boundary region of free spins between them. Consequently, a transition over

TABLE 4 Characteristic properties of the first excitations for 8,555 realizations of systems of size $n = 4 \times 4 \times 4$ with different ground-state energies E_0. $\overline{N}_1$ denotes the mean number of first excited states, $\overline{N}_{cl1}$ the mean number of clusters, $\overline{N}_s$ the mean number of saddle states, $\overline{N}_{cls}$ the mean number of saddle clusters, $\overline{N}_m$ the mean number of metastable states, and $\overline{N}_{clm}$ the mean number of metastable clusters.

E_0	$\overline{N}_1$	$\overline{N}_{cl1}$	$\overline{N}_s$	$\overline{N}_{cls}$	$\overline{N}_m$	$\overline{N}_{clm}$
-100	384825	61.00 ± 8.51	191598	0.70 ± 0.45	351	7.20 ± 3.56
-104	102879	47.64 ± 15.76	50016	0.98 ± 0.68	543	6.31 ± 3.58
-108	25469	33.48 ± 14.37	11079	0.99 ± 0.90	375	4.59 ± 2.73
-112	6653	23.02 ± 10.81	2199	0.54 ± 0.77	151	2.57 ± 1.98
-116	1895	17.01 ± 7.76	375	0.30 ± 0.62	42	1.25 ± 1.28
-120	775	14.18 ± 5.07	81	0.16 ± 0.41	14	0.59 ± 0.94
-124	395	12.81 ± 4.11	22	0.15 ± 0.44	2	0.14 ± 0.36
-128	322	13.42 ± 4.75	13	0.14 ± 0.38	0.43	0.14 ± 0.37
	17634	26.5 ± 14.4	7326	0.68 ± 0.83	225	3.14 ± 2.73

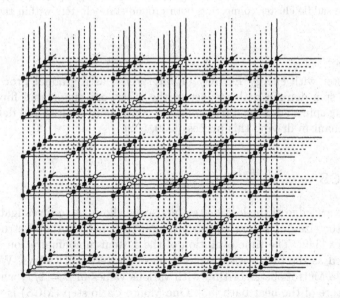

FIGURE 3 Two spin domains of a $\pm J$ spin glass with $n = 6 \times 6 \times 6$, marked by full circles and shaded triangles. All spins that are free in either of the ground-state clusters are marked by empty circles.

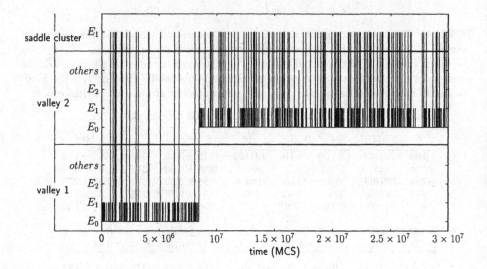

FIGURE 4 An individual Monte Carlo run through the landscape vs. time, at inverse temperature $\beta = 2.5$ for one system of size $n = 4 \times 4 \times 4$ (fig. 2). The process starts from an arbitrary state within the ground-state cluster on the left in figure 2. The vertical axis shows different energies belonging to valleys 1 and 2, respectively, and the energy E_1 of the saddle cluster connecting both ground-state clusters within the first excitation.

the saddle cluster with energy E_1 may be interpreted as a gradual process of reversal of one spin domain with respect to the other, one single spin flip at a time. By flipping spins, additional free spins are continually created and deleted: that is the mechanism driving this process [230, 512].

5 DYNAMICS

The complete knowledge of the low-energy landscape allows us to investigate the influence that the size of clusters and valleys and their neighborhood structure has on dynamics [349]. The time evolution of the system in configuration space can be described as the progressive exploration of clusters and valleys. We use the Monte Carlo Metropolis algorithm with various values of $\beta = 1/\tau$, where τ is the temperature of the heat bath [53]. One Monte Carlo step (MCS) is taken as the time unit. An individual run through the landscape is shown in figure 4. We start from an arbitrary state in the leftmost ground-state cluster of figure 2.

At first, the system walks in the valley, sometimes touching the saddle cluster in the first excitation. After an escape time t_{esc} of the order of 10^7 MCS, the system leaves the first valley and goes through the saddle cluster to the second

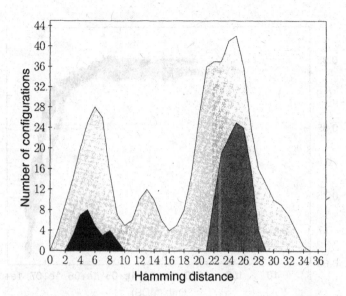

FIGURE 5 The transition profile of the saddle cluster with energy E_1 in figure 2 illustrated by the number of configurations vs. Hamming distance from a reference state (see text). The shaded area marks all configuration in the saddle cluster. States having connections with valley 1 (dark) and 2 (middle) are shown in black and grey respectively.

one. This transition is governed by the internal structure of the saddle cluster, shown as its transition profile in figure 5. First, all pairs of configurations are checked to find the largest Hamming distance h_d (the number of spin values differing between the two configurations). Then, using one of these states as the reference state, the h_d values of all configurations in the saddle cluster with respect to the reference state are calculated. Two sets of states are marked, one consisting of states connected by a single spin flip with the first valley and the other with the second valley. These sets denote the input and the output areas for a transition from one valley to the other. Considering a transition as a walk between these sets, it is clearly slowed down by the small numbers of states in between.

Quantitatively, the random walk can be described by the spin correlation function

$$q(t) = \frac{1}{n}\left\langle \sum_{i=1}^{n} s_i^G(0)s_i(t) \right\rangle, \tag{7}$$

where $s_i^G(0)$ is the ith spin of the starting configuration chosen arbitrarily from the ground states of valley 1 or 2. The brackets denote the average of 100 runs starting from the same state (fig. 6).

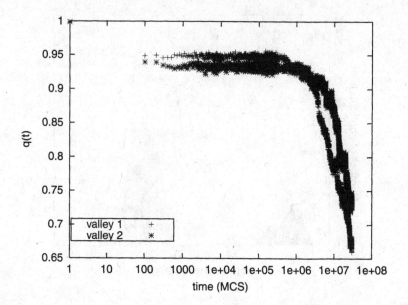

FIGURE 6   The spin correlation function vs. time for the system of size $n = 4 \times 4 \times 4$ (fig. 2). The starting configuration is selected from the set of ground states of valley 1 and 2. The Monte Carlo process is run at inverse temperature $\beta = 2.5$.

The spin correlation function $q(t)$ vs. time is characterized by a plateau with the value q_{pl} followed by a temperature-dependent decay. It should be noted that such a plateau is typical for supercooled liquids, where the dynamical process is called α-relaxation. To examine the correlation between the structure of the landscape and the dynamics, we compare q_{pl} with the size of the valley, keeping in mind that the spin correlation within the valley can be calculated using the mean Hamming distance $\bar{h}_d$ of all pairs of states by

$$q_{pl}^{(Ham)} = 1 - 2\bar{h}_d/n. \tag{8}$$

We find an agreement between q_{pl} and $q_{pl}^{(Ham)}$ (table 5), where the average in eq. (8) approximated by the average over all states in the corresponding ground-state clusters.

The plateau thus reflects the dynamics within the valley. The subsequent decay of $q(t)$ shows the escape from the valley. The escape time t_{esc} depends on the temperature and can be fitted by $t_{esc} \sim exp(\beta \, \Delta E_{eff})$. We found $\Delta E_{eff} = 4.24 \pm 0.08$ for valley number 1 and $\Delta E_{eff} = 4.46 \pm 0.09$ for valley number 2. The effective energy barrier is larger than the real one, which is $\Delta E = E_1 - E_0 = 4$ in our example. Moreover, ΔE_{eff} is larger for valley 2 than for 1. This reflects the fact that the system can leave the saddle cluster more easily in the direction of 2, as there are more exit connections (see fig. 5).

6 SUMMARY

Due to the physical complexity of the spin glass problem, advanced methods of combinatorial optimization are required. In recent years, powerful numerical algorithms have become available, enabling us to study model systems of small and moderate sizes from the microscopic point of view. For example, it is possible to determine the internal structure of an energy landscape in a high-dimensional configuration space. Understanding the slow dynamics of glassy systems is a current challenge of solid state physics. Spin glasses are good candidates for modeling glassy behavior.

In this chapter, we have discussed the $\pm J$ spin glass model, and shown the correlation between the microscopic structure of the energy landscape and the dynamical behavior. The characteristic shape of the correlation function may be attributed to the restricted connectivity of clusters and valleys in the energy landscape and to their internal profiles. Finding better algorithms for NP-hard problems remains an ongoing challenge. Our hope is that with the development of improved algorithms, the restriction to small system sizes can be eased, and the ground-state behavior of $\pm J$ spin glasses can be analyzed with improved confidence [534].

ACKNOWLEDGMENTS

The authors wish to thank A. Heuer, S. Boettcher, and A. K. Hartmann for valuable discussions. Thanks to A. K. Hartmann also for permission to use his data concerning the ground state of a three-dimensional system with $n = 6 \times 6 \times 6$. This work has been supported by Graduiertenkolleg "Struktur- und Korrelationseffekte in Festkörpern."

TABLE 5   The values of q_{pl} obtained from simulations (fig. 6) and calculation (eq. (8))

	Figure 6	Equation (8)
q_{pl} (1)	0.947 ± 0.004	0.936
q_{pl} (2)	0.932 ± 0.004	0.924
Δq_{pl}	0.015 ± 0.004	0.012

Part 3: Identifying the Threshold

CHAPTER 7

The Satisfiability Threshold Conjecture: Techniques Behind Upper Bound Improvements

Lefteris M. Kirousis
Yannis C. Stamatiou
Michele Zito

1 INTRODUCTION

One of the most challenging problems in probability and complexity theory is to establish and determine the *satisfiability threshold*, or phase transition, for random k-SAT instances: Boolean formulas consisting of clauses with exactly k literals. As the previous part of the volume has explored, empirical observations suggest that there exists a critical ratio of the number of clauses to the number of variables, such that almost all randomly generated formulas with a higher ratio are unsatisfiable while almost all randomly generated formulas with a lower ratio are satisfiable. The statement that such a crossover point really exists is called the *satisfiability threshold conjecture*. Experiments hint at such a direction, but as far as theoretical work is concerned, progress has been difficult. In an important advance, Friedgut [177] showed that the phase transition is a sharp one, though without proving that it takes place at a "fixed" ratio for large formulas. Otherwise, rigorous proofs have focused on providing successively better upper and lower bounds for the value of the (conjectured) threshold. In this chapter,

Computational Complexity and Statistical Physics, edited by
Allon G. Percus, Gabriel Istrate, and Cristopher Moore, Oxford University Press.

our goal is to review the series of improvements of upper bounds for 3-SAT and the techniques leading to these. We give only a passing reference to the improvements of the lower bounds as they rely on significantly different techniques, one of which is discussed in the next chapter.

Let ϕ be a random k-SAT formula constructed by selecting, uniformly and with replacement, m clauses from the set of all possible clauses with k literals (no variable repetitions allowed within a clause) over n variables. It has been experimentally observed that as the numbers n, m of variables and clauses tend to infinity while the ratio or *clause density* m/n is fixed to a constant α, the property of satisfiability exhibits a phase transition. For the case of 3-SAT, when α is greater than a number that has been experimentally determined to be approximately 4.27, then almost all random 3-SAT formulas are unsatisfiable; that is, the fraction of unsatisfiable formulas tends to 1. The opposite is true when $\alpha < 4.27$. Analogous phenomena have been observed for k-SAT with $k > 3$, and the experimentally determined threshold point increases with k. The experiments that led to these conclusions were initiated by the work of Cheeseman et al. [91]. For detailed numerical results see Crawford and Auton [110] and Mitchell et al. [400]. For $k = 2$, it has been rigorously established, independently by Chvátal and Reed [94], Goerdt [201, 202], and Fernandez de la Vega [156], that a transition from almost certain satisfiability to almost certain unsatisfiability takes place at a clause-to-variable ratio equal to 1.

For $k \geq 3$, finding the exact value of the threshold point where this transition occurs—or even proving that such a threshold exists—is still an open problem. The following is known. Friedgut [177] has shown that for k-SAT the transition is sharp, so that in the large n limit, the probability of satisfiability changes from arbitrarily close to 1 to arbitrarily close to 0, as the density α moves along arbitrarily short intervals. However, it is not known whether these intervals converge to a fixed point. Also, Istrate et al. [273] have shown that the transition is *first order*: as α moves along these intervals of asymptotically zero length, the value of a certain combinatorial parameter of the random formula jumps from zero to a nonzero multiple of n. Such parameters are called *order parameters* in statistical physics. The specific one used in 3-SAT is the size of the formula's *spine*, defined as the set of all literals l for which a subformula $\psi \subseteq \phi$ can be found, so that l is FALSE in every truth assignment satisfying ψ. Furthermore, recent theoretical work in statistical physics [395] has supplied additional and almost conclusive evidence—though not a formal proof in the mathematical sense—for the existence of the threshold point. Some of this has been discussed in chapter 4.

Apart from the results above, much effort has been put into rigorously establishing upper and lower bounds for the region where the k-SAT transition occurs. These efforts have resulted in interesting and novel probabilistic techniques. In this chapter we will mainly concentrate on presenting the upper-bound results and the techniques that lead to them (see also the review by Dubois [129]).

2 GENERATING RANDOM 3-SAT FORMULAS

Let Ω denote the set of all $2^3 \binom{n}{3}$ possible 3-SAT clauses. A random 3-SAT formula ϕ on $m = \alpha n$ clauses can be formed using one of the following frequently employed probability models:

1. Model $\mathcal{G}_{m,m}$: select the m clauses of ϕ by drawing them uniformly at random, independently of one another and with replacement, from Ω;
2. Model $\mathcal{G}_m$: as above, but with no replacement;
3. Model $\mathcal{G}_p$: place each clause of Ω in ϕ independently of the others and with probability p; and
4. Model $\mathcal{G}_{3,\alpha n}$: fill each of the $3\alpha n$ possible literal positions (αn clauses each having 3 literals) with literals chosen uniformly at random, independently and with replacement, from the set of $2n$ possible literals over the n variables. Note that this model allows the formation of clauses containing variable repetitions.

All of these models are variations on the *fixed clause length* model introduced by Franco and Paull [167]. That model was an adaptation of the classical model for random graphs introduced by Erdős and Rényi in a series of seminal papers published starting in 1959 [147, 148] (see the book of Bollobás [62] for the historical development of the field of random graphs).

The fixed clause length model is in sharp contrast to the *variable clause length* model introduced by Goldberg et al. [204] in order to study the average time complexity of satisfiability algorithms. In the variable clause length model, each of a fixed number of clauses is formed by placing every possible literal in the clause with some probability, and independently of the others. This model has the disadvantage of inducing, on the set of all Boolean formulas with given n, a probability distribution that favors easy instances. The fixed clause length model does not have this feature, since it allows the manipulation of the instance hardness by means of the clause density parameter (clause-to-variable ratio) α.

Each of these models has its own distinct advantages and disadvantages. The model $\mathcal{G}_{m,m}$ usually leads to tighter results than $\mathcal{G}_p$. On the other hand, the latter has the important property of independence for events involving "nonintersecting" sets of clauses, events that may be dependent in $\mathcal{G}_{m,m}$. Finally, as we will see later, $\mathcal{G}_{3,\alpha n}$ enables one to study as well as manipulate individual literal appearances in a formula. This fact leads to a finer description of the formula than the detail that the other models can achieve. As we discuss in section 6, this may lead to better upper bound values, as one usually applies the techniques we will examine on a more limited and well-defined set of formulas. However, it can be shown that if a threshold exists in any one of the models above, it exists in all of them and its value is equal in all of them, even though the bounds obtained by a given method may differ from model to model.

In the sections where we examine the rigorous techniques that have been used in order to bound the satisfiability threshold from above, we will see examples of the advantages and disadvantages mentioned and how they are exploited or circumvented, respectively. Unless stated otherwise, we will assume throughout this chapter that we work with the model $\mathcal{G}_{m,m}$.

3   THE SUCCESSIVE THRESHOLD APPROXIMATIONS

For the purposes of this chapter, we will accept the satisfiability threshold conjecture, and denote the k-SAT transition point in the large n limit by α_k. The basic mathematical tool employed for bounding α_k from above is a probabilistic technique known as the *first moment method*. This method makes use of *Markov's inequality*: let X be a nonnegative integer random variable and let $\mathbf{E}[X]$ be the expectation of X, then $\mathbf{Pr}[X \geq 1] \leq \mathbf{E}[X]$. In our case Markov's inequality is applied on a sequence of random variables $X = X_n, n = 0, 1, \ldots$, that depend on certain *control parameters*. If one finds a condition on the control parameters that forces $\mathbf{E}[X]$ to approach zero as n approaches infinity, then the probability of X being nonzero also vanishes in the large n limit as long as that condition holds. Despite its simplicity, the first moment method is a powerful tool that quickly provides us with a condition (though most often not the tightest possible) for proving that asymptotically a random variable is almost certainly zero,

The connection of the first moment method with the satisfiability threshold conjecture was observed by a number of researchers, including Franco and Paull [167], Simon et al. [475] and Chvátal and Szemerédi [95]. Let ϕ be a random 3-SAT formula on n variables generated according to $\mathcal{G}_{m,m}$ and let $\mathcal{A}_n(\phi)$—or simply $\mathcal{A}_n$ if ϕ is implied by the context—be the random set consisting of the truth assignments that satisfy ϕ. The probability that a truth assignment satisfies a single clause is $7/8$, so given 2^n possible truth assignments, $\mathbf{E}[|\mathcal{A}_n|] = 2^n (7/8)^m$. Since $\mathbf{Pr}[\phi$ is satisfiable$] = \mathbf{Pr}[|\mathcal{A}_n| \geq 1]$, from Markov's inequality it follows that

$$\mathbf{Pr}[\phi \text{ is satisfiable}] \leq 2^n \left(\frac{7}{8}\right)^{\alpha n}. \tag{1}$$

If by α_M we denote the exact solution of the equation $2(7/8)^\alpha = 1$ (so that $\alpha_M = \log 2/\log(8/7) \approx 5.19$), then we observe that under the condition $\alpha > \alpha_M$ the right-hand side of eq. (1) tends to zero. This establishes the value α_M as an upper bound for the critical value α_3.

It is perhaps instructive at this point to provide the Markov inequality computations for model $\mathcal{G}_p$ as an example of the difference in accuracy that can be obtained using various random models. In $\mathcal{G}_p$, the probability that a truth assignment satisfies a random formula is the probability that none of the $\binom{n}{3}$ clauses violated by the assignment are part of the formula, or $(1 - p)^{\binom{n}{3}}$. Let us set $p = (6\alpha)/(8n^2)s$, so that for large n the mean number of clauses in ϕ is αn. Note that for such a choice of selection probability, it holds that if the event "ϕ is satisfiable" has a vanishingly small probability in the $\mathcal{G}_p$ model, the probability of this event is also small in $\mathcal{G}_m$ and $\mathcal{G}_{m,m}$ for $m = \alpha n$, as well as in $\mathcal{G}_{3,\alpha n}$. By Markov's inequality in $\mathcal{G}_p$ we have

$$\mathbf{Pr}[\phi \text{ is satisfiable}] \leq \mathbf{E}[|\mathcal{A}_n|] = 2^n \left(1 - p(n)\right)^{\binom{n}{3}} = 2^n \left(1 - \frac{6\alpha}{8n^2}\right)^{\binom{n}{3}}, \qquad (2)$$

so asymptotically, $\mathbf{Pr}[\phi \text{ is satisfiable}] \leq 2^n e^{-\alpha n/8}$. This leads to the inequality $\alpha_3 < 8 \log 2 \approx 5.545$, a weaker one than for $\mathcal{G}_{m,m}$. Equations (1) and (2) provide a simple demonstration of a frequently occurring tradeoff among the various probabilistic models: accuracy of results vs. ease of handling complicated situations, such as the computation of the probability of conjunctions of events.

The first observation that the inequality $\alpha_3 < 5.19$ is not the best one possible came from Broder, Frieze, and Upfal [81], who pointed out that the condition $\alpha > \alpha_M - 10^{-7}$ is sufficient to guarantee that $\mathbf{Pr}[\phi \text{ is satisfiable}]$ tends to zero. El Maftouhi and Fernandez de la Vega [145] obtained a further improvement, by showing that the condition can be relaxed to $\alpha > 5.08$. Then Kamath et al. [292] obtained the improved condition $\alpha > 4.758$ using a numerical computation while also giving an analytical proof of the condition $\alpha > 4.87$. Using a refinement of Markov's inequality based on the definition of a restricted class of satisfying truth assignments, Kirousis, Kranakis, and Krizanc [323] proved an upper bound value $\alpha > 4.667$. Using the same class of satisfying truth assignments, after more accurate but lengthier computations, Dubois and Boufkhad [130] independently obtained the upper bound 4.642. Also, Kirousis et al. [324] give the bound 4.602 by what they call "the method of local maxima." Later, Janson, Stamatiou, and Vamvakari [278] lowered this value to 4.596 through two different approaches: by viewing a formula as a physical spin system and taking advantage of techniques from statistical physics to compute an asymptotic expression for its energy, and by obtaining an improved upper bound to the Rogers-Szegő polynomials. In Zito's doctoral thesis [538] the upper bound was further improved to about 4.58 while Kaporis et al. [293] obtained the value 4.571 using a new upper bound for the q-binomial coefficients obtained in Kirousis et al. [325]. Finally, Dubois, Boufkhad, and Mandler [132, 134] gave an upper bound of 4.506 using an approach involving formulas with a "typical" number of appearances of signed occurrences of their variables.

For general k, Franco and Paull [167] used the first moment method and derived an upper bound for the value of the satisfiability threshold of k-SAT equal to $2^k \log 2$, while the same derivation was also observed by Simon et al. [475] and Chvátal and Szemerédi [95]. Kirousis, Kranakis, Krizanc and Stamatiou [324] and, independently, Dubois and Boufkhad [130] gave techniques that improved this general upper bound without, however, improving the leading term that in both approaches is equal to $2^k \log 2$.

On the lower bound side, Chao and Franco [88, 89] were the first to analyze the asymptotic behavior of algorithms that apply a heuristic in order to iteratively assign a truth value to all the variables of a formula. If the heuristic is sure to succeed at a given value of α, this clearly provides a lower bound on the threshold. One of the algorithms they analyzed applied the *unit clause* heuristic defined in chapter 3. Using a technique relying on differential equations in order to model the workings of their algorithm, they showed that the algorithm succeeds with positive probability (but *not* necessarily with high probability) for clause-to-variable ratio less than 2.9. Following this, the first lower bound for 3-SAT was established by Franco [166], who analyzed an algorithm that satisfies only literals whose complements do not appear in the formula (pure literals). He showed that for $\alpha < 1$, the algorithm succeeds almost certainly—meaning with probability approaching one in the large n limit—in finding a satisfying truth assignment to all the variables. Broder, Frieze and Upfal [81] then showed that the pure literal heuristic actually succeeds almost certainly in satisfying a formula if the ratio is smaller than 1.63. Frieze and Suen [182] improved the lower bound to 3.003 by analyzing the *generalized unit clause* heuristic (GUC) with limited backtracking and showing that it succeeds almost certainly for ratios lower than 3.003, as discussed in chapter 3. Finally, using the differential equations method developed by Wormald [529] for approximating the evolution of discrete random processes, Achlioptas [2] and Achlioptas and Sorkin [5] reached the values 3.143 and 3.26, respectively. They developed a framework for a special class of algorithms called *myopic*, and showed that no algorithm in this class can succeed almost certainly in satisfying formulas with clause-to-variable ratios larger than 3.26. Recently, Kaporis, Kirousis, and Lalas [294, 295] analyzed a simple greedy heuristic using the methodology discussed in chapter 8, where the literal that is selected to be satisfied at each step is the one with the maximum number of occurrences in the formula. They obtained the lower bound of 3.42. This was the first time that a heuristic making use of information related to the number of appearances of literals in a random formula (*degree sequence*) has been analyzed. With a little more complicated greedy heuristics that at each step satisfy a literal with a large degree but whose negation has a small degree, a lower bound of more than 3.52 can be attained. This is currently the best value.

The best currently known *general* lower bound for k-SAT, for any fixed value of k, is given by a recent result by Achlioptas and Moore [3] who showed that $\alpha_k \geq 2^k \log 2/2 - c$, for some constant $c > 0$ independent of k. This result essentially bridged the asymptotic gap between the $2^k \log 2$ general upper bound and the $1.817(2^k/k)$ previously best general lower bound obtained by Frieze and Suen [182]. Moreover, Frieze and Wormald [183] showed that α_k is asymptotic to $2^k \log 2$ if k is a function of n and $k - \log_2 n \to \infty$. Both results are the first successful efforts (to the best of our knowledge) in applying the *second moment method* in order to prove a lower bound to the satisfiability threshold, something that previously was feasible only through the probabilistic analysis of satisfiability algorithms relying on specific heuristics for random formulas, as discussed above. Finally, using a technique known in physics as the *replica method*, Monasson and Zecchina predicted [405] that the asymptotic (in k) expression for the threshold is equal to $2^k \log 2$, although their approach was not a rigorous one.

4 UPPER BOUND APPROACHES BASED ON THE HARMONIC MEAN

The first moment method is simple to apply, but does not lead to the best possible upper bounds for k-SAT. For values of the clause-to-variable ratio smaller than α_M as defined in the previous section, the expected number of satisfying truth assignments of a random formula tends to infinity, even though the empirical evidence suggests that most such formulas have no satisfying truth assignment at all. This is due to the fact that there exist very rare formulas that are satisfiable and have a large number of satisfying assignments. El Maftouhi and Fernandez de la Vega [145] and, independently, Kamath et al. [292], studied this situation in detail. They resorted to the harmonic mean formula, first introduced (or formalized) by Aldous [13] to address the problem.

Aldous's result. Let $(B_i : i \in I)$ be a finite family of events in a probability space. For a permutation π of I, call (B_i) *invariant under π* if

$$\mathbf{Pr}[B_{i_1} \cap B_{i_2} \ldots \cap B_{i_r}] = \mathbf{Pr}[B_{\pi(i_1)} \cap B_{\pi(i_2)} \ldots \cap B_{\pi(i_r)}]$$

for all $\alpha \geq 1$ and $i_1, \ldots, i_r \in I$. Call the family (B_i) *transitive invariant* if for each $i_1, i_2 \in I$ there exists π such that $\pi(i_1) = i_2$ and (B_i) is invariant under π. In particular, transitive invariance implies that $\mathbf{Pr}[B_i] = p$ is actually independent of i.

Let N be the random variable counting the number of B_i's that occur. Then, if $(B_i : i \in I)$ is a transitive invariant family of events (with $p = \mathbf{Pr}[B_i]$ independent of i),

$$\mathbf{Pr}[\bigcup_{i \in I} B_i] = p \cdot |I| \cdot \mathbf{E}[N^{-1}|B_j]$$

for any $j \in I$. The method gives a new expression for $\mathbf{Pr}[\phi$ is satisfiable], if one interprets B_i as the event "assignment A_i satisfies ϕ." (Note that $|\mathcal{A}_n|$, the number of truth assignments satisfying ϕ, is denoted by $|\mathcal{M}od(\mathcal{F})|$ in El Maftouhi and Fernandez de la Vega [145] and $\#F$ in Kamath et al. [292].) Let T_i be the set of formulas satisfied by the ith truth assignment when these assignments are placed in reverse lexicographic order, so that T_1 consists of those formulas satisfied by all variables set to TRUE. In that case, letting $j = 1$ without loss of generality, the following is a restatement of Aldous's result in the context of 3-SAT formulas:

$$\mathbf{Pr}[\phi \text{ is satisfiable}] = \mathbf{E}[|\mathcal{A}_n|] \times \sum_{\psi \in T_1} \frac{1}{|\mathcal{A}_n(\psi)| \cdot |T_1|} \ .$$

Notice that an expression equivalent to the equation above is the following (this is the one proven explicitly in El Maftouhi and Fernandez de la Vega [145, eq. (1)]):

$$\mathbf{E}_{\phi \in \mathrm{SAT}}[|\mathcal{A}_n|] = \frac{|T_1|}{\sum_{\psi \in T_1} \frac{1}{|\mathcal{A}_n(\psi)|}}$$

where $\mathbf{E}_{\phi \in \mathrm{SAT}}[|\mathcal{A}_n|]$ is the expectation of $|\mathcal{A}_n|$ with respect to all satisfiable formulas on n variables and m clauses. El Maftouhi and Fernandez de la Vega [145] prove that it is possible to define a class of formulas $T_1^* \subseteq T_1$ of size at least $(1 - 2^{-\delta n})|T_1|$, where δ is a constant, such that each formula in T_1^* has at least $2^{\delta n}$ satisfying truth assignments. As we will see in subsection 4.1, this implies $\mathbf{E}_{\phi \in \mathrm{SAT}}[|\mathcal{A}_n|] \geq 2^{\delta n - 1}$. Therefore, the probability that a random 3-SAT formula is satisfiable is at most

$$2^n \left(\frac{7}{8}\right)^{\alpha n} 2^{1-\delta n} .$$

The authors set $\alpha = 5.08$, and using a simple random experiment they find a class of formulas T_1^* satisfying the conditions above when $\delta = 0.02137$. The satisfiability probability goes to zero asymptotically for these values, establishing the improved bound $\alpha_3 < 5.08$. The different quality of the bounds derived in El Maftouhi and Fernandez de la Vega [145] and Kamath et al. [292] is due not only to the use of coarse upper bounds rather than exact asymptotics in El Maftouhi and Fernandez de la Vega [145] for estimating the proportion of "interesting" formulas with a particular structure—it can in fact be proven that the difference between the two is vanishingly small—but also to the different experiment used to count this proportion. In the following sections we report, briefly, the results in the two papers. The careful reader will be able to pick up the similarities and the differences in the two approaches.

4.1 ACCOUNTING FOR RARE FORMULAS WITH MANY SATISFYING TRUTH ASSIGNMENTS: DISPENSABLE VARIABLES

El Maftouhi and Fernandez de la Vega [145] define the subset T_1^* of T_1 such that $|T_1^*| \geq (1 - 2^{-\delta n})|T_1|$ and all formulas in T_1^* have at least $2^{\delta n}$ satisfying assignments. Notice that this can be rewritten as $\mathbf{Pr}[|\mathcal{A}_n| \geq 2^{\delta n} \mid \phi \in T_1] \geq 1 - 2^{-\delta n}$. Since $|\mathcal{A}_n| \geq 1$ for any $\phi \in T_1$, one can write:

$$\sum_{\psi \in T_1} \frac{1}{|\mathcal{A}_n(\psi)|} = \sum_{\psi \in T_1^*} \frac{1}{|\mathcal{A}_n(\psi)|} + \sum_{\psi \in T_1 \setminus T_1^*} \frac{1}{|\mathcal{A}_n(\psi)|}$$

$$\leq \frac{|T_1^*|}{2^{\delta n}} + (|T_1| - |T_1^*|) \leq \frac{|T_1|}{2^{\delta n}} + \frac{|T_1|}{2^{\delta n}} .$$

Therefore $\mathbf{E}_{\phi \in \mathrm{SAT}}[|\mathcal{A}_n|] \geq \frac{|T_1|}{2|T_1|/2^{\delta n}} = 2^{\delta n - 1}$.

In order to describe how T_1^* is defined, let $\mathcal{C}_i = \mathcal{C}_i(\psi)$ be the set of clauses in ψ containing exactly i positive literals. We first estimate $|\mathcal{C}_i|$ under the assumption that $\psi \in T_1$. Notice that no formula in T_1 can contain a clause with only negated variables, therefore, $|\mathcal{C}_0| = 0$. Furthermore, for formulas in T_1 with n variables and αn clauses, it is fairly easy to compute the asymptotic distribution of the formulas with $|C_i| = m_i$ for each $i \in \{1, 2, 3\}$ (where $\alpha n = m_1 + m_2 + m_3$):

$$\mathbf{Pr}[m_1, m_2, m_3] = \frac{\binom{\alpha n}{m_1, m_2, m_3}\left[n\binom{n}{2}\right]^{m_1+m_2}\binom{n}{3}^{m_3}}{\left[\binom{2n}{3} - \binom{n}{3}\right]^{\alpha n}}.$$

Using Stirling's approximation for the various factorials involved and setting $\gamma_i = m_i/n$, it is easy to prove that $\mathbf{Pr}[m_1, m_2, m_3]^{1/n}$ is asymptotic to $(6\alpha/7)^\alpha / (2\gamma_1)^{\gamma_1}(2\gamma_2)^{\gamma_2}(6\gamma_3)^{\gamma_3}$, assuming all m_i's tend to infinity. Considered as a function of γ_1, γ_2 and γ_3, this expression reaches its maximum (equal to one) for $\gamma_1 = \gamma_2 = \frac{3\alpha}{7}$ and $\gamma_3 = \frac{\alpha}{7}$. Now let $T_1^* \subseteq T_1$ be the set of all those formulas in T_1 with $\gamma_1 \leq 2.37$, $\gamma_2 \leq 2.37$ and $\gamma_3 \leq 0.87$ (recall that $\alpha = 5.08$ for all formulas in T_1). It may be shown from the asymptotic probability expression [145] that these inequalities hold with probability greater than $1 - 2^{-0.02137n}$, implying that $|T_1^*| > (1 - 2^{-0.02137n})|T_1|$.

To prove that 5.08 is an upper bound to the satisfiability threshold, following the reasoning given earlier, it remains to demonstrate that the formulas in T_1^* have, with sufficiently high probability, at least $2^{0.02137n}$ satisfying truth assignments. To this end, the authors introduce the notion of *dispensable variables*. Given a formula ϕ, a truth assignment A that satisfies ϕ and a set D consisting of certain variables taking the values dictated by A, we call D a set of *dispensable* variables if its elements can be set in any arbitrary way and still result in the truth assignment satisfying ϕ. Let $D(\phi)$ be the set of dispensable variables in ϕ with respect to the assignment that sets all variables to TRUE. Clearly $|\mathcal{A}_n(\phi)| \geq 2^{|D(\phi)|}$, so it is then sufficient to show that for all $\phi \in T_1^*$, $|D(\phi)| \geq 0.02137n$ with high probability. The authors do this by analyzing the size of the set of dispensable variables returned by the following "greedy" algorithm:

1. Take all clauses in C_1, and call I_1 the set of all positive literals in these clauses. These are known as *isolated* literals. Let $n_1 = |I_1|$.
2. Take all clauses in C_2 whose two positive literals are both absent from I_1, and for each such clause select at random one of its two positive literals. Call I_2 the set of all such literals. Set $J_2 = I_1 \cup I_2$. Let $n_2 = |I_2|$, so that $|J_2| = n_1 + n_2$.
3. Take all clauses in C_3 whose three positive literals are all absent from J_2, and for each such clause select at random one literal. Call I_3 the set of all such literals. Set $J_3 = J_2 \cup I_3$. Let $n_3 = |I_3|$, so that $|J_3| = n_1 + n_2 + n_3$.

One may readily verify that all variables *not* represented in J_3 form a set of dispensable variables, so $|J_3|$ needs to be bounded from above. For the range of values of γ_i that defines T_1^*, an estimate on $|J_3|$ is obtained by finding upper bounds on: n_1; n_2 conditioned on n_1; and n_3 conditioned on n_2 and n_1.

In order to estimate n_1, the authors resort to the *occupancy problem*. In this problem, one throws μn balls (μ is a constant) uniformly at random into n boxes and asks for the distribution of the random variable Y that counts the number of *non-empty* boxes. Then for any $\epsilon > 0$, $r = r(\epsilon) = (1 + \epsilon)(1 - e^{-\frac{\mu}{1+\epsilon}})$ and $s = s(\epsilon) = 1 - r(\epsilon)$, the following is established:

$$\frac{1}{n} \log \mathbf{Pr}[Y \geq \lfloor r(\epsilon)n \rfloor] \leq (1 - o(1)) \log \frac{(s+\epsilon)^{s+\epsilon}(1+\epsilon)^{\mu-1-\epsilon}}{s^s}.$$

As n_1 can be viewed as the number of non-empty boxes that result from the random placement of $\gamma_1 n$ balls into n boxes, we have that $Y = n_1$ and $\mu = \gamma_1$. Setting $\epsilon = 0.062$ and exponentiating both sides of the inequality above, we obtain for $\gamma_1 \leq 2.37$ that

$$\mathbf{Pr}[n_1 \leq 0.94800n] \geq 1 - e^{-0.01513n}. \tag{3}$$

Now define m_2' as the number of clauses in $\mathcal{C}_2$ (clauses with two positive literals) identified by the greedy algorithm as having both of their positive literals absent from I_1. This is binomially distributed, with number of trials $\gamma_2 n$ and success probability $(1 - n_1/n)(1 - (n_1 - 1)/n)$, where success means "absent from I_1." Conditioning on $n_1 \leq 0.94800n$, we can use the Chernoff bound on the upper tail of the binomial distribution, $\mathbf{Pr}[B(m,p) \geq \beta mp] \leq (e^{\beta-1}/\beta^\beta)^{mp}$, setting $\beta = 3.84$ and $mp = 0.006408n$ to obtain for $\gamma_2 \leq 2.37$ that $\mathbf{Pr}[m_2' \leq 0.02461n|n_1 \leq 0.94800n] \geq 1 - e^{-0.01490n}$. Since $n_2 \leq m_2'$,

$$\mathbf{Pr}[n_2 \leq 0.02461n|n_1 \leq 0.94800n] \geq 1 - e^{-0.01490n}. \tag{4}$$

Similarly, define m_3' as the number of clauses in $\mathcal{C}_3$ identified by the greedy algorithm as having all of their literals absent from J_2. Again bounding the tail of the relevant binomial distribution, we obtain for $\gamma_3 \leq 0.87$ that $\mathbf{Pr}[m_3' \leq 0.00356n|n_1 + n_2 \leq 0.97261n] \geq 1 - e^{-0.0153n}$. Since $n_3 \leq m_3'$,

$$\mathbf{Pr}[n_3 \leq 0.00356n|n_1 + n_2 \leq 0.97261n] \geq 1 - e^{-0.0153n}. \tag{5}$$

Finally, multiplying together (3), (4) and (5) we may verify that for sufficiently large n, $\mathbf{Pr}[n_1 + n_2 + n_3 \leq 0.97617n] \geq 1 - e^{-0.01481n} = 1 - 2^{-0.02137n}$. Thus, with probability at least $1 - 2^{-0.02137n}$, $|D(\phi)| \geq 0.02383n > 0.02137n$ for $\phi \in T_1^*$.

4.2 SHARPER ESTIMATE OF OCCUPANCY PROBABILITIES: INDEPENDENT VARIABLES

Kamath et al. [292] performed a similar investigation of the structure of the typical $\phi \in T_1$. A variable x is said to *cover* a clause C if x occurs unnegated in C—that is, as a positive literal. For instance, in the formula below (which does not belong to T_1), represented by the sequence of sets of literals forming individual clauses in it,

$\phi(x_1, x_2, x_3, x_4, x_5) =$

C_3 $\{x_1, x_2, x_3\}, \{x_2, x_3, x_4\}, \{x_3, x_4, x_5\},$

C_2 $\{\overline{x}_2, x_3, x_4\}, \{x_1, x_4, \overline{x}_5\}, \{x_1, \overline{x}_2, x_5\}, \{x_1, \overline{x}_2, x_4\}, \{x_1, \overline{x}_3, x_5\}, \{x_3, x_4, \overline{x}_5\},$
      $\{x_2, x_4, \overline{x}_5\}, \{x_2, \overline{x}_3, x_4\}, \{\overline{x}_3, x_4, x_5\}, \{x_1, x_3, \overline{x}_4\},$

C_1 $\{x_2, \overline{x}_3, \overline{x}_5\}, \{\overline{x}_1, x_4, \overline{x}_5\}, \{\overline{x}_1, \overline{x}_3, x_5\}, \{\overline{x}_1, \overline{x}_2, x_5\}, \{\overline{x}_1, x_3, \overline{x}_4\}, \{\overline{x}_3, x_4, \overline{x}_5\},$
      $\{\overline{x}_2, x_4, \overline{x}_5\}, \{\overline{x}_1, x_4, \overline{x}_5\}, \{x_1, \overline{x}_3, \overline{x}_5\},$

C_0 $\{\overline{x}_1, \overline{x}_2, \overline{x}_3\}, \{\overline{x}_1, \overline{x}_4, \overline{x}_5\}, \{\overline{x}_2, \overline{x}_3, \overline{x}_4\}, \{\overline{x}_1, \overline{x}_2, \overline{x}_4\}, \{\overline{x}_1, \overline{x}_3, \overline{x}_5\}$

the variable x_1 covers the clauses:

$$\{x_1, x_2, x_3\}, \{x_1, x_4, \overline{x}_5\}, \{x_1, \overline{x}_2, x_5\}, \{x_1, \overline{x}_2, x_4\}, \{x_1, \overline{x}_3, x_5\}, \{x_1, x_3, \overline{x}_4\},$$
$$\{x_1, \overline{x}_3, \overline{x}_5\}.$$

A set of variables V covers a set of clauses if every one of these clauses is covered by at least one variable in V. Such a variable set is called a *cover*. Obviously, a random formula $\phi \in T_1$ has a trivial cover, namely the set consisting of all the n variables. However, it is possible that there exists a smaller cover than the trivial one. For instance, Figure 1 shows a formula in T_1 with the sets of clauses in C_1, C_2, C_3, the set X containing all variables, a set I_1 of variables covering C_1, a second set of variables (from the set R of remaining variables) needed to cover the uncovered portions of C_2 and C_3, and the remaining *independent set I*. The set $X \setminus I$ in the figure is an example of a cover for all the clauses of the formula that is smaller than the trivial one. Therefore, setting the variables in $X \setminus I$ to TRUE is sufficient to satisfy the formula. Since all the variables in the independent set I can be set arbitrarily, if ϕ has a cover of size s then $|\mathcal{A}_n(\phi)| \geq 2^{n-s}$.

To estimate $\mathbf{E}_{\phi \in \mathrm{SAT}}[|\mathcal{A}_n|] = |T_1| / \sum_{\psi \in T_1} \frac{1}{|\mathcal{A}_n(\psi)|}$, we partition T_1 into "slices" containing formulas with minimal cover of size s, and use:

$$\sum_{\psi \in T_1} \frac{1}{|\mathcal{A}_n(\psi)| \cdot |T_1|} = \sum_{s=1}^{n} \sum_{\psi \in T_1 : |\mathrm{cover}(\psi)| = s} \frac{1}{|\mathcal{A}_n(\psi)| \cdot |T_1|}$$

$$\leq \sum_{s=1}^{n} \sum_{\psi \in T_1 : |\mathrm{cover}(\psi)| = s} \frac{\mathbf{Pr}[|\mathrm{cover}(\psi)| = s]}{|\mathcal{A}_n(\psi)|}$$

$$\leq \sum_{s=1}^{n} \sum_{\psi \in T_1 : |\mathrm{cover}(\psi)| = s} 2^{s-n} \mathbf{Pr}[|\mathrm{cover}(\psi)| = s].$$

The problem thus reduces to one of estimating, as accurately as possible, the probability that an arbitrary formula in T_1 has minimal cover size s. As in subsection 4.1, this is done using asymptotic expressions for binomial tails and occupancy probabilities.

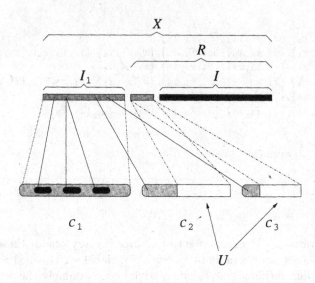

FIGURE 1 A set X of variables and the clauses in C_1, C_2, C_3 that it covers. Broken lines denote covered regions; solid lines represent some of the specific clauses covered by variables in I_1.

1. First fix the size m_1 of C_1, the set of clauses containing a single unnegated variable. The probability $\mathbf{Pr}[u]$ that this number is within u times its mean can be shown to be close to 1 using bounds on binomial tails.
2. Then determine the set I_1 of isolated variables ($X \setminus R$ in fig. 1). Conditioning on m_1 being within u of its mean, the probability $\mathbf{Pr}[v|u]$ that the size n_1 of I_1 is within v times its mean is estimated using occupancy asymptotics (again we are throwing clauses "into" variables: the empty bins correspond to the variables in $X \setminus I_1$).
3. Next, compute the number of clauses in $C_2 \cup C_3$ that are *not* covered by I_1, conditioned on n_1 and m_1. The *set* of these clauses is denoted by U in figure 1. The probability that this number is within w times its mean, $\mathbf{Pr}[w|u, v]$, is again a binomial tail.
4. Finally, bound the size of the variable set needed to cover U.

The improvement in Kamath et al. [292] comes from estimating the size of U, and from adding to the cover only those variables needed to cover U rather than the whole set $I_2 \cup I_3$ (as done in the other paper, in the second and third selection step).

5 SPECIAL CLASSES OF SATISFYING TRUTH ASSIGNMENTS

The next improvements resulted from a different kind of exploitation of rare formulas with many satisfying truth assignments.

Since the main disadvantage of Markov's inequality can be attributed to rare formulas having a large number of satisfying truth assignments, a plausible approach for improvement is to use in the inequality not the expected cardinality of the set of satisfying truth assignments of a random formula, but the expected cardinality of a smaller set. Of course one needs to prove that the expectation of the new random set still provides an upper bound on the probability that ϕ is satisfiable.

This idea was introduced by Kirousis, Kranakis, and Krizanc [323] ("single flips") and independently by Dubois and Boufkhad [130]. In this section we describe these approaches and several others derived from them, and we report the resulting improvements on the estimate of α_3. With regard to the techniques described in sections 5.1 and 5.2 in particular, we should point out in advance the following important difference between them. The technique described in section 5.1 approximates α_3 by computing an upper bound on the expected cardinality of a special class of satisfying truth assignments, employing a simple correlation inequality that bounds from above the probability that some dependent events hold simultaneously. On the other hand, the technique described in section 5.2 results in a slightly better approximation of α_3 by computing *exactly* the expected cardinality of the same class of satisfying truth assignments. The former method, however, is much simpler to apply and easily generalizes to k-SAT random formulas for any $k > 3$ while the latter is complicated and cannot be readily applied to k-SAT in general.

The following formula

$$\phi(x_1, x_2, x_3, x_4) = \{x_1, \overline{x}_2, x_4\}, \{\overline{x}_1, \overline{x}_2, \overline{x}_4\}, \{x_1, \overline{x}_3, \overline{x}_4\}, \{\overline{x}_1, x_2, \overline{x}_3\}$$

whose satisfying assignments are

	x_1	x_2	x_3	x_4
A_1	FALSE	FALSE	FALSE	FALSE
A_2	FALSE	FALSE	FALSE	TRUE
A_3	FALSE	FALSE	TRUE	FALSE
A_4	FALSE	TRUE	FALSE	TRUE
A_5	TRUE	FALSE	FALSE	FALSE
A_6	TRUE	FALSE	FALSE	TRUE
A_7	TRUE	TRUE	FALSE	FALSE
A_8	TRUE	TRUE	TRUE	FALSE

will be used as an example in the next subsection, to convey a better understanding of the various results.

5.1   SINGLE FLIPS

The strategy described here was introduced in Kirousis et al. [323]. In what follows it may be convenient to identify classes of truth assignments on n variables with sets of lexicographically ordered sequences over a two-letter alphabet (say the numbers 0 and 1 with $0 < 1$).

Given a random formula ϕ, the set $\mathcal{A}_n^1$ is defined as the class of truth assignments A such that the following two conditions hold: (1) A satisfies ϕ, and (2) any assignment obtained from A by changing exactly one FALSE value to TRUE does not satisfy ϕ. Such a change is called a *single flip* and will be denoted by sf. The truth assignment that results from the flip will be denoted by A^{sf}. The class $\mathcal{A}_n^1$ contains the elements of $\mathcal{A}_n$ that are *local maxima* with respect to single flips. In other words, a truth assignment belongs to $\mathcal{A}_n^1$ if it satisfies ϕ and if no other possible satisfying truth assignment can be obtained from it by changing a single FALSE value to TRUE—by performing all possible single flips in isolation.

In our example formula, the first truth assignment A_1 does not belong to $\mathcal{A}_n^1$: if we change the value assigned to x_1 the resulting truth assignment is still in $\mathcal{A}_n$. However, the fourth truth assignment A_4 does belong to $\mathcal{A}_n^1$: changing the value assigned to x_1 produces the assignment $A' \notin \mathcal{A}_n$ where all variables except x_3 are set to TRUE, and changing the value of x_3 similarly leads to the ϕ not being satisfied. Since all possible transformations changing a FALSE value result in truth assignments that do not satisfy ϕ, the assignment A_4 is in $\mathcal{A}_n^1$. It is easy to verify that the set $\mathcal{A}_n^1$ for ϕ is formed by the assignments A_3, A_4, A_6, and A_8.c

Since $\mathcal{A}_n^1 \subseteq \mathcal{A}_n$, $\mathbf{E}[|\mathcal{A}_n^1|] \leq \mathbf{E}[|\mathcal{A}_n|]$. Thus, to relax Markov's inequality we need only establish that $\mathbf{Pr}[\phi \text{ is satisfiable}] \leq \mathbf{E}[|\mathcal{A}_n^1|]$. This can easily be seen as follows. Let I_ϕ be the random indicator for the property "ϕ is satisfiable." Clearly $I_\phi \leq |\mathcal{A}_n^1|$. If we now write

$$\mathbf{Pr}[\phi \text{ is satisfiable}] = \sum_\phi I_\phi \mathbf{Pr}[\phi]$$

then the desired inequality follows immediately. To exploit this technique one then needs to prove a bound on $\mathbf{E}[|\mathcal{A}_n^1|]$, asymptotically smaller than $2^n(7/8)^{\alpha n}$.

Using a correlation inequality to compute the probability that a single flip results in an assignment not satisfying ϕ, it can be proven that in the random formula model $\mathcal{G}_{m,m}$, the expected size of class $\mathcal{A}_n^1$ is at most $(7/8)^{\alpha n}(2 - e^{-3\alpha/7} + o(1))^n$. Therefore, the unique positive solution of the equation $(7/8)^\alpha(2 - e^{-3\alpha/7}) = 1$ gives an upper bound for the satisfiability threshold critical value α_3. This solution is approximately 4.667.

If instead one uses the $\mathcal{G}_p$ ensemble, one avoids having to compute probabilities of conjunctions of *dependent* events. Applying Markov's inequality then leads to the solution of the equation $e^{-\alpha/8}(2 - e^{-3\alpha/7}) = 1$. This expression, however, gives an upper bound equal to 5.07—worse than the bound given for $\mathcal{G}_{m,m}$.

5.2   THE SET OF NEGATIVELY PRIME SOLUTIONS (NPS)

Independently from Kirousis et al., Dubois and Boufkhad introduced a class of satisfying truth assignments that they called *negatively prime solutions* (NPS) [130]. This class turns out to coincide with the class $\mathcal{A}_n^1$ described in subsection 5.1.

Dubois and Boufkhad proved the following exact expression for the expected cardinality of NPS for k-SAT:

$$\mathbf{E}[|\text{NPS}|] = \sum_{i=0}^{n} \sum_{j=i}^{m} 2^{-km} 2^i \binom{n}{i} \binom{m}{j} S_2(j,i) i! \left(\frac{k}{n}\right)^j (2^k - 1 - k)^{m-j} \qquad (6)$$

where $S_2(j,i)$ are the *Stirling numbers of the second kind* that count the number of ways of partitioning a set of j elements into i *non-empty* subsets. By way of a series of asymptotic manipulations, the authors then arrived at a closed-form upper bound for (6), showing that it converges to 0 for values of the clause-to-variable ratio greater than 4.642.

5.3   RESTRICTING FURTHER THE CLASS OF SATISFYING TRUTH ASSIGNMENTS: DOUBLE FLIPS

Kirousis et al. [324] define as a *double flip* the change of exactly two variables x_i and x_j, with $i < j$, where x_i is changed from FALSE to TRUE and x_j from TRUE to FALSE. Notice that the restriction $i < j$ implies that a double flip always leads to a lexicographically "greater" assignment. Let A^{df} denote the truth assignment that results from A after the application of the double flip df. Let $\mathcal{A}_n^{2\sharp}$ be the set of truth assignments A that have the following three properties: (1) A satisfies ϕ, (2) for all possible single flips sf of A, A^{sf} does not satisfy ϕ, and (3) for all possible double flips df of A, A^{df} does not satisfy ϕ.

It is proven in [324] that the following inequality holds:

$\mathbf{Pr}[\phi \text{ is satisfiable}]$

$$\leq \mathbf{E}[|\mathcal{A}_n^{2\sharp}|] = (7/8)^{\alpha n} \sum_{A \in S} \mathbf{Pr}[A \in \mathcal{A}_n^1 \mid A \in \mathcal{A}_n] \mathbf{Pr}[A \in \mathcal{A}_n^{2\sharp} \mid A \in \mathcal{A}_n^1], \quad (7)$$

where S is the set of all 2^n possible truth assignments on the n variables. As before, to get an upper bound on α_3 it suffices to find the smallest possible value for the clause-to-variable ratio for which the right-hand side of this inequality tends to 0. In what follows, $sf(A)$ denotes the total number of possible single flips of A (which is simply the number of variables assigned the value FALSE in A) and $df(A)$ denotes the total number of possible double flips.

It can be proven that $\mathbf{Pr}[A \in \mathcal{A}_n^1 \mid A \in \mathcal{A}_n]$ may be bounded from above by an expression of the form $P^{sf(A)}$, where P depends only on the clause-to-variable ratio α (this is in fact the expression used to derive the improved upper bound in subsection 5.1). The hard part is computing the second probability, involving the realization of the double flip events conditional upon the realization of the single flip events. As it turns out, the computation of this probability in the model $\mathcal{G}_{m,m}$ involves dependencies among events of a very complicated nature. In $\mathcal{G}_p$, on the other hand, there at least are no dependencies ensuing from the fundamental requirement of $\mathcal{G}_{m,m}$ that the size of the formula is fixed. The remaining dependencies are those arising from the fact that some of the double flip events involve double flips sharing a particular FALSE variable. The computation of an upper bound to this probability was made possible by the use of a version of Suen's correlation inequality [492] proven by Janson [277]. The bound has the form $Q^{df(A)}$ with Q dependent on n and α. The reader may consult [324] for the derivation of this bound. Inequality (7) may then be rewritten as follows:

$$\mathbf{Pr}[\phi \text{ is satisfiable}] \leq 3m^{1/2}(7/8)^{\alpha n} \sum_{A \in S} P^{sf(A)} Q^{df(A)}, \tag{8}$$

with the polynomial factor $3m^{1/2}$ arising due to the change from $\mathcal{G}_{m,m}$ to $\mathcal{G}_p$ (see Bollobás [62]). To complete the derivation of the improved bound, the authors noted the following combinatorial identity, which can be proven by induction on n:

$$\sum_{A \in S} P^{sf(A)} Q^{df(A)} = \sum_{i=0}^{n} \begin{bmatrix} n \\ i \end{bmatrix}_Q P^i, \tag{9}$$

where $\begin{bmatrix} n \\ i \end{bmatrix}_q = \frac{(q^n-1)(q^n-q)\cdots(q^n-q^{i-1})}{(q^i-1)(q^i-q)\cdots(q^i-q^{i-1})}$ are the q-binomial or Gaussian coefficients [192] for $0 \leq i \leq n$ and $q \neq 1$. The right-hand expression in eq. (9) is also known as the Rogers-Szegő polynomial $F_{n,Q}(P)$, and leads to the inequality

$$\sum_{A \in S} P^{sf(A)} Q^{df(A)} \leq \prod_{i=0}^{n-1} (1 + PQ^{i/2}), \qquad 0 \leq P^2 \leq Q \leq 1. \tag{10}$$

Equation (8) thus becomes

$$\mathbf{Pr}[\phi \text{ is satisfiable}] \leq 3m^{1/2}(7/8)^{\alpha n} \prod_{i=0}^{n-1} (1 + PQ^{i/2}). \tag{11}$$

The product on the right-hand side can be estimated by making use of hypergeometric series (see also Gasper and Rahman [192]) and an inequality derived in Kirousis [322], ultimately leading to the bound $\alpha_3 < 4.602$.

5.4 OCCUPANCY BOUNDS AND Q-BINOMIAL COEFFICIENTS

As we have just seen, a key step in the improvement of the upper bound to 4.602 is the derivation of an upper bound to the sum in eq. (8) through its connection with the q-binomial coefficients. This bound can be improved further: two approaches have been given by Janson, Stamatiou, and Vamvakari [278]. Both result in the same value, namely $\alpha_3 < 4.596$. However, these approaches are interesting in their own right. The first approach links the problem of determining upper bounds to the satisfiability threshold with the study of *Ising spin systems* in statistical mechanics, while the second approach links the problem with the branch of mathematics dealing with q-hypergeometric series and their generating functions.

More specifically, in the first approach the sum in eq. (8) is written as

$$\sum_{\varepsilon_1,\ldots,\varepsilon_n \in \{0,1\}} \exp\left(a \sum_{i=1}^{n} \varepsilon_i + \frac{b}{n} \sum_{1 \le i < j \le n} \varepsilon_i(1 - \varepsilon_j) \right), \tag{12}$$

with the outer sum ranging over all the 2^n sequences $\varepsilon_1, \ldots, \varepsilon_n$ of 0's and 1's, each of them coding a truth assignment A with 0 and 1 representing TRUE and FALSE respectively. This sum is indeed equal to the sum of eq. (8) when $a = \log P$ and $b = n \log Q$.

The expression in eq. (12) enables the application of an optimization technique common in statistical physics, resulting in an asymptotic expression for the sum. This particular form of the sum is precisely the *partition function* $Z = \sum_{\varepsilon_1,\ldots,\varepsilon_n} \exp(-\beta H)$ for a system with n spin sites, each having a spin value $\varepsilon_i \in \{0, 1\}$ and with an energy function equal to $H = -a \sum_{i=1}^{n} \varepsilon_i - \frac{b}{n} \sum_{1 \le i < j \le n} \varepsilon_i(1 - \varepsilon_j)$, in units where the inverse temperature $\beta = 1$. The first term in H corresponds to an external field acting on all the spins of the system, the second to an interaction acting between arbitrary pairs of spin sites with the left site having spin 1 and the right site having spin 0. The energy function can easily be rewritten into a more conventional form: $H = \sum_{i=1}^{n}(-a - b + b\frac{i}{n})\varepsilon_i + \frac{b}{n} \sum_{i<j} \varepsilon_i\varepsilon_j$, or, substituting $\varepsilon_i = (1 + s_i)/2$ to have more traditional (and symmetric) spins with values ± 1,

$$H = -\frac{a}{2}n - \frac{b}{8}(n - 1) + \sum_{i=1}^{n}\left(-\frac{a}{2} + b\frac{i - (n+1)/2}{2n}\right)s_i + \frac{b}{4n} \sum_{i<j} s_i s_j \,.$$

The value H may be interpreted in statistical physics as the energy function for a mean-field Ising model with an inhomogeneous (linear) external field. Ultimately, this leads to an asymptotic expression for the partition function, and an estimate for the sum in eq. (8), resulting in $\alpha_3 < 4.596$.

In the second approach, a sharp upper bound is derived for the sum in eq. (8). Recall that it is equal to the Rogers-Szegő polynomial $F_{n,Q}(P)$. Then, using the *Eulerian* generating function and a technique described in Lemma 8.1

of Odlyzko [417], the following upper bound is obtained (see Janson et al. [278]) for any t, $0 < t < \min(1, 1/P)$:

$$F_{n,Q}(P) \leq \frac{t^{-n}}{(1-t)(1-tP)} \exp\left[-\frac{1}{\log Q}\left(\mathrm{Li}_2(tP) + \mathrm{Li}_2(t) + \mathrm{Li}_2(Q^n) - \mathrm{Li}_2(Q)\right)\right]$$

where $\mathrm{Li}_2(y) = \mathrm{dilog}(1 - y) = \mathrm{Polylog}(2, y) = \sum_{i \geq 1} \frac{y^i}{i^2}$ is the *dilogarithm* function. By finding the value of t that optimizes this upper bound and plugging it into eq. (8), we obtain precisely the same upper bound as in the first approach, $\alpha_3 < 4.596$.

5.5 BALLS AND BINS

The calculations described in the previous sections are useful, but limited. The idea of single flips, although leading to a good improvement of the upper bound, only takes into account a very limited range of locality around a given satisfying assignment. The results described in subsection 5.3 exploit wider locality ranges, but because of the complexity of the resulting numerical expressions, the authors were forced to use weak bounds on $\mathbf{Pr}[A \in \mathcal{A}_n^1 \mid A \in \mathcal{A}_n]$ and the overall $\mathbf{E}[|\mathcal{A}_n^{2\sharp}|]$. Finally, the two approaches in subsection 5.4 reached the limits of what can be exploited from the upper bound shown in eq. (8). In order to obtain further improvements, one must step back from the derivation of eq. (8) and attempt to realize improvements on the probabilities involved.

Kaporis et al. [293] have achieved an improved bound of 4.571 using sharp estimates for certain probabilities related to the classical occupancy problem that we have seen in subsection 4.1. For a given satisfying assignment A, the probability that no single flip satisfies ϕ is best computed (up to polynomial factors) by noticing the following "structural" condition imposed on ϕ by the event in question: for each variable x set to FALSE in A, the formula must contain a *critical* clause $\{\overline{x}, \ell_1, \ell_2\}$ with $A(\ell_1) = A(\ell_2) = $ FALSE.

If a satisfying assignment A sets j variables to FALSE, no single flip satisfies ϕ when: (1) some $l \geq j$ clauses out of $m = \alpha n$ are critical (the remaining ones being consistent with A^{sf}), and (2) these l clauses can be seen as a sequence of balls that are dropped into j distinct bins (corresponding to different single flips) in a way that leaves no bin completely empty.

Asymptotic estimates on the occupancy probability that the l critical clauses indeed cover all possible j single flips [292], as well as a change of models from $\mathcal{G}_{m,m}$ to $\mathcal{G}_m$ in order to be able to formulate our problem in the balls and bins framework, lead to a sharper bound of $\mathbf{Pr}[A \in \mathcal{A}_n^1 \mid A \in \mathcal{A}_n]$. Note that Zito [538] has performed a similar analysis using coupon collector probabilities instead, deriving a bound of about 4.58. The analysis in Kaporis et al. [293] improves on the previous results for another reason as well: the overall bound on $\mathbf{E}[|\mathcal{A}_n^{2\sharp}|]$ is tightened by means of a more direct estimate of the q-binomial coefficient involved. Using simple generating function inequalities (and elementary calculus) it is possible to bound the term $\begin{bmatrix} n \\ i \end{bmatrix}_Q$ directly and avoid the use of Rogers-Szegő polynomials.

Finally, to obtain their results, the authors had to establish a relationship between the probabilities of the event, $A \in \mathcal{A}_n^{2\sharp} \mid A \in \mathcal{A}_n^1$, in the models $\mathcal{G}_{m,m}$ and $\mathcal{G}_p$. Using results described in Bollobás [62, ch. II], it is easy to prove the desired relationship for *unconditional* events when the average length of a random formula constructed in the model $\mathcal{G}_p$ equals m. However, the conditioning here may bias the expected length of the formula to higher values. The authors have shown how to adjust p appropriately so as to obtain the bound above.

6 TYPICAL FORMULAS

In the last section, all the techniques have worked by placing restrictions on the *assignments* (*semantics*, in some sense) satisfying a formula and for which the expectation, which is required by the first moment method, is computed. As mentioned above, the application of this technique to the kinds of assignment restrictions described in subsections 5.1, 5.2, and 5.3 results provably in no further upper bound improvement.

Dubois, Boufkhad, and Mandler [132, 134] consider random formulas with the special characteristic that the numbers of appearances of their literals fall within certain ranges that are "typical" for randomly generated formulas. In this way, they are able to disallow the rare formulas that seem to prevent Markov's inequality from giving an upper bound close to the experimentally determined value. By computing the expected number of negative prime solutions for these formulas only, making use of the model $\mathcal{G}_{3,\alpha n}$, they achieve an upper bound improvement. In contrast to the semantic methods that rely on restricting the set of truth assignments taking part in the application of the first moment method, this approach can be characterized as *syntactic*: it focuses on restricting the *form* or *syntax* of the set of formulas participating in the first moment method calculations. However, the approach also limits the possible truth assignments using the restricted sets defined in Sections 5.1 and 5.2. Without going into detail, Dubois, Boufkhad and Mandler give an expression for the expected number of negative prime solutions for these formulas. In so doing, they obtain $\alpha_3 < 4.506$, the best upper bound to date for the location of the random 3-SAT threshold.

ACKNOWLEDGMENTS

We would like to thank the editors as well as the anonymous referee for their invaluable contribution to the improvement of the presentation in the chapter. This work has been supported by the University of Patras Research Committee (Project C. Carathéodory no. 20445) and by the EU within the 6th Framework Programme under contract 001907 (DELIS).

CHAPTER 8

Proving Conditional Randomness using the Principle of Deferred Decisions

Alexis C. Kaporis
Lefteris M. Kirousis
Yannis C. Stamatiou

1   INTRODUCTION

In order to prove that a certain property holds asymptotically for a restricted class of objects such as formulas or graphs, one may apply a heuristic on a random element of the class, and then prove by probabilistic analysis that the heuristic succeeds with high probability. This method has been used to establish lower bounds on thresholds for desirable properties such as satisfiability and colorability: lower bounds for the 3-SAT threshold were discussed briefly in the previous chapter. The probabilistic analysis depends on analyzing the mean trajectory of the heuristic—as we have seen in chapter 3—and in parallel, showing that in the asymptotic limit the trajectory's properties are strongly concentrated about their mean. However, the mean trajectory analysis requires that certain random characteristics of the heuristic's starting sample are retained throughout the trajectory.

We propose a methodology in this chapter to determine the conditional that should be imposed on a random object, such as a conjunctive normal form (CNF)

Computational Complexity and Statistical Physics, edited by
Allon G. Percus, Gabriel Istrate, and Cristopher Moore, Oxford University Press.

formula or a graph, so that conditional randomness is retained when we run a given algorithm. The methodology is based on the principle of deferred decisions. The essential idea is to consider information about the object as being stored in "small pieces," in separate registers. The contents of the registers pertaining to the conditional are exposed, while the rest remain unexposed. Having separate registers for different types of information prevents exposing information unnecessarily. We use this methodology to prove various randomness invariance results, one of which answers a question posed by Molloy [402].

2   PRINCIPLE OF DEFERRED DECISIONS

Let $G \in \mathcal{G}_{n,m}$ be a graph chosen uniformly at random, conditional on its number of vertices n and number of edges m. All G with n vertices and m edges are thus equiprobable. Intuitively, if we delete from G a vertex v chosen uniformly at random and also delete all edges incident on v, the new graph should be random conditional on the new number of vertices, $n - 1$, and the new number of edges m', where m' is a random variable. In other words, given m', the new graph is equiprobable among all graphs with $n - 1$ vertices and m' edges. Note that here and in what follows, "random" will mean "uniformly random," that is, equiprobable, on conditionals that will be either explicit or clear from the context.

Knuth [332, Lecture 3] has introduced a method, known as the *principle of deferred decisions*, by which randomness claims such as the one above can be verified. In the specific example of vertex deletion from a $\mathcal{G}_{n,m}$ graph, it works as follows. Consider $n+m$ cards facing down, or more precisely, $n+m$ registers with unexposed content. The first n of them correspond to the vertices of the graph and the remaining m to its edges. The register of a vertex v contains pointers to the registers of the edges incident on v. The register of an edge e contains pointers to the registers of the two endpoints of e. That the registers are unexposed means that the pointers can be specified randomly. To delete a random vertex, do the following: point randomly to a vertex register; expose its contents; expose all edge registers pointed to by this vertex register; delete the exposed vertex register and the exposed edge registers; nullify pointers in other vertex registers that point to deleted edge registers (without exposing these vertex registers). The registers that have not been deleted remain unexposed and, therefore, they can be filled in randomly. The only conditional, that is, the only exposed information about the graph, is the new number of vertex registers and the new number of edge registers.

The principle of deferred decisions states that conditional randomness is retained as long as no new information about the current contents of unexposed registers can be determined, at any given update step, from information exposed up until that step. The method can be applied in more complicated situations. Consider a random graph conditional on (i) the number of vertices, (ii) the number of edges, and (iii) for each $i = 0, \ldots, n - 1$, the number of vertices of degree i (the degree sequence). We claim that upon deleting a random vertex of degree i (for any i) and its i incident edges, the new graph is random conditional on the same type of information. Indeed, it suffices to augment the argument of the previous paragraph with the additional assumption that for each vertex v there is an *exposed* degree register containing an integer equal to its degree. This degree register needs to be exposed so that the algorithm may choose, at random, a vertex of a given degree. After a deletion step, the contents of the remaining vertices are updated. After the update, no information about the new values of the unexposed registers can be determined from what still is, or previously was, exposed. Therefore, the new graph is random given the number of its vertices, the number of its edges, and its degree sequence.

Notice that keeping a register unexposed is not in itself sufficient to guarantee that its contents stay random. Randomness is destroyed if one could even *implicitly* infer additional information about the current contents of an unexposed register from the combined knowledge of the current and previous contents of exposed registers. Therefore, in all cases, a proof is necessary that no new information about the current values of unexposed registers can be implicitly revealed. On the other hand, it is permissible for a given update step to implicitly reveal information about previous contents—subsequently overwritten—of an unexposed register. This does not destroy randomness, in that it is the *updated* structure that must be proven random. Since revealing past secret information causes no harm as long as no current secret information is revealed, it is convenient to imagine an omniscient "intermediary": an agent independent of the deleting algorithm who updates all necessary registers in total confidence (see, in this respect, the "card model" in Achlioptas [1]). Randomness is retained even if the actions (updates) of the intermediary combined with all exposed information implicitly yield some information about past values of unexposed registers, as long as no information about their current contents is revealed. Of course, this construct of "intermediary" is not a formal notion, but simply a convenient way to describe the updating mechanism.

Notice also that one should not assume that *all* previously unexposed information that is going to be overwritten is necessarily exposed at an update. Doing so might make it possible to infer additional implicit information about the updated contents of unexposed registers. In general, only part of the information to be overwritten needs to be known in order to carry out the update, and thus is implicitly revealed. The construct of the omniscient intermediary operating in secrecy frees us from having to make explicit exactly what secret information (to be overwritten) is implicitly revealed at an update. We simply need to make sure that no updated secret information is implicitly revealed after the update.

We illustrate these points by a further example. Consider a random graph conditional on (i) the number of vertices, (ii) the number of edges, (iii) the number of vertices of degree 1, and (iv) the number of vertices of degree 0 (isolated vertices). We claim that upon deleting a random vertex of degree 1 and its incident edge, the new graph is random conditional on the same type of information. This randomness claim is an immediate consequence of a more general theorem proven in Pittel et al. [436] (see also Broder et al. [81]), where the degrees of the vertices to be deleted are allowed to take values up to an arbitrary fixed integer k, assuming that the degree sequence of the graph is given up to k (we have seen in the previous example that this is true if we allow the degrees to range up to $n - 1$). The proof in Pittel et al. [436] depends upon counting all possibilities. However, the result can also be proven using the principle of deferred decisions. Assume that for each vertex v there is an *exposed* degree register that contains a three-valued parameter, indicating whether the degree of that vertex is 0, 1, or ≥ 2. In contrast to the case where the whole degree sequence was known, updating these registers after a deletion step presupposes knowledge of *unexposed* information. For instance, to update the degree register of a vertex that had degree at least 2 before the deletion, and which lost an incoming edge because of the deletion, we need to know whether its degree was previously exactly 2 or strictly more. However, it is easy to see that no information about the *updated* value of the unexposed registers is revealed by the combined knowledge of what currently is and previously was exposed: if an updated degree register ends up with the value ≥ 2, beyond this information we still have no knowledge of its actual degree. Therefore, randomness is retained. The omniscient intermediary secretly carries out the updating, using unexposed information. Even though the intermediary might reveal implicit information about the past values of registers, an observer cannot obtain any knowledge about the current contents of any unexposed register from what is and was unexposed.

On the other hand, the fact that additional current information is implicitly revealed is sometimes hard to notice. The subtlety of implicit disclosure can be illustrated by the following example. Let a *B&W graph* be a graph whose *edges* are either black or white. Call a vertex *all-white* if all the edges incident on it are white. Let the w-degree of a vertex v be the number of all-white vertices that v is connected with (see figs. 1 through 3). Notice that a black edge incident on v does not count towards the w-degree of v, while a white edge incident on v may or may not count towards the w-degree of v. Suppose we are given a random B&W graph G conditional on the number of vertices, and for each vertex v, the w-degree of v as well as the number of black edges and the number of white edges incident on v. All other characteristics of G are assumed to be random. Formally, given a fixed integer n and a fixed array of integers $d_{w,i,j}$ where $w, i, j = 0, \ldots, n - 1$, then G is chosen with equal probability among all B&W graphs such that $d_{w,i,j}$ is the number of vertices in the graph with w-degree w, i incident white edges and j incident black edges. We assume that the values of the array are such that there is at least one such graph. Suppose now that we delete from G a vertex v, chosen at random among all vertices with a specified w-degree (say 0). Suppose we also delete all edges, black and white, incident on v. Is the new graph random conditional on the same type of information?

Prima facie, one may think that the answer to this question is yes. Indeed, suppose that the exposed registers give for each vertex its w-degree, as well as the number of black edges and the number of white edges incident on it. All other information about the graph is assumed to be unexposed, that iss, random. After the deletion of a vertex v as previously described, and the subsequent deletion of all edges incident on v, all registers are updated. We may be tempted to conclude that the same type of information about the graph is known before and after the deletion, leading to an affirmative answer to the question. Unfortunately, this argument is erroneous. To see why, observe what happens if, after the deletion of v, the exposed w-degree of another vertex u increases. Using the combined knowledge of the current and previous contents of exposed registers, we can infer that in the new graph there exists at least one vertex v' that has just become all-white (as the result of the deletion of a black edge joining v with v'). Additionally, we learn that u is connected with at least one of these newly-all-white vertices. However, this last type of information is not supplied by the currently exposed registers, which give only the w-degree of u and the number of black and white edges incident on it. They do not specify a subset of the all-white vertices connected with u. The fact that we have implicit access to that information means that randomness cannot be retained in the new graph.

We now show a specific case of this. Consider the list of degree parameters (w-degree, number of incident white edges and number of incident black edges) given in figure 1(a) for each vertex of a random B&W graph. Then, by an easy case analysis we may verify that the only graphs having these degree parameters are the two depicted in figure 1(b) and (c). These two graphs are equiprobable, and any information about them other than what is in the upper table is assumed to be stored in unexposed registers. Suppose now that we delete the vertex v_5 from the random graph. Then the resulting graph, depending on which the original one was, will have the degree parameters given either in figure 2(a) or in figure 3(a). Suppose the resulting graph has the degree parameters of figure 2(a), so that the original graph was the one in figure 1(b)—examining this case will be sufficient for the purposes of demonstration. Again, by an easy case analysis we can verify that the only graphs having these degree parameters are the two depicted in figure 2(b) and 2(c). (If the original graph was the one in fig. 1(c), then the only possible graph having the degree parameters of fig. 3(a) is the one depicted in fig. 3(b)—we do not examine that case here.)

If deleting vertex v_5 did not destroy randomness, then both graphs in figure 2(b) and 2(c) should be equiprobable. However, from the *combined* knowledge of the tables in figure 1(a) and figure 2(a), we can easily infer that the graph in figure 2(c) is impossible. This is so because combining the information in the last columns of the tables in figure 1(a) and 2(a) we find that the newly all-white vertex is v_6 (it is the only vertex that previously had, but no longer has, an incident black edge). Also, from the combined information in the third and fourth rows of the second columns of these tables we see that both v_3 and v_4 are adjacent to v_6, as their w-degree has increased. Continuing with an easy case analysis,

(a)

vertices	w-degree	# of white incident edges	# of black incident edges
v_1	0	2	0
v_2	1	1	1
v_3	1	2	2
v_4	0	1	1
v_5	0	0	1
v_6	0	2	1

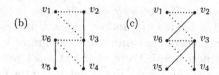

FIGURE 1   Original B&W graph: (a) exposed register values; (b) and (c) the two possible graphs corresponding to these values. Solid lines represent black edges and dashed lines represent white edges.

we conclude that the only graph that has the degree parameters of figure 2(a) and was obtained from a graph that has the degree parameters of figure 1(a) by deleting v_5 is the graph in figure 2(b). In other words, the combined knowledge of the two tables—the one before the deletion and the one after—reveals additional information that cannot be obtained exclusively from the current table, after the deletion. This proves that randomness is not retained. It is instructive to note that if no information were given about the w-degree of the vertices and we dealt only with information about the ordinary degrees (even if they were categorized by the number of incident white edges and the number of incident black edges) then randomness would be retained. That is true because combined knowledge of the two consecutive tables would not then be enough for us to infer additional unexposed information about the resulting graph.

The execution of an algorithmic step on the graph, such as the deletion of a vertex and the edges incident on it, can thus implicitly but subtly expose additional information about the current values of unexposed registers. In section 4, we describe more fully the methodology that is helpful in checking whether any implicit exposure of additional information has taken place as the result of the application of an algorithmic step. As we have seen here, the basic idea is to store information about the random structure in registers, in sufficiently "small pieces." The payoff of doing so is that implicit disclosure of information can be detected easily. Again, we do not require that updates be performed only on the basis of exposed information: unexposed information can be made available to the omniscient "intermediary" doing the updating. But there must be no way for us to infer this information.

vertices	w-degree	# of white incident edges	# of black incident edges
v_1	0	2	0
v_2	1	1	1
v_3	2	2	2
v_4	1	1	1
—	-	-	-
v_6	0	2	0

(a)

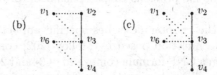

FIGURE 2   B&W graph from figure 1(b) with vertex v_5 deleted: (a) exposed register values; (b) and (c) the two possible graphs corresponding to these values.

vertices	w-degree	# of white incident edges	# of black incident edges
v_1	0	2	0
v_2	1	1	1
v_3	1	2	1
v_4	0	1	1
—	-	-	-
v_6	0	2	1

(a)

(b)

FIGURE 3   B&W graph from figure 1(c) with vertex v_5 deleted: (a) exposed register values; (b) the only possible graph corresponding to these values.

One might say that a safer way to prove conditional randomness claims is by rigorous counting arguments, rather than through the principle of deferred decisions. In complicated situations, however, counting arguments are practically impossible. As we will see from specific applications, our methodology makes it easy to specify what the *a priori* exposed information should be in order to retain randomness throughout the execution of an algorithm, given the type of operations that the algorithm allows. Such considerations have attracted much attention lately, in view of the increased interest in the probabilistic analysis of heuristics on random Boolean formulas and graphs. This has been discussed in chapter 3 (see also Molloy [403] for an overview of satisfiability and colorability thresholds). The probabilistic analysis involves analyzing the mean path of the heuristic [529], while showing that randomness is retained throughout the course of the heuristic. It is in situations like this where our methodology is particularly useful. This approach can ultimately be used to obtain lower bounds on threshold locations: indeed, the best lower bound to date on the satisfiability threshold [294, 295], mentioned in the previous chapter, has been proven using the principle of deferred decisions.

The rest of the chapter describes specific applications of this nature. We answer, notably, a question posed by Molloy [402] concerning a Davis-Putnam heuristic acting on a CNF formula comprised of 3- and 2-clauses, when the literals to be satisfied are selected on the basis of how often they appear in each of the two types of clauses. Using the principle of deferred decisions, we show what characteristics must be conditional in order to retain randomness throughout the procedure (theorem 4.2 in section 4), and conjecture that this is the minimal set of conditionals needed.

3   TERMINOLOGY AND NOTATION

Our results can be applied in various contexts related to random graphs or formulas. However, for concreteness, we first present them in the context of random formulas comprised only of 3- and 2-clauses. We introduce below the related terminology and notation.

Let V be a set of variables of cardinality n. Let L be the set of literals of V, that is, elements of V and their negations. A k-clause is a disjunction of exactly k literals from L. Let ϕ be a Boolean formula in conjunctive normal form (CNF), comprised of 3- and 2-clauses. Let m be the total number of clauses of the formula. Let C_3 and C_2 denote the collections of 3-clauses and 2-clauses of ϕ, respectively, and let c_3, c_2, and l be the respective cardinalities of the sets C_3, C_2, and L. Clearly $c_3 + c_2 = m$, and $l = 2n$. (Note that the notation used here is slightly different from that of chapter 3: there, C_3 and C_2 were the numbers of 3- and 2-clauses, and c_3 and c_2 were the respective *densities*. Note also that C_3 and C_2 are distinct from $\mathcal{C}_3$ and $\mathcal{C}_2$ from the previous chapter, where they denoted the collections of clauses containing exactly 3 and 2 *positive* literals, respectively.)

For $i = 0, 1, \ldots, 3c_3 + 2c_2$, let D_i be the set of literals in L that have exactly i occurrences in ϕ. The elements of D_i are said to have *degree i*. Literals whose negation is in D_0 are called *pure*. Notice that according to our terminology, a literal in L whose variable does not appear at all in the formula is pure.

Let D_1^3 and D_1^2 be the sets of literals that have exactly one occurrence in ϕ, in a 3-clause and 2-clause, respectively. D_1 is then the disjoint union of D_1^3

and D_1^2. Let also $D_{1\times 1}^2$ be the subset of D_1^2 comprised of literals that appear in a 2-clause whose second literal also belongs to D_1^2.

Let d_i, d_1^3, d_1^2, and $d_{1\times 1}^2$ be the respective cardinalities of D_i, D_1^3, D_1^2, and $D_{1\times 1}^2$. Obviously, $d_1 = d_1^3 + d_1^2$.

Consider the collection of formulas comprised of 3- and 2-clauses that have given, fixed values for the parameters l, c_3, c_2, d_0, d_1^3, d_1^2, and $d_{1\times 1}^2$. Make this collection into a probability space by assigning to each one of its elements the same probability (we assume that the values of the parameters are such that this space is not empty). An element of this space is called a random $\{3, 2\}$-CNF formula conditional on the values of l, c_3, c_2, d_0, d_1^3, d_1^2, and $d_{1\times 1}^2$. One could define random graphs similarly, conditional, for instance, on the number of edges and vertices, as we did in the previous section. Such formulas and graphs are called *conditionally random objects*.

We will consider algorithms on random $\{3, 2\}$-CNF formulas that only apply steps of the following three types (one step may comprise several constituent sub-steps):

- *Set a pure literal.* Select at random a pure literal, set it to TRUE and delete all clauses where it appears.
- *Set a degree-one literal from a 3-clause.* Select at random a literal in D_1^3, set it to FALSE, delete it from the 3-clause where it appears and delete all clauses where its negation appears.
- *Set a degree-one literal from a 2-clause.* Select at random a literal in D_1^2, set it to FALSE, delete it from the 2-clause where it appears and delete all clauses where its negation appears. This can create a 1-clause. As long as there are 1-clauses, choose one at random, set its literal to TRUE, delete all clauses where it appears and delete its negation from any clause in which it appears. Ignore (delete) any empty and thus trivially unsatisfiable clause that may occur during this step. This last provision is simply a technicality introduced to study the randomness of the formula independently of its satisfiability. Of course, when such a step is used as a subroutine of an algorithm for satisfiability, the occurrence of an empty clause is an indication to stop immediately and report unsatisfiability.

4   RESULTS

Theorem 4.1. *Let ϕ be a random $\{3, 2\}$-CNF formula conditional on the values of the parameters l, c_3, c_2, d_0, d_1^3, d_1^2, and $d_{1\times 1}^2$. If any algorithmic step like the ones described above is applied to ϕ, then the formula obtained is a random $\{3, 2\}$-CNF formula conditional on the new values of the parameters l, c_3, c_2, d_0, d_1^3, d_1^2, and $d_{1\times 1}^2$.*

Proof Notice that no algorithmic step differentiates between degree-one literals appearing in 2-clauses on the basis of the degree of the other literal in the 2-clause. Still, according to the statement of the theorem, randomness is preserved if it is conditional not only on d_1^2 but also on $d_{1\times1}^2$. The reason for this will become clear later in this proof.

We first introduce some general notions, in more formal terms than before. An object such as a formula or a graph can be modeled by a data structure. Let us think of a data structure as a collection of registers containing information about the object. For example, a data structure modeling a graph includes a register for each vertex, with pointers to the registers of the edges incident on the vertex. It also includes a register for each edge, with pointers to the registers of the vertices on which the edge is incident. A data structure modeling a formula includes a register for each literal, with pointers to the registers of the literal appearing in the clause. It also includes a register for each clause (more information about the registers of a formula is given below).

Registers are partitioned into groups. The elements of each group contain various types of information for the same part of the modeled object. For example, for each vertex of a graph, we may have several registers in one group: one with pointers to the edges incident on the vertex, another with the degree of this vertex, etc. For the present purposes, we refer to the registers belonging to the same group as sub-registers of the group. We also imagine, for each group, a head register with pointers to its sub-registers. When a sub-register of a group contains a pointer to another group, it is assumed to point to the head register of that other group. Intuitively, the reason for storing different types of information in separate sub-registers is to avoid exposing all information about a part of the modeled object when it is necessary to expose only a "small piece" of it.

A *data structure with unexposed information* is a data structure whose (sub-)registers are partitioned into two categories, called *unexposed* and *exposed* registers. The partitioning is done according to rules given in the definition of the structure. These rules are based on the type of contents of the registers. The head registers of the groups are always exposed. Intuitively, one may think of such a structure as modeling an object whose characteristics stored in the unexposed registers are random, conditional on the information stored in the exposed registers. The same group may contain both exposed and unexposed sub-registers. For example, although the specific edges where a vertex appears may not be exposed, its degree may be exposed.

In general, given a conditionally random object, we associate with it a data structure as above. An algorithmic step that deletes an element of the object (such as the deletion of a vertex or the assignment of a variable) corresponds to the deletion of the group of registers associated with the deleted element of the object. After the deletion, all registers are updated.

Definition 1. *An algorithmic step is called randomness preserving if, after the corresponding deletions and updates of registers, no information about the contents of unexposed registers can be inferred from what currently is and previously was exposed, beyond what can be inferred from what is currently exposed. In other words, no additional information is implicitly revealed by knowing both past and current exposed information.*

To prove a randomness claim such as the theorem under consideration, it suffices to find a data structure with unexposed information that models the conditionally random object in the claim, and then to show (i) that the algorithmic steps are randomness preserving and (ii) that the information in the conditional is exactly the information that can be extracted from the exposed registers of the structure.

We describe below a structure $\mathcal{S}$, with unexposed information, that models a random $\{3, 2\}$-CNF formula conditional on the parameters l, c_3, c_2, d_0, d_1^3, d_1^2, and $d_{1\times 1}^2$.

- For each literal t in L, the structure $\mathcal{S}$ contains a group of sub-registers collectively called literal sub-registers. These contain information about the degree of the literal, its occurrences in the formula and its negation. The information that is assumed exposed is (i) the degree and (ii) the position in the formula of literals with a single occurrence that happens to be in a 2-clause. All other information is unexposed. More formally, one of these sub-registers contains two bits of information indicating whether t belongs to D_0 (t does not appear in the formula), D_1^3 (t has degree 1 and appears in a 3-clause), D_1^2 (t has degree 1 and appears in a 2-clause) or none of these (t has degree at least 2). This sub-register is exposed. Also, we assume that there are sub-registers containing pointers to the positions of all occurrences of t in the formula (to the heads of all clause sub-registers where t appears; see below). These sub-registers are exposed if t is in D_1^2 and unexposed otherwise. The reason for exposing the position in the formula of literals in D_1^2 will become apparent later. Finally, we assume that there is an unexposed sub-register pointing to the head of the literal sub-register of the negation of t. It is important to notice that because the pointer to the negation of a literal is unexposed, each literal is paired with its logical negation randomly.

- For each clause in the formula, the structure $\mathcal{S}$ contains a group of sub-registers collectively called clause sub-registers. These contain information about the type of the clause (3-clause or 2-clause) and pointers to the heads of literal sub-registers corresponding to the literals that appear in the clause. The information about the type of the clause is exposed, while the pointers to the literal registers are unexposed.

It is straightforward to verify that after the application of any of the algorithmic steps, no information about an unexposed register can be deduced from what is and previously was exposed. Under these circumstances, the randomness of the structure S is preserved under an algorithmic step. The need for having the positions of literals in D_1^2 exposed can be seen in the event that under an update step, exactly one 3-clause shrinks to a 2-clause and exactly one literal moves from D_1^3 to D_1^2. In that case, the information about the type of each clause and the degree of each literal is sufficient to allow us to infer the position of this literal.

Now the theorem follows because the information that can be extracted from S consists only of the values of the following parameters: l (the number of groups of literal sub-registers), c_2 (the number of groups of clause sub-registers for 2-clauses), c_3 (the number of groups of clause sub-registers for 3-clauses), d_0, d_1^3, d_1^2, and $d_{1 \times 1}^2$. The value of $d_{1 \times 1}^2$ can be obtained from S because the positions in the formula of literals in D_1^2 are exposed. All other information that can be extracted from S can be expressed in terms of the values of the parameters l, c_3, c_2, d_0, d_1^3, d_1^2, and $d_{1 \times 1}^2$, only. (One can immediately see, for instance, that the number of 2-clauses where both positions are filled with literals of degree at least 2 or the number of 2-clauses where one position contains a literal of degree at least two and the other a literal of degree exactly one can be expressed in terms of the values of the parameters l, c_3, c_2, d_0, d_1^3, d_1^2, and $d_{1 \times 1}^2$). This completes the proof of Theorem 4.1.

We now come to the generalization of the previous result to arbitrary degrees, where algorithms making use of the overall number of occurrences of literals in 3-clauses and 2-clauses, separately, are allowed. To preserve randomness in this case, a conditional given by a number of integer parameters—as in the previous theorem—is not enough. We have to assume that the positions of all literals appearing in 2-clauses are known, regardless of their degree: this information is revealed when a 3-clause shrinks to a 2-clause and the exposed degree information of literals is updated. However, no information about negations of literals or identification of literals need be revealed, nor does information on the positions of literals appearing in 3-clauses. In other words, we have to assume that the *pattern* in which literals are paired in 2-clauses is conditional, though the pattern need not reveal the pairing of literals of opposite logical sign. This is still a severe restriction on the randomness of the formula. Below, we formalize the notion of pattern.

Fix an even integer $2n$ representing the number of literals of a formula, and an integer c_3 representing the number of 3-clauses in a formula. A *pattern for 2-clauses and degree sets that is transparent with respect to negations* (pattern, in short) is a set of unordered pairs $\underline{C}_2$ of integers from $\{1, \ldots, 2n\}$, representing the collection of 2-clauses of the formula, together with a collection of sets $D_i^3 \subseteq \{1, \ldots, 2n\}, i = 0, \ldots, c_3$, such that $\sum_i i |D_i^3| = 3c_3$, representing the collection of sets of literals whose number of occurrences in 3-clauses is i.

Now fix a pattern P as described above. A random formula ϕ conditional on P is constructed as follows: randomly choose c_3 unordered triplets from $1, \ldots, 2n$

so that all integers in each D_i^3 appear in exactly i such triplets; denote this set by $\underline{C}_3$; randomly select a one-to-one and onto mapping neg : $\{1, \ldots, n\} \to \{n+1, \ldots, 2n\}$ representing the negations; in the tuples of $\underline{C}_3$ and of $\underline{C}_2$, replace each $k = 1, \ldots, n$ with variable x_k and each neg(k), $k = 1, \ldots, n$, with its negation $\overline{x_k}$, and denote by C_3 and C_2, respectively, the sets of clauses thus obtained; let the formula ϕ be the one that has as 3-clauses and as 2-clauses the sets C_3 and C_2, respectively. Notice that since the negation function "neg" was random, a literal and its negation may appear in the same clause. If we wish to avoid this, "neg" may instead be a random one-to-one and onto mapping made conditional on the fact that for no $i = 1, \ldots, n$ can both i and neg(i) appear in the same tuple of either $\underline{C}_3$ or $\underline{C}_2$. Based on the method of proof of the previous theorem, one can obtain the following result that answers an open question posed by Molloy [402].

Theorem 4.2. *Let ϕ be a random $\{3, 2\}$-CNF formula conditional on a given pattern P, as described above. For arbitrary i and j, choose at random a literal t with i occurrences in 3-clauses and j occurrences in 2-clauses. Assign to t the value* TRUE *and perform the necessary deletions and shrinking of clauses accompanied by repeated setting to* TRUE *of literals in 1-clauses, as long as 1-clauses exist. The new formula is then random, conditional on its new pattern P'.*

Proof Again, we introduce a structure S that contains groups of sub-registers corresponding to literals and to clauses. This time, the exposed degree sub-registers of a literal t contain two integers: one giving the number of occurrences of t in 3-clauses and the other giving the number of occurrences of t in 2-clauses. Furthermore, the group of literal sub-registers of t contains information on which 3-clauses and which 2-clauses include t. The information regarding 3-clauses is unexposed. The information regarding 2-clauses, however, must be exposed because after an algorithmic step, it can be inferred from the knowledge of the previous and current values of the registers giving the type of each clause (3-clause or 2-clause) and the degrees of the literals. One may readily confirm that nothing can then be inferred about the unexposed registers after the application of an algorithmic step. It is also immediately apparent that the information that can be extracted from such a structure S is given by the pattern P.

Note that if the algorithm does not make use of the number of occurrences of literals separately in 3-clauses and 2-clauses, but only needs the total number of occurrences of a literal in the formula, then the conditional does not have to include the pairing of literals in 2-clauses. It is sufficient in this case for the conditional to contain the total degree sequence, the number of 3-clauses, and the number of 2-clauses.

Finally, as a further application, let us see what information must be placed in the conditional for an algorithm deleting vertices of a specified w-degree from a random B&W graph, as discussed in section 2.

Given a B&W-graph G, let $W \subseteq G$ be the subgraph comprised of the vertices of G, with vertices marked according to whether or not they are all-white (in the sense of G), and all white edges with at least one endpoint incident on an all-white vertex. Call W the subgraph of w-degree *witnesses*. Without giving details, one can again define a notion of random B&W graphs conditional on the number of vertices, the total number of edges and the precise subgraph of w-degree witnesses. Note that to construct the rest of the graph from this information, one can arbitrarily place edges between vertices that are not all-white and then arbitrarily color them black or white, taking care that at least one black edge is incident on each vertex that is not all-white.

Then the following theorem holds. We omit its easy proof, as the notion of B&W graphs was introduced only for illustrative purposes.

Theorem 4.3. *If we delete a random vertex of a specified arbitrary w-degree from a B&W graph that is random conditional on the number of vertices, the total number of edges and the subgraph of w-degree witnesses, then the new graph is random conditional on the new number of vertices, the new total number of edges and the new subgraph of w-degree witnesses.*

An analogous result holds if the deleted vertex has specified numbers of white and black edges incident on it (the conditional in the latter case must be augmented to contain the sequence d_{ij} giving the number of vertices with i white and j black edges incident on them).

We conclude this chapter by the following

Informal Conjecture. The conditionals of theorems 4.1, 4.2, and 4.3 contain the least information possible. With weaker conditionals, randomness would not be retained.

ACKNOWLEDGMENTS

This research has been supported by the University of Patras, Research Committee (Project C. Carathéodory no. 2445). We would like to thank an anonymous referee and the editors for their comments that led to substantial improvements. The second author thanks D. Achlioptas and M. Molloy for several discussions about probabilistic arguments, and acknowledges partial support by the EU within the 6th Framework Programme under contract 001907 (DELIS).

CHAPTER 9

The Phase Transition in the Random HornSAT Problem

Demetrios D. Demopoulos
Moshe Y. Vardi

1 INTRODUCTION

This chapter presents a study of the satisfiability of random Horn formulas and a search for a phase transition. In the past decade, phase transitions or *sharp thresholds*, have been studied intensively in combinatorial problems. Although the idea of thresholds in a combinatorial context was introduced as early as 1960 [147], in recent years it has been a major subject of research in the communities of theoretical computer science, artificial intelligence, and statistical physics. As is apparent throughout this volume, phase transitions have been observed in numerous combinatorial problems, both for the probability that an instance of a problem has a solution and for the computational cost of solving an instance. In a few cases (2-SAT, 3-XORSAT, 1-in-k SAT) the existence and location of these phase transitions have also been formally proven [7, 94, 101, 131, 156, 202].

The problem at the center of this research is that of *3-satisfiability* (3-SAT). An instance of 3-SAT consists of a conjunction of clauses, where each clause is a disjunction of three literals. The goal is to find a truth assignment that satisfies

Computational Complexity and Statistical Physics, edited by
Allon G. Percus, Gabriel Istrate, and Cristopher Moore, Oxford University Press.

all clauses. The *density* of a 3-SAT instance is the ratio of the number of clauses to the number of Boolean variables. We call the number of variables the *size* of the instance. Experimental studies [110, 395, 397, 466, 469] have shown that there is a shift in the probability of satisfiability of random 3-SAT instances, from 1 to 0, located at around density 4.27 (this is also called the *crossover point*). So far, in spite of much progress in obtaining rigorous bounds on the threshold location, highlighted in the previous chapters, there is no mathematical proof of a phase transition taking place at that density [1, 132, 177]. Experimental studies also show a peak of the computational complexity around the crossover point. In Kirkpatrick and Selman [319], finite-size scaling techniques were used to suggest a phase transition at the crossover point. Later, in Coafra et al. [96], experiments showed that a phase transition of the running time from polynomial in the instance size to exponential is solver-dependent, and for several different solvers this transition occurs at a density lower than the crossover point. This phenomenon has been further discussed in chapter 3. A restriction on all the experimental studies is imposed by the inherent difficulty of the problem, especially around the crossover point. We can only study instances of limited size (usually up to a few hundred) before the problems get too hard to be solved in reasonable time using available computational resources.

A problem similar to random 3-SAT is that of the satisfiability of random Horn formulas, also called random HornSAT. A Horn formula in conjunctive normal form (CNF) is a conjunction of Horn clauses; each Horn clause is a disjunction of literals of which *at most one* can be positive. Unlike 3-SAT, HornSAT is a tractable problem. The complexity of the HornSAT is linear in the size of the formula [128]. The linear complexity of HornSAT allows us to study experimentally the satisfiability of the problem for much bigger input sizes than those used in similar research on other problems like 3-SAT or 3-Colorability [96, 110, 253, 469].

An additional motivation for studying random HornSAT comes from the fact that Horn formulas are related to several other areas of computer science and mathematics [375]. In particular, Horn formulas are connected to automata theory, as the transition relation, the starting state, and the set of final states of an automaton can be described using Horn clauses. For example, if we consider automata on binary trees (see definition below), then Horn clauses of length three can be used to describe its transition relation, while Horn clauses of length one can describe the starting state and the set of the final states of the automaton (we elaborate on that later). Then, the question about the emptiness of the language of the automaton can be translated to a question about the satisfiability of the formula. There is also a close relation between knowledge-based systems and Horn formulas, though we do not consider that relation in this chapter. Finally, there is a correspondence between Horn formulas and hypergraphs that we use to show how results on random hypergraphs relate to our research on random Horn formulas.

The probability of satisfiability of random Horn formulas generated according to a variable clause length model has been studied by Istrate [271], who showed that random Horn formulas have a *coarse* rather than a sharp satisfiability threshold, meaning that the problem does not have a phase transition. The variable clause length distribution model used by Istrate is ideally suited to studying Horn formulas in connection with knowledge-based systems [375].

Motivated by the connection between the *automata emptiness* problem and Horn satisfiability, we study the satisfiability of two types of random Horn formulas in conjunctive normal form (CNF) that are generated according to a variation of the *fixed* clause length distribution model mentioned in chapter 7. We consider the 1-3-HornSAT, where formulas consist of clauses of length one and three only, and 1-2-HornSAT, where formulas consist of clauses of length one and two only. We are looking to identify regions in the problems' space where instances are almost surely satisfiable or almost surely unsatisfiable. We are also interested in finding if the problems exhibit a sharp threshold.

The random 1-2-HornSAT problem is related to the random 1-3-HornSAT problem in the same way that random 2-SAT is related to random 3-SAT. That is, as some algorithm searches for a satisfying truth assignment for a random 1-3-Horn formula by assigning truth values to the variables, a random 1-2-Horn formula is created as a subformula of the original formula. This is a result of 3-clauses being shortened to 2-clauses by a subtitution of truth values. The relation between random 2-SAT and random 3-SAT has been exploited by Achlioptas [1] to improve on the lower bound for the threshold of random 3-SAT. In this work, Achlioptas uses differential equations to analyze the execution of a broad family of SAT algorithms. In general, one can try to analyze phase transitions using differential equations (cf. Istrate [272]). The 1-2-HornSAT problem can also be analyzed with the help of random graphs [62]. We show how results on random digraph connectivity, presented by Karp [300], can be used to model the satisfiability of random 1-2-Horn formulas. These results can be used to show that there is no phase transition for 1-2-HornSAT and are matched by our experimental data.

Our experimental investigation of 1-3-HornSAT shows that there are regions where a random 1-3-Horn formula is almost surely satisfiable and regions where it is almost surely unsatisfiable. Analysis of the satisfiability percentile window and finite-size scaling methods [485] suggest that there is a sharp threshold line between these two regions. Just as 1-2-HornSAT can be analyzed using random digraphs, 1-3-HornSAT can be analyzed using random hypergraphs. We show that some recent results on random hypergraphs [116] fit our experimental data well. Unlike the data analysis, however, the hypergraph-based model suggests that the transition from the satisfiable to unsatisfiable regions is a steep function rather than a step function. It is, therefore, not clear if the problem exhibits a phase transition, in spite of our having made use of experimental data for instances of large size.

Our results here also relate to those of Kolaitis and Raffill [339], who carried out a search for a phase transition in another NP-complete problem, that of AC-matching. The similarity between their work and ours is that the experimental data provide evidence that both problems have a slowly emerging phase transition. The difference is that in our case, because of the linear complexity of Horn satisfiability, we are able to test instances of Horn satisfiability of much larger size than the instances of AC-matching in Kolaitis and Raffill [339], or for that matter of most NP-complete problems such as 3-SAT and 3-COL.

2   PRELIMINARIES AND FINITE AUTOMATA

Before discussing our main results on thresholds in HornSAT, let us review some definitions related to combinatorial phase transitions, and show explicitly the relationship between HornSAT and finite automata. Let X be a finite set and $|X| = n$. Let A be a random subset of X constructed by a random procedure according to the probability space $\Omega(n, m)$ with measure:

$$\Pr_{\Omega(n,m)}(A) = \begin{cases} 1/\binom{n}{m^*} & \text{if } |A| = m^*; \\ 0 & \text{otherwise}, \end{cases}$$

where m is an integer and

$$m^* = \begin{cases} 0 & \text{if } m < 0; \\ m & \text{if } 0 \le m \le n; \\ n & \text{if } m > n. \end{cases}$$

The random procedure consists of selecting m^* elements of X without replacement. A (set) property Q of X is a subset of 2^X, the power set of X consisting of all subsets of X. Q is increasing if $A \in Q$ and $A \subseteq B \subseteq X$ implies $B \in Q$. Q is non-trivial if $\varnothing \notin Q$ and $X \in Q$. A property sequence Q consists of a sequence of sets $\{X_n : n \ge 1\}$ such that $|X_n| < |X_{n+1}|$ and a family $\{Q_n : n \ge 1\}$ where each Q_n is a property of X_n. Q is increasing if Q_n is increasing for every $n \ge 1$, and Q is non-trivial if Q_n is non-trivial for every $n \ge 1$.

Let Q_n be an increasing non-trivial property sequence, and $\theta : \mathbb{N} \to \mathbb{R}^+$ a strictly positive function. We say that θ is a threshold for Q if for every $f : \mathbb{N} \to \mathbb{N}$:

1. If $\lim_{n \to \infty} f(n)/\theta(n) = 0$ then $\lim_{n \to \infty} \Pr_{\Omega(n,f(n))}(Q_n) = 0$
2. If $\lim_{n \to \infty} f(n)/\theta(n) = \infty$ then $\lim_{n \to \infty} \Pr_{\Omega(n,f(n))}(Q_n) = 1$.

θ is a sharp threshold Q if for every $f : \mathbb{N} \to \mathbb{N}^+$:

1. If $\sup_{n \to \infty} f(n)/\theta(n) < 1$ then $\lim_{n \to \infty} \Pr_{\Omega(n,f(n))}(Q_n) = 0$
2. If $\inf_{n \to \infty} f(n)/\theta(n) > 1$ then $\lim_{n \to \infty} \Pr_{\Omega(n,f(n))}(Q_n) = 1$.

We say that Q exhibits a phase transition if it has a sharp threshold. Our interest is in satisfiability of Horn formulas. Thus, in our framework X_n is the set of Horn clauses over a set with n Boolean variables. A set of Horn clauses is a Horn formula.

Our main motivation for studying the satisfiability of Horn formulas is that, unlike 3-SAT, this problem is tractable. Therefore, we will have numerical data for instances of much larger size to help us answer questions similar to those previously asked about 3-SAT.

Apart from that, it is also of interest to us that Horn formulas can be used to describe finite automata. A finite automaton A is a 5-tuple $A = (S, \Sigma, \delta, s, F)$, where S is a finite set of states, Σ is an alphabet, δ is a transition relation, s is a starting state, and $F \subseteq S$ is the set of final (accepting) states.

In a word automaton, δ is a function from $S \times \Sigma$ to 2^S. In a binary-tree automaton δ is a function from $S \times \Sigma$ to $2^{S \times S}$. Intuitively, for word automata δ provides a set of successor states, while for binary-tree automata δ provides a set of successor state pairs. A run of an automaton on a word $a = a_1 a_2 \cdots a_n$ is a sequence of states $s_0 s_1 \cdots s_n$ such that $s_0 = s$ and $(s_{i-1}, a_i, s_i) \in \delta$. A run is successful if $s_n \in F$: in this case we say that A accepts the word a. A run of an automaton on a binary tree t labeled with letters from Σ is a binary tree r labeled with states from S such that $\text{root}(r) = s$ and for a node i of t, $(r(i), t(i), r(\text{left-child-of-}i), r(\text{right-child-of-}i)) \in \delta$. Thus, each pair in $\delta(r(i), t(i))$ is a possible labeling of the children of i. A run is successful if for all leaves l of r, $r(l) \in F$: in this case we say that A accepts the tree t. The language $L(A)$ of a word automaton A is the set of all words a for which there is a successful run of A on a. Likewise, the language $L(A)$ of a tree automaton A is the set of all trees t for which there is a successful run of A on t. An important question in automata theory that is also of great practical importance in the field of formal verification [510] is, given an automaton A, is $L(A)$ non-empty? We can show how the problem of non-emptiness of automata languages translates to Horn satisfiability.

Consider first a word automaton $A = (S, \Sigma, \delta, s_0, F)$. Construct a Horn formula ϕ_A over the set S of variables as follows:

- create a clause $(\overline{s_0})$
- for each $s_i \in F$ create a clause (s_i)
- for each element (s_i, a, s_j) of δ create a clause $(\overline{s_j}, s_i)$,

where $(s_i, \ldots, s_k)$ represents the clause $s_i \vee \cdots \vee s_k$ and $\overline{s_j}$ is the negation of s_j.

Theorem 2.1. *Let A be a word automaton and ϕ_A the Horn formula constructed as described above. Then $L(A)$ is non-empty if and only if ϕ_A is unsatisfiable.*

Proof ($\Rightarrow$) Assume that $L(A)$ is non-empty, i.e., there is a path $\pi = s_{i_0} s_{i_1} \cdots s_{i_m}$ in A such that $s_{i_0} = s_0$ and $s_{i_m} = s_k$ where s_k is a final state. Since s_k is a final state, (s_k) is a clause in ϕ_A. Also $(\overline{s_k}, s_{i_{m-1}})$ is a clause in ϕ_A. For ϕ_A to be satisfiable s_k must be TRUE, and consequently $s_{i_{m-1}}$ must be TRUE. By induction on the length of the path π we can show that for ϕ_A to be satisfiable s_0 must be TRUE, which is a contradiction.

($\Leftarrow$) Assume that ϕ_A is unsatisfiable. It then must have positive-unit resolution refutation [239], i.e., a proof by contradiction where in each step one of the resolvents must be a positive literal and the last resolution step is with the clause $(\overline{s_0})$. Let (s_i) be the first positive literal resolvent in the proof. By construction, s_i is a final state of A. By induction on the length of the refutation, we can construct a path in A from s_0 to s_i, Therefore, $L(A)$ is non-empty.

Similarly to the word automata case, we can show how to construct a Horn formula from a binary-tree automaton. Let $A = (S, \Sigma, \delta, s_0, F)$ be a binary-tree automaton. Then we can construct a Horn formula ϕ_A using the construction above with the only difference that since δ in this case is a function from $S \times \{\alpha\}$ to $S \times S$, for each element (s_i, α, s_j, s_k) of δ we create a clause $(\overline{s_j}, \overline{s_k}, s_i)$. It is not difficult to see that also in this case we have:

Theorem 2.2. *Let A be a binary-tree automaton and ϕ_A the Horn formula constructed as described above. Then $L(A)$ is non-empty if and only if ϕ_A is unsatisfiable.*

Motivated by the connection between tree automata and Horn formulas described in theorem 2.2, we study the satisfiability of two types of random Horn formulas. More precisely, let $H^{1,2}_{n,d_1,d_2}$ denote a random formula in CNF over a set of variables $X = \{x_1, \ldots, x_n\}$, containing:

- a single negative literal chosen uniformly among the n possible negative literals;
- $d_1 n$ positive literals that are chosen uniformly, independently and without replacement among all $n - 1$ possible positive literals (the negation of the single negative literal already chosen is not allowed); and
- $d_2 n$ clauses of length two that contain one positive and one negative literal chosen uniformly, independently and without replacement, among all $n(n-1)$ possible clauses of that type.

We call the number of variables n the *size* of the instance.

Let also $H^{1,3}_{n,d_1,d_3}$ denote a random formula in CNF over the set of variables $X = \{x_1, \ldots, x_n\}$, containing:

- a single negative literal chosen uniformly among the n possible negative literals;
- $d_1 n$ positive literals that are chosen uniformly, independently and without replacement among all $n - 1$ possible positive literals (the negation of the single negative literal already chosen is not allowed); and
- $d_3 n$ clauses of length three that contain one positive and two negative literals chosen uniformly, independently and with replacement among all $n(n - 1)(n - 2)/2$ possible clauses of that type.

The sampling spaces $H^{1,3}$ and $H^{1,2}$ are slightly different: we sample with replacement in the first, and without replacement in the second. Here we explain why. Assume that we sample dn clauses out of N uniformly at random with replacement. Let us consider the (asymptotic) expected number of distinct clauses we get. Each one of the N clauses will be chosen with probability $1 - (1 - 1/N)^{dn}$. The expected number of distinct chosen clauses is $N(1 - (1 - 1/N)^{dn})$. Notice that $N(1 - (1 - 1/N)^{dn}) \simeq dn - O((dn)^2/N)$. In the case of a random $H^{1,3}_{n,d_1,d_3}$ formula $N = n(n - 1)(n - 2)/2$ and clearly the expected number of distinct clauses we sample is asymptotically equivalent to dn; thus we sample with replacement for experimental ease. In the case of a random $H^{1,2}_{n,d_1,d_2}$ formula, we sample without replacement to ensure that we do not have many repetitions among the chosen clauses.

3   1-2-HORNSAT

In this section we present our results on the probability of satisfiability of random 1-2-Horn formulas. We first present an experimental investigation of the satisfiability on the $d_1 \times d_2$ quadrant. We then discuss the relation between random 1-2-Horn formulas and random digraphs and show that our data agree with analytical results on graph reachability presented in Karp [300].

To study the probability of satisfiability of $H^{1,2}_{n,d_1,d_2}$ random formulas in the $d_1 \times d_2$ quadrant, we have generated and solved 1200 random instances of size 20000 per data point. Figure 1 shows the average satisfiability probability versus the two input parameters d_1 and d_2 (a) and the corresponding contour plot (b).

The satisfiability plot in figure 1 suggests that the problem does not have a phase transition. This can also be observed if we fix the value of one of the input parameters. In figure 2 we show the satisfiability plot for random 1-2-HornSAT for various instance sizes ranging from 500 to 32000, and for fixed $d_1 = 0.1$. We now explain why random 1-2-HornSAT does not have a phase transition, based on known results on random digraphs.

There are two most frequently used models of random digraphs. The first one, $\mathcal{G}_{n,m}$ consists of all digraphs on n vertices having m edges; all digraphs have equal probability. The second model, $\mathcal{G}_{n,p}$ with $0 < p < 1$, consists of all digraphs on n vertices in which the edges are chosen independently with probability p.

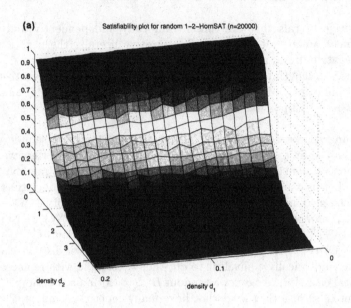

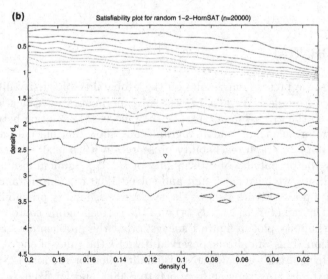

FIGURE 1   Satisfiability probability of a random 1-2-Horn formula of size 20,000 (a) and the corresponding contour plot (b). The contour plot contains 25 lines that separate consecutive percentage intervals $[0\% - 4\%), [4\% - 8\%), \ldots, [96\% - 100\%]$.

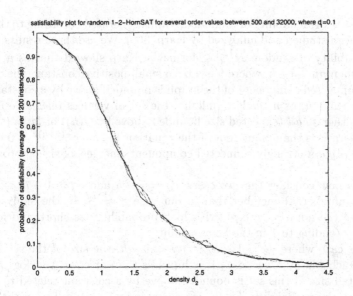

satisfiability plot for random 1–2–HornSAT for several order values between 500 and 32000, where $d_1 = 0.1$

FIGURE 2 Satisfiability probability of random 1-2-Horn formulas when $d_1 = 0.1$

It is known that in most investigations the two models are interchangeable, provided certain conditions are met. In what follows, we will take advantage of this equivalence in order to show how our experimental results relate to analytical results on random digraphs [300].

We will first show that there is a relation between the satisfiability of a random $H_{n,d_1,d_2}^{1,2}$ formula and the vertex reachability of a random digraph $\mathcal{G}_{n,m=d_2n}$. Let $\phi \in H_{n,d_1,d_2}^{1,2}$, $(\overline{x_0})$ be the unique single negative literal in ϕ, and F be the set of all variables that appear as single positive literals in ϕ. Obviously $|F| = d_1n$. Construct a graph G_ϕ such that for every variable x_i in ϕ there is a corresponding node v_i in G_ϕ and for each clause $(\overline{x_i}, x_j)$ of ϕ there is a directed edge in G_ϕ from v_i to v_j. G_ϕ is a random digraph from the $\mathcal{G}_{n,m=d_2n}$ model.

It is not difficult to see that ϕ is unsatisfiable if and only if the node v_0 in G_ϕ is reachable from a node v_i such that $x_i \in F$. In other words, the probability of unsatisfiability of a random $H_{n,d_1,d_2}^{1,2}$ formula ϕ is equal to the probability that a vertex of the random digraph $\mathcal{G}_{n,m=d_2n}$ is reachable from a set of vertices of size d_1n. (A vertex is reachable from a set of vertices if it is reachable by at least one of the vertices of the set.)

As mentioned above, the $\mathcal{G}_{n,m}$ and $\mathcal{G}_{n,p}$ models can be used interchangeably, when $m \simeq \binom{n}{2}p$ [62]. Therefore, the relation we have established between the satisfiability of a random $H_{n,d_1,d_2}^{1,2}$ formula ϕ and the vertex reachability of a random digraph $\mathcal{G}_{n,m=d_2n}$ also holds between ϕ and a random digraph $\mathcal{G}_{n,p=d_2/n}$.

The vertex reachability of random digraphs generated according to the model $\mathcal{G}_{n,p}$ has been studied and analyzed by Karp [300]. We use those results to study the satisfiability of random $H_{n,d_1,d_2}^{1,2}$ formulas. Karp showed that as n tends to infinity, when $np < 1 - h$, where h is a fixed small positive constant, the expected size of a connected component of the graph is bounded above by a constant $C(h)$. When $np > 1 + h$, as n tends to infinity, the set of vertices reachable from one vertex is either *small* (expected size bounded above by $C(h)$) or *large* (size close to Θn, where Θ is the unique root of the equation $1 - x - e^{-(1+h)x} = 0$ in $[0,1]$). Moreover, a *giant* strongly connected component emerges, of size approximately $\Theta^2 n$.

Let us now consider the two cases, $d_2 = 1 - h$ and $d_2 = 1 + h$, where h is a positive number. Remember that in our case $p = \frac{d_2}{n}$. In the analysis below we will use the notation *w.h.p.* (with high probability) as shorthand for "with probability tending to 1 in the large n limit."

In the case where $d_2 = 1 - h$, or $np < 1 - h$, the size of the set $X(v_i)$ of vertices reachable by a vertex v_i is w.h.p. less than or equal to $3 \log n h^{-2}$, and the expected size of this set is bounded above by a constant related to h. Thus we get that the probability that v_0 is reachable by v_i w.h.p. lies in the interval $[0, 3 \log n / n (1 - d_2)^2]$, and its expected value is bounded above by a constant. The expected probability that v_0 is reachable by a set of $d_1 n$ vertices should increase with d_1. The plots in figures 1 and 2 show that when $d_2 < 1$, the probability of satisfiability of ϕ (1 minus the probability that v_0 is reachable by a set of $d_1 n$ vertices in G_ϕ), decreases as we increase d_2 and/or d_1.

When $d_2 = 1 + h$, or $np > 1 + h$, we know that the set $X(v_i)$ of vertices reachable by a vertex v_i is w.h.p. either in the interval $[0, 3 \log n / (1 - d_2)^2]$ or around Θn. We also know that the probability that $X(v_i)$ is small tends to $1 - \Theta$. Therefore, w.h.p. at least one of the $d_1 n$ vertices will have a large reachable set. That is, the probability that v_0 is reachable by a set of $d_1 n$ vertices is bounded below from Θ. Notice that Θ increases with d_2. Again, the plots in figures 1 and 2 show that when $d_2 > 1$, the probability of satisfiability of ϕ decreases as d_2 increases. So the experimental observations are in agreement with the expectations based on the digraph reachability analysis.

Going back to digraphs' reachability, Karp's results show that for each vertex the set of its reachable vertices is very small up to the point where $np = 1$. We can observe the same behavior in 1-2-HornSAT if we change our distribution model by setting $d_1 = c/n$ for some constant c. By doing so, we are adjusting our model to fit the reachability analysis done by Karp that is based on a single starting vertex in the digraph. The result of this modification is that d_1 is no longer a factor on the probability of satisfiability of ϕ, which now depends solely on d_2. See figure 3, where we show the satisfiability plot in that case, and contrast with the picture that emerges when d_1 is a constant (shown in fig. 2). While before the satisfiability probability was steadily decreasing as we increased d_2, now the satisfiability probability is practically 1 until d_2 becomes larger than one. In both

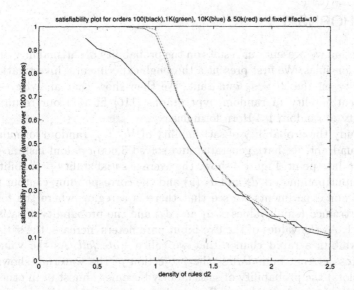

FIGURE 3   Satisfiability probability of random 1-2-Horn formulas when $d_1 = 10/n$ for sizes 100 (lower curve), 1,000, 10,000, and 50,000 (upper curve).

cases, however, the reachability analysis and the experimental data show that the satisfiability of random 1-2-Horn formulas is a problem that lacks a phase transition.

Remark 1. *The probability of satisfiability of 1-2-Horn can in fact be calculated exactly. Using the combinatorics of labeled trees, one can calculate exactly the probability $P(k)$ that a given vertex v has an out-tree of size k, not including itself, in a random digraph with mean out-degree d_2. This is*

$$P(k) = \frac{e^{-(k+1)d_2} d_2^k}{k!}(k+1)^{k-1}.$$

The probability of satisfiability is then

$$\mathbf{Pr}[\text{SAT}] = \sum_{k=0}^{\infty} P(k)(1 - d_1)^k.$$

Numerical computation indicates a close fit with our experimental results.

4   1-3-HORNSAT

In this section we present our results on the probability of satisfiability of random 1-3-Horn formulas. We first present a thorough experimental investigation of the satisfiability on the $d_1 \times d_3$ quadrant. We then show that analytic results on vertex identifiability in random hypergraphs [116] fit well our results on the satisfiablity of random 1-3-Horn formulas.c

To study the probability of satisfiability of $H^{1,3}_{n,d_1,d_3}$ random formulas in the $d_1 \times d_3$ quadrant, we have generated and solved 3,600 random instances of size 20,000 per data point. Figure 4 shows the average satisfiability probability versus the two input parameters d_1 and d_3 (a) and the corresponding contour plot (b).

From our experiments we see that there is a region where the formula is underconstrained (small values of d_1 and d_3) and the probability of satisfiability is almost 1. As the values of the two input parameters increase, the satisfiability terrain exhibits a rapid change that we call a *waterfall*. As the values of d_1 and d_3 cross certain boundaries (the projection of the waterfall shown in the contour plots) the probability of satisfiability becomes almost 0. In other words, the transition appears to be similar to those observed in other combinatorial problems such as 3-SAT and 3-COL.

There is a significant difference though, between these previously studied transitions and the one we observe in 1-3-HornSAT. In cases such as 3-SAT or 3-COL there are two input parameters describing a random instance; the size and the constrainedness of the instance. The constrainedness is defined as the ratio of clauses to variables for satisfiability, and edges to vertices for graph coloring. In random 1-3-HornSAT, there are three parameters: the size of the instance and the two densities, namely d_1 and d_3. By taking a cut along the three-dimensional surface shown in figure 4(a), we can study the problem as if it had only two input parameters.

We have taken two straight-line cuts of the surface. For the first cut, we fix d_1 to be 0.1, we let d_3 take values in the range $[1, 5.5]$ with step 0.1, and we choose instance sizes 500, 1,000, 2,500, 5,000, 10,000, 20,000, and 40,000. See figure 5(a), where we plot the probability of satisfiability along this cut. This plot reveals a quick change on the probability of satisfiability as the input parameter d_3 passes through a critical value, around 3. One technique that has been used to support experimental evidence of a phase transition is finite-size scaling. This is a technique coming from statistical mechanics that has been used in studying the phase transitions of several NP-complete problems, such as k-SAT and AC-matching [319, 339]. The technique uses data from finite-size instances to extrapolate to infinite-size instances. The transformation is based on a rescaling according to a power law of the form $d' = d - d_c/d_c n^r$, where d is the density, d' is the rescaled parameter, d_c is the critical value, n is the instance size and r is a scaling exponent. As a result, a function $f(d, n)$ is transformed to a function $f(d')$. We apply finite-size scaling to our data to observe the sharpness of the transition, following the procedure presented by Kolaitis et al. [339]. Our

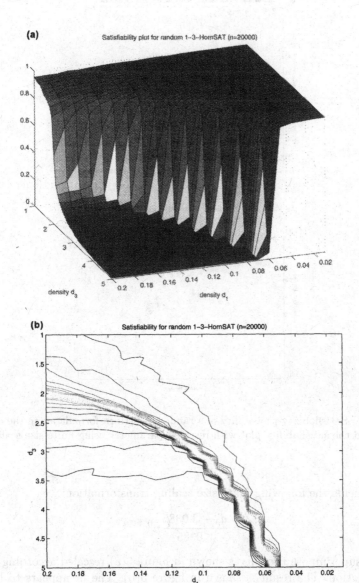

FIGURE 4 Satisfiability probability of a random 1-3-Horn formula of size 20,000 (a) and the corresponding contour plot (b). The contour plot contains 25 lines that separate consecutive percentage intervals $[0\% - 4\%), [4\% - 8\%), \ldots, [96\% - 100\%]$.

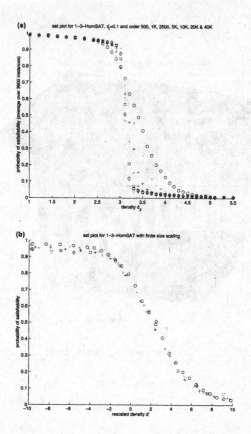

FIGURE 5 Satisfiability probability of a random 1-3-Horn formula along the $d_1 = 0.1$ cut (a) and the satisfiability plot with rescaled parameter using finite-size scaling (b).

analysis yields the following finite-size scaling transformation:

$$d' = \frac{d_3 - 3.0385}{3.0385} n^{0.4859}.$$

We then superimpose the curves shown in figure 5(a) rescaled according to this transformation. The result is shown in figure 5(b). The fit appears to be very good around zero, where curves collapse to a single universal curve, although as we move away it gets weaker. In the plot, the universal curve seems to be monotonic with limits $\lim_{d' \to -\infty} f(d') = 1$ and $\lim_{d' \to \infty} f(d') = 0$. This evidence would seem to suggest a phase transition near $d_3 = 3$ for $d_1 = 0.1$.

We repeat the same experiment and analysis with the second cut, a straight line cut along the diagonal of the $d_1 \times d_3$ quadrant. In this case our formal parameter is an integer i. An instance with input parameter value i corresponds to an instance with densities $d_1 = i/200$ and $d_3 = i/10 + 1$. Here, by making d_1 and d_3 dependent, we effectively reduce the number of input parameters of the problem from three, (d_1, d_3, n), to two, (i, n). We let i take values in the range $[1, 40]$ in increments of 1, and use instance sizes 500, 1,000, 2,500, 5,000, 10,000, 20,000, and 40,000. In figure 6(a) we plot the probability of satisfiability along this cut. This plot, as the one for the previous cut, reveals a quick change on the probability of satisfiability as the input parameter i passes through a critical value (around 19). We again use finite-size scaling on these data, looking for further support of a phase transition. For this cut, the analysis yields the following transformation:

$$i' = \frac{i - 19.1901}{19.1901} n^{0.2889}.$$

In figure 6(b) we superimpose the curves shown in the same figure (a) using the transformation above. As with the previous cut, the fit seems quite good, especially around zero, and the universal curve seems to have limits 1 and 0 in the infinities.

In an attempt to find further evidence of a phase transition, we perform the following experiment for the cut used to produce the data in figure 5 ($d_1 = 0.1$). For several instance sizes between 500 and 200,000 and for density d_3 taking values in the range $[2.7, 3.8]$ in increments of 0.02, we generate and solve 1,200 instances. We record for each different instance size the values of density d_3 for which the average probability of satisfiability is 0.1, 0.2, 0.8, and 0.9, respectively. (We actually used linear regression on the two closest points to compute the density for each satisfiability percentage.) The idea behind this experiment is that if the problem has a sharp threshold, then as the size of the instances increases the window between the 10th and 90th probability percentiles, as well as that between the 20th and the 80th probability percentiles, should shrink and become zero at the large n limit. In figure 7 we plot these windows. Indeed, they do get smaller as the instance size increases.

Similar analysis has been performed in the past for k-SAT. The *width* of the satisfiability phase transition, namely the amount by which the number of clauses of a random instance needs to be increased so that the probability of satisfiability drops from $1 - \epsilon$ to ϵ, is thought to grow as $\Theta(n^{1-\frac{1}{\nu}})$, with the exponent ν for $2 \leq k \leq 6$ estimated in Kirkpatrick and Selman [319], Kirkpatrick et al. [321], and Monasson et al. [406, 407]. Notice that the window that we estimate is equal to the normalized width (divided by the instance size). It was also conjectured that as k gets large, ν tends to 1. However, Wilson has proven [527] that for all $k \geq 3$, $\nu \geq 2$, so the transition width is at least $\Theta(n^{1/2})$. Our experiments suggest that the window of the satisfiability transition for 1-3-HornSAT shrinks as fast as $n^{-1/2}$, thus the transition width grows as $n^{1/2}$. We believe that the analysis in Wilson [527] can be applied in the case of 1-3-HornSAT, and can complement our experimental findings.

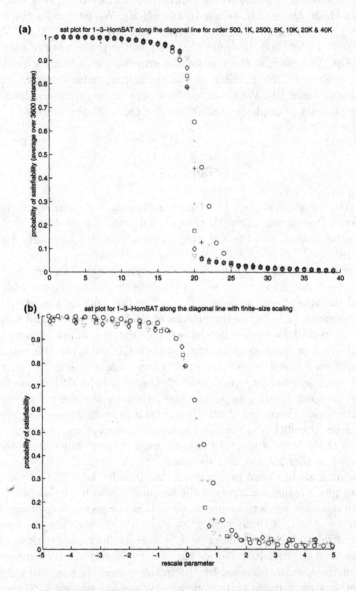

FIGURE 6   Satisfiability probability of a random 1-3-Horn formula along the diagonal cut (a) and the satisfiability plot with rescaled parameter using finite-size scaling (b).

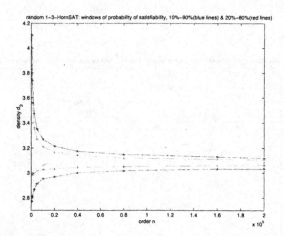

FIGURE 7   Windows of satisfiability probability of random 1-3-Horn formulas along the $d_1 = 0.1$ cut. The outer two curves show the 10%–90% probability window, and the inner two curves show the 20%–80% probability window.

Although figure 7 shows that these windows indeed shrink as the instance size increases, it is not at all clear whether in the limit they would go to zero. A further curve-fitting analysis is more revealing. In figure 8 we plot the size of the 10%–90% probability of satisfiability window (a) and the 20%–80% probability of satisfiability window (b) as a function of the instance size. Using MATLAB to do curve fitting on our data, we find that both windows decrease almost as fast as $1/\sqrt{n}$. The correlation coefficient r^2 is almost 0.999, giving high confidence for the validity of the fit. This analysis suggests that, indeed, the two windows should be zero at the limit, evidence supporting the existence of a phase transition for 1-3-HornSAT.

In the rest of this section we will discuss the connection between random Horn formulas and random hypergraphs. We will see that recent results on random hypergraphs provide a good fit for our experimental data on random 1-3-HornSAT presented so far. On the other hand, these results suggest that the transition is steep, but not the step function needed for a sharp threshold.

There is a one-to-one correspondence between random Horn formulas and random directed hypergraphs. Let ϕ be a $H_{n,d_1,d_3}^{1,3}$ random formula. We can represent ϕ with the following hypergraph G_ϕ:

- represent each variable x_i in ϕ with a node v_i in G_ϕ;
- represent each unit clause $\{x_k\}$ as a hyperedge in G_ϕ over v_k (hyperedges over vertices are also called *patches* [116] or *loops* [135]); and
- represent each clause $\{x_j, \overline{x_k}, \overline{x_l}\}$ as a directed hyperedge in G_ϕ over the set $\{v_j, v_k, v_l\}$.

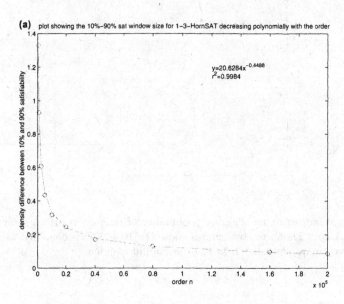

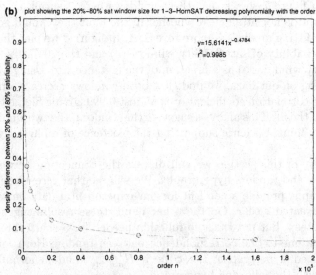

FIGURE 8   Plot of the 10%–90% satisfiability probability window (a) and of the 20%–80% satisfiability probability window (b) as a function of the intance size n.

Note that we omit the single negative literal appearing in ϕ.

In a recent development, Darling and Norris [116] proved certain results on vertex identifiability in random undirected hypergraphs. A vertex v of a hypergraph is *identifiable in one step* if there is a hyperedge over v. A vertex v is *identifiable in n steps* if there is a hyperedge over a set S, such that $v \in S$ and all other elements of S are identifiable in fewer than n steps. Finally, a vertex v is *identifiable* if it is identifiable in n steps for some positive n. We now establish the equivalence between the satisfiability of ϕ and the *identifiability* of vertex v_k of G_ϕ, where $\{\overline{x}_k\}$ is the unique single negative literal clause of ϕ.

First, we introduce a simple algorithm for deciding whether a Horn formula is satisfiable or not, presented by Dowling and Gallier [128] (see also Beeri and Bernstein [41]). The algorithm runs in time $O(n^2)$ where n is the number of variables in the formula. Dowling and Gallier actually describe in their work how to improve this algorithm to run in linear time, though for our purposes and for the sake of simplicity we use the simple quadratic algorithm.

Algorithm A.

```
begin
        let φ = {c₁,...,cₘ}
        consistent:=TRUE; change:=TRUE;
        set each variable xᵢ to be FALSE;
        for each variable xᵢ such that {xᵢ} is a clause in φ
            set xᵢ to TRUE
        endfor;
        while (change and consistent) do
            change:=FALSE;
            for each clause cⱼ in φ do
                if (cⱼ is of the form (x̄₁,...,x̄q)
                    and all x₁,...,xq are set to TRUE) then
                        consistent:=FALSE;
            else
                if cⱼ is of the form {x₁,x̄₂,...,x̄q}
                    and all x₂,...,xq are set to TRUE
                    and x₁ is set to FALSE
                        then set x₁ to TRUE; change:=TRUE; φ := φ − cⱼ
                endif
            endif
            endfor
        endwhile
end
```

If Algorithm A terminates with consistent:=TRUE, then a satisfying truth assignment has been found. Otherwise, the formula ϕ is unsatisfiable.

Given a formula ϕ, its corresponding directed hypergraph G_ϕ, and a variable x_i, we prove the following relation between the truth value that Algorithm A assigns to x_i and the identifiability of vertex v_i of G_ϕ:

Lemma 1. *Algorithm A running on ϕ assigns the value* TRUE *to x_i if and only if the vertex v_i of G_ϕ is identifiable.*

Proof It is easy to show the equivalence by induction on the number of steps required to identify v_k (equivalently the number of iterations of the **while** loop of Algorithm A needed to set the value of x_k to TRUE).

Base Case: If v_k is identifiable in one step, then $\{x_k\}$ is a clause in ϕ and Algorithm A immediately assigns the value TRUE to it, and vice versa.

Inductive Hypothesis: A vertex is identifiable in $n-1$ steps if and only if the corresponding variable is set to TRUE by Algorithm A in no more than $n-1$ iterations of the **while** loop.

Inductive Step: A vertex v_j that is identifiable in n steps corresponds to a variable that appears in a clause of the form $\{v_j, \overline{v_{i_1}}, \ldots, \overline{v_{i_q}}\}$, and since all of $x_{i_1}, \ldots, x_{i_q}$ are already set to TRUE, A will set x_j to TRUE in the nth iteration of the **while** loop. Conversely, if x_j is set to TRUE in the nth iteraton of the **while** loop of Algorithm A, then we derive that it appears in a clause of the form $\{x_j, \overline{x_{i_1}}, \ldots, \overline{x_{i_q}}\}$, where all of $x_{i_1}, \ldots, x_{i_q}$ are already set to TRUE. But this implies that all $v_{i_1}, \ldots, v_{i_q}$ are identifiable in $n-1$ steps; therefore v_j is identifiable in n steps.

As an immediate result of this lemma we obtain:

Corollary 4.1. *Let ϕ be a $H_{n,d_1,d_3}^{1,3}$ random formula and $\{\overline{x_k}\}$ be the unique single negative literal clause of ϕ. Let G_ϕ be the directed hypergraph corresponding to ϕ. The formula ϕ is satisfiable if and only if the vertex v_k of G_ϕ is not identifiable.*

Darling and Norris [116] studied the vertex identifiability in random undirected hypergraphs. Although Horn formulas correspond to directed hypergraphs, we have decided to use the results of Darling and Norris in an effort to approximate the satisfiability of Horn formulas. The authors use the notion of a *Poisson random hypergraph*. A Poisson random hypergraph on a set V of n vertices with non-negative parameters $\{\beta_k\}_{k=0}^{\infty}$ is a random hypergraph Λ, where the numbers $\Lambda(A)$ of hyperedges of Λ over sets $A \subseteq V$ of vertices are independent random variables, depending only on $|A|$, such that $\Lambda(A)$ has distribution $\text{Poisson}(n\beta_k/\binom{n}{k})$ where $|A| = k$. Thus, the number of hyperedges of size k is $\text{Poisson}(n\beta_k)$, distributed uniformly at random among all vertex sets of size k. (Recall that the distribution function of $\text{Poisson}(\lambda)$ is $f(x) = \exp(-\lambda)\lambda^x/x!$. The expectation of $\text{Poisson}(\lambda)$ is λ.) Note that this model allows for more than one edge over a set $A \subseteq V$; for our purposes we only care whether $\Lambda(A) = 0$ or not.

One of the key results they proved is the following:

Theorem 4.1 (Darling and Norris [116]). *Let $\beta = (\beta_j : j \in \mathbb{Z})$ be a sequence of non-negative parameters. Let $\beta(t) = \Sigma_{j\geq 0}\beta_j t^j$ and $\beta'(t)$ the derivative of $\beta(t)$. Let $z^* = \inf\{t \in [0,1) : \beta'(t) + \log(1-t) < 0\}$; if the infimum is not well-defined then let $z^* = 1$. Denote by ζ the number of zeros of $\beta'(t) + \log(1-t)$ in $[0, z^*)$.*

Assume that $z^* < 1$ and $\zeta = 0$. For $n \in \mathbb{N}$, let V^n be a set of n vertices and let G^n be a Poisson(β) hypergraph on V^n. Then, as $n \to \infty$ the number V^{n*} of identifiable vertices satisfies the following limit w.h.p.: $V^{n*}/n \to z^*$.

If we ignore the direction of the hyperedges then the random hypergraph G_ϕ representing an $H_{n,d_1,d_3}^{1,3}$ random formula corresponds to a Poisson(β) hypergraph G^n. To see that, notice that the hyperedges in G_ϕ are distributed uniformly at random among all possible 1- and 3-sets of vertices, just as in a Poisson random hypergraph with only two non-zero parameters, β_1 and β_3. To find the values of these parameters, we set equal the probabilities that a hyperedge exists in the two hypergraphs G_ϕ and G^n. In G_ϕ, the probability that a variable x_i is selected as a positive unit literal is d_1. In G^n, the probability that there are zero hyperedges on x_i is e^{β_1}. From this we get $\beta_1 = -\log(1 - d_1)$. In G_ϕ, the probability that a 3-clause is selected (ignoring directions) is $nd_3/\binom{n}{3}$. In G^n, the probability that there are zero edges on the three variables in that clause is $e^{-n\beta_3/\binom{n}{3}} \simeq 1 - n\beta_3/\binom{n}{3}$ (as $n \to \infty$). From this we get $\beta_3 = d_3$.

Note that ignoring the direction of the hyperedges is equivalent to adding to the formula, for each clause $(x \vee \overline{y} \vee \overline{z})$, two more clauses $(\overline{x} \vee y \vee \overline{z})$ and $\overline{x} \vee \overline{y} \vee z$. Therefore, we expect that the probability of satisfiability we get from the hypergraph model should be lower than the actual probability as it is measured by our experiments. This is indeed the case, as will be apparent in figure 10.

We used MATLAB to compute z^* for the hypergraph G^n on the quadrant $d_1 \times d_3$ (the Darling-Norris Theorem does not provide us with an explicit result for z^*). From corollary 1, we get that the probability of satisfiability of ϕ is 1 minus the probability that v_k is identifiable in G^n, which, by theorem 4.1, is $1 - z^*$. In figure 9(a) we plot the satisfiability probability of ϕ against the input parameters d_1 and d_3. A contour plot is given in figure 9(b).

Comparing the results derived by this model (fig. 9) and the results obtained by our experiments (fig. 4), we see that the results from the hypergraph analysis provide a very good fit of the experimental data. This is also clear in figure 10 where we plot the 50 percent satisfiability line according to the model above (the rough curve) and according to our experimental data (smoother curve).

Finally, we used our model to estimate the probability of satisfiability along the same two cuts that we presented earlier (the $d_1 = 0.1$ and the diagonal cut). See figure 11 for the probability estimation along the two cuts according to the hypergraph-based model, and compare with our experimental findings shown in figure 5(a) and figure 6(a). For both cuts, the estimated probability has a steep drop that happens at the exact same point that the respective drop is observed in the experimental data. In table 1 we give the raw data that correspond to the

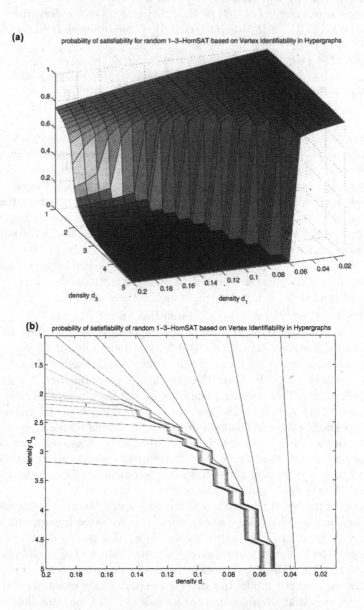

FIGURE 9   Satisfiability probability of a random 1-3-Horn formula according to the vertex-identifiability model (a) and the corresponding contour plot (b).

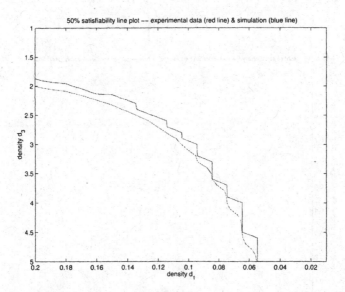

FIGURE 10 50% satisfiability line, according to the model derived through hypergraphs (rough line) and according to our experimental data (smoother line).

plots in figure 11. Notice that, despite the very steep transition, the estimated curve is not a step function as we would expect from our data and the limiting curve under finite-size scaling analysis (figs. 5(b) and 6(b)). Should this be an accurate model for the 1-3-HornSAT, the probability of satisfiability is not a step function at the limit, so the threshold function is not in fact a constant function.

5 CONCLUSIONS

We have set out to investigate the existence of a phase transition for the satisfiability of the random 1-3-HornSAT problem. This is a problem that is similar to 3-SAT, but its polynomial complexity allows us to collect data for much higher instance sizes.

We first showed, through our experimental findings and an analysis based on known results from digraphs' reachability, that the 1-2-HornSAT is a problem lacking a phase transition.

On the other hand, our experiments provide evidence that the 1-3-HornSAT has a phase transition. By thoroughly sampling the $d_1 \times d_3$ quadrant, solving a large number of random instances of large size, we document a waterfall-like probability of satisfiability surface. In addition, by taking cuts of this surface, we are able to observe a quick transition from a satisfiable to an unsatisfiable

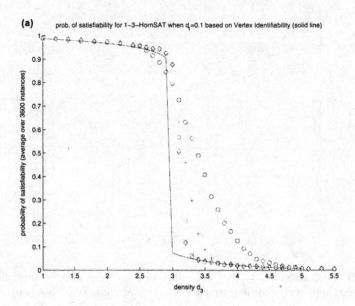

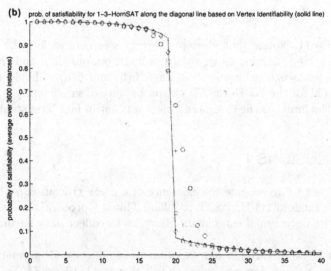

FIGURE 11   Satisfiability probability of a random 1-3-Horn formula according to the vertex-identifiability model, along the $d_1 = 0.1$ cut (a) and the diagonal cut (b). The solid line corresponds to the model; experimental data points are shown for comparison.

TABLE 1   Data for the satisfiability probability of a random 1-3-Horn formula according to the vertex-identifiability model, along the $d_1 = 0.1$ and the diagonal cut.

$d_1 = 0.1$ cut		diagonal cut	
d_3	sat. prob.	input parameter i	sat. prob.
1	0.98775	1	0.99997
1.1	0.98619	2	0.99988
1.2	0.98455	3	0.9997
1.3	0.98282	4	0.99941
1.4	0.98098	5	0.99899
1.5	0.97903	6	0.99841
1.6	0.97694	7	0.99764
1.7	0.9747	8	0.99664
1.8	0.9723	9	0.99537
1.9	0.96969	10	0.99376
2	0.96685	11	0.99175
2.1	0.96372	12	0.98924
2.2	0.96026	13	0.98611
2.3	0.95637	14	0.98217
2.4	0.95194	15	0.97717
2.5	0.94679	16	0.97069
2.6	0.94062	17	0.96202
2.7	0.9329	18	0.94968
2.8	0.92244	19	0.9294
2.9	0.90522	20	0.072832
3	0.072832	21	0.063411
3.1	0.063588	22	0.055476
3.2	0.055745	23	0.048727
3.3	0.049039	24	0.042943
3.4	0.043267	25	0.038
3.5	0.038272	26	0.0335
3.6	0.033928	27	0.029856
3.7	0.030137	28	0.026559
3.8	0.026815	29	0.023665
3.9	0.023896	30	0.021117
4	0.021324	31	0.018868
4.1	0.019052	32	0.016878
4.2	0.017041	33	0.015114
4.3	0.015257	34	0.013547
4.4	0.013672	35	0.012153
4.5	0.012262	36	0.01091
4.6	0.011006	37	0.0098016
4.7	0.0098849	38	0.0088112
4.8	0.0088836	39	0.0079255
4.9	0.0079881	40	0.0071324

region. Finite-size scaling applied along these cuts suggests that there is a phase transition, and analysis of the transition window provides further evidence of this.

We then used recent results on random hypergraphs to generate a model for our experimental data. By comparing the waterfall-like probability surface against the estimated probability according to this model, we see that the hypergraph-based model fits our experimental data well. This suggests that further analysis based on hypergraphs could provide a rigorous analysis of the conjectured phase transition for the 1-3-HornSAT. Such a development would be very significant since very few phase transitions have been proven analytically (2-SAT, 3-XORSAT, 1-in-k SAT) [7, 94, 101, 131, 156, 202]. Interestingly, in spite of how well this model fits our data, when calculating the estimated probability along the two cuts we see that in the limit of large instance sizes, the probability of satisfiability is a very steep function, but does not have a discontinuity. This last result conflicts with our experimental findings and demonstrates the difficulty of using numerics to show a phase transition, even for tractable problems such as 1-3-HornSAT.

ACKNOWLEDGMENTS

We thank one of the anonymous referees for pointing out the exact calculation of the 1-2-HornSAT satisfiability probability, given in Remark 1.

Part 4: Extensions and Applications

CHAPTER 10

Phase Transitions for Quantum Search Algorithms

Tad Hogg

1 INTRODUCTION

Phase transitions have long been studied empirically in various combinatorial searches and theoretically in simplified models [91, 264, 301, 490]. The analogy with statistical physics [397], explored throughout this volume, shows how the many local choices made during search relate to global properties such as the resulting search cost. These studies have led to a better understanding of typical search behaviors [514] and improved search methods [195, 247, 261, 432, 433].

Among the current research questions in this field are the range of algorithms exhibiting the transition behavior and the algorithm-independent problem properties associated with the difficult instances concentrated near the transition. Towards this end, the present chapter examines quantum computer [123, 126, 158, 486] algorithms for nondeterministic polynomial (NP) combinatorial search problems [191].

As with many conventional methods, they exhibit the easy-hard-easy pattern of computational cost as the degree of constraint in the problems varies. We

Computational Complexity and Statistical Physics, edited by
Allon G. Percus, Gabriel Istrate, and Cristopher Moore, Oxford University Press.

describe how properties of the search space affect the algorithms and identify an additional structural property, the *energy gap*, motivated by one quantum algorithm but applicable to a variety of techniques, both quantum and classical. Thus, the study of quantum search algorithms not only extends the range of algorithms exhibiting phase transitions, but also helps identify underlying structural properties.

Specifically, the next two sections describe a class of hard search problems and the form of quantum search algorithms proposed to date. The remainder of the chapter presents algorithm behaviors, relevant problem structure, and an approximate asymptotic analysis of their cost scaling. The final section discusses various open issues in designing and evaluating quantum algorithms, and relating their behavior to problem structure.

2   RANDOM SATISFIABILITY PROBLEMS

The k-satisfiability (k-SAT) problem, as discussed earlier in this volume, consists of n Boolean variables and m clauses. A clause is a logical **OR** of k variables, each of which may be negated. A solution is an assignment, that is, a value for each variable, TRUE or FALSE, satisfying all the clauses. An assignment is said to conflict with any clause it does not satisfy. Thus, a possible 2-SAT problem instance with 3 variables and 2 clauses might be $(x_1$ **OR** $\overline{x_2})$ **AND** $(x_2$ **OR** $x_3)$, which has 4 solutions: for example, $x_1 = $ FALSE, $x_2 = $ FALSE and $x_3 = $ TRUE is one of them. When $k \geq 3$, k-SAT is NP-complete [191], and so it is among the most difficult NP problems.

Evaluating a search algorithm's average cost requires defining a problem ensemble, meaning a class of problem instances and probabilities for their selection.

The random k-SAT ensemble with given n and m consists of instances whose m clauses are selected uniformly at random. Specifically, for each clause, a set of k variables is selected randomly from among the $\binom{n}{k}$ possibilities. Then each selected variable is negated with probability $1/2$ to produce the clause. Each of the m clauses is, therefore, selected, with replacement, uniformly from among the $N_{\text{clauses}} = \binom{n}{k}2^k$ possible clauses. This ensemble (called $\mathcal{G}_{m,m}$ in ch. 7) has a high concentration of hard instances when the *clause-to-variable ratio* $\alpha \equiv m/n$ is near a critical value where the fraction of solvable instances exhibits a phase transition [91, 254, 319], dropping abruptly from near 1 to near 0. For random 3-SAT this transition is at $\alpha \approx 4.27$.

The quantum algorithms discussed here are incomplete methods: failure to find a solution does not guarantee no solution exists. Thus, for empirical evaluation, we restrict attention to instances with at least one solution (determined via exhaustive classical search of the randomly generated instances). This restriction gives the random solvable k-SAT ensemble.

3   QUANTUM SEARCH METHODS

The quantum algorithms considered in this chapter consist of a series of trials, each operating on vectors of size 2^n, whose components consist of a complex number, or *amplitude*, for each assignment or *search state*. These vectors are often described as *superpositions* of all search states. After each trial, observing or measuring the superposition randomly produces a single assignment, with probabilities equal to the squared magnitudes of the final amplitudes. This result is then tested with a conventional computer, a rapid operation for NP problems. Trials repeat until a solution is found.

A trial starts with equal amplitudes, so that the initial vector has components $\psi_s^{(0)} = 2^{-n/2}$, for s ranging over all 2^n assignments. The trial consists of a prespecified number of steps j, unlike classical algorithms which can halt as soon as they find a solution. For step $h = 0, \ldots, j$ we define the *step fraction* $f = h/j$ as the fraction of steps completed.

A single step performs a matrix multiplication on the superposition, giving new values for the amplitudes. Quantum algorithms generally perform this operation in two parts, corresponding to multiplying by two matrices. First, the phases of the amplitudes are adjusted based on properties of the problem instance to be solved. Typically, this adjustment to the amplitude ψ_s associated with search state s depends on the *state cost* $c(s)$. In the case of k-SAT, $c(s)$ is the number of the m clauses conflicting with s. The phase adjustment has the form $e^{-i\rho(f,c(s))}$ where the *cost phase function* ρ is a real-valued function of the step fraction and the state cost.

The second part of a step mixes amplitudes among different states, corresponding to multiplication by a unitary matrix with nonzero off-diagonal terms. Usually, the mixing matrix for step h is taken to have the form $U^{(h)} = WT^{(h)}W$ with W and T defined as follows. W is the Walsh-transform, a $2^n \times 2^n$ matrix with elements $W_{rs} = 2^{-n/2}(-1)^{|r \wedge s|}$ where $|r \wedge s|$ is the number of 1's the two assignments have in common when viewed as bit-vectors ($0 \leq r, s \leq 2^{n-1}$). For instance, when $n = 1$, $W = \frac{1}{\sqrt{2}} \begin{pmatrix} 1 & 1 \\ 1 & -1 \end{pmatrix}$. The matrix $T^{(h)}$ is diagonal with $T_{ss}^{(h)} = e^{-i\tau(f,|s|)}$, where $|s|$ denotes the number of 1-bits in bit-vector s and the *mixing phase function* τ is another real-valued function. With these definitions, $U_{rs}^{(h)}$ depends only on the Hamming distance $d(r,s)$ between the states [250], that is, the number of variables they assign different values. Thus, the mixing matrix has the form $U_{rs}^{(h)} = u_{d(r,s)}^{(h)}$ with the u_d values determined by the choice of τ.

Combining these operations, a single step is[1]

$$\psi_r^{(h)} = \sum_s u_{d(r,s)}^{(h)} e^{-i\rho(f,c(s))} \psi_s^{(h-1)} . \tag{1}$$

Although the vectors have exponentially many components, quantum computers perform these operations efficiently [76, 215, 255].

Observing the final superposition gives an assignment having c conflicts with probability

$$P_{\mathrm{conf}}^{(j)}(c) = \sum_{s|c(s)=c} |\psi_s^{(j)}|^2$$

with the sum over all assignments with c conflicts. In particular, $P_{\mathrm{soln}}(j) = P_{\mathrm{conf}}^{(j)}(0)$ is the probability to find a solution in a single trial. For good performance, the choices for ρ and τ should ensure a large value of $P_{\mathrm{soln}}(j)$ after only a modest number of steps j.

Optimization versions of the search considered in this chapter amount to finding states with the minimum cost. This use of "cost" associated with individual states should not be confused with the *search cost*, that is, the number of elementary computation steps required to find a solution. For the probabilistic algorithms considered here, we focus on the expected value of the search cost and usually consider its median value over an ensemble of problems rather than its value for any single instance. This corresponds to the notion of average complexity in chapter 3.

As a reminder concerning notation, to compare function growth rates [208], $F = O(G)$ indicates F grows no faster than G as a function of n when $n \to \infty$. Conversely, $F = \Omega(G)$ means F grows at least as fast as G, and $F = \Theta(G)$ means both functions grow at the same rate.

The remainder of this section describes choices for ρ and τ giving unstructured search and then techniques exploiting problem structure.

3.1 UNSTRUCTURED SEARCH

In thec notation introduced above, Grover's unstructured search [215] has phase functions

$$\rho(f,c) = \begin{cases} \pi & \text{if } c = 0 \\ 0 & \text{otherwise} \end{cases}$$

$$\tau(f,b) = \begin{cases} \pi & \text{if } b = 0 \\ 0 & \text{otherwise} \end{cases}$$

---

[1] To provide a more direct connection with the Hamiltonian formulation of this algorithm, in section 3.4, the values of ρ and τ used here are $-\pi$ times the values introduced previously in the literature [250].

With these choices, which are the same for all steps (i.e., independent of f), the phase adjustment simply multiplies amplitudes for solutions by -1, leaving the others unchanged, and the mixing is a diffusion matrix: $u_d = -\delta_{d0} + 2^{1-n}$ where δ_{ab} is one if $a = b$ and zero otherwise.

The probability of finding a solution after j steps is [76]

$$P_{\text{soln}}(j) = \sin((2j + 1)\theta)^2 \qquad (2)$$

where $\theta \equiv \sin^{-1}\sqrt{S/2^n}$ and S is the number of solutions. For hard, solvable random k-SAT, θ is exponentially small. Thus the solution probability $P_{\text{soln}} \approx 1$ when the number of steps is $j \approx \pi/(4\theta) \approx \frac{\pi}{4}\sqrt{2^n/S}$, that is, after exponentially many steps for hard problems.

The corresponding unstructured classical algorithm, generate-and-test, requires on average $\frac{1}{2}2^n/S$ state tests to find a solution. The quantum method thus achieves a square-root speedup, the best possible improvement for unstructured search [76].

In practice, the number of solutions S and hence the best choice for the number of steps j are not known *a priori*. Moreover, since the solution probability $P_{\text{soln}}(j)$ oscillates with the number of steps j, picking j too large not only increases the search cost for each trial but can also result in a smaller value of P_{soln} and hence require more trials. One approach to this difficulty selects j differently for each trial as follows [76]: Starting with $J = 1$,

• perform a single trial with j selected uniformly at random between 0 and $J-1$
• if a solution is found, stop. Otherwise, set $J = \min(2^{n/2}, 6J/5)$ and repeat.

To evaluate the expected search cost, from eq. (2) the probability of obtaining a solution is [76]

$$p_{\text{random}}(J) = \frac{1}{J}\sum_{j=0}^{J-1} P_{\text{soln}}(j) = \frac{1}{2} - \frac{\sin(4J\theta)}{4J\sin(2\theta)} \qquad (3)$$

which approaches $1/2$ as J increases. The trial with a given J takes $(J-1)/2$ steps, on average. With probability $1 - p_{\text{random}}(J)$ the trial is not successful. Thus the expected search cost for all trials starting with J is

$$\text{cost}(J) = \frac{J-1}{2} + (1 - p_{\text{random}}(J))\,\text{cost}(\min(2^{n/2}, 6J/5))\,. \qquad (4)$$

When $J \geq 2^{n/2}$, further iterations have $J = 2^{n/2}$ so eq. (4) gives

$$\text{cost}(2^{n/2}) = \frac{2^{n/2} - 1}{2\,p_{\text{random}}(2^{n/2})}\,.$$

This condition and eq. (4) enable us to compute the full expected search cost, cost(1), recursively. This technique, allowing a different number of steps j for each trial, increases the expected search cost by at most a factor of 4 [76] compared to having prior knowledge of S.

3.2   SINGLE-STEP SEARCH

Unstructured search ignores all properties of search states except whether they are solutions. Fortunately, many problems allow efficient evaluation of additional useful properties. An extreme example is a problem with one solution and an efficient method giving the distance of any state to that solution. We then take this distance value to be the cost associated with that state.

Such problems are quite easy: classical methods can solve them with only a linear number of state evaluations. For instance, pick a random state, compute its distance to the solution, and then evaluate the distance for each of its n neighboring states (assignments that differ from the original by changing a single variable's value). Selecting the value for each variable giving the smaller distance then directly constructs a solution.

For a corresponding quantum search [249], take cost phase $\rho(f, c) = \frac{\pi}{2}c$ and mixing phase $\tau(f, b) = \frac{\pi}{2}b$ and let σ be the solution. The mixing matrix value is then $u_d = (e^{-i\pi/4}/\sqrt{2})^n i^d$. Since we take $c(s) = d(\sigma, s)$ and the initial state has equal amplitudes, a single step gives

$$\psi_r^{(1)} = 2^{-n} e^{-in\pi/4} \sum_s i^{d(r,s)} (-i)^{d(\sigma,s)}. \tag{5}$$

For $r = \sigma$, all terms in the sum are 1, that is, all contributions to the amplitude associated with the solution add in phase, giving $\psi_\sigma^{(1)} = e^{-in\pi/4}$ with absolute value 1. The amplitudes for the remaining states are zero. Thus, this quantum algorithm finds a solution, with probability 1, in just a single step.

This algorithm can work well even with some error in estimating the distances. As an extreme example, if the estimated solution distance has an error of any multiple of four, the quantum algorithm's behavior is unchanged. By contrast, such errors would change conventional algorithms based on comparing the distance values for neighboring states.

This scenario is not applicable to hard search problems, which lack efficient methods to determine distance to solution from most states. Nevertheless the single-step method illustrates how quantum computers exploit problem structure, in this case a strong correlation between easily computed measures (e.g., number of conflicts) and the distance to desired states. As described below, this observation, combined with the typical properties of problem structure, gives qualitative insight into why quantum algorithms show the phase transition behavior.

3.3   USING PROBLEM STRUCTURE

The previous two subsections described algorithms for two extremes: first the case in which a solution can be recognized when it is found, but no other information is available, and second the case of perfect information in which it is easy to determine the distance to the solution from any state. Typical NP search problems are between these extremes. Readily computed information about search states gives some information on the location of solutions, but is not always accurate. Using such information in the quantum algorithm of eq. (1) is conceptually straightforward. For a problem ensemble, such as random k-SAT, pick the phase functions $\rho(f, c)$ and $\tau(f, b)$ and number of steps j to minimize the search cost for typical instances.

We thus have another example of a common situation with heuristics: tuning algorithm parameters with respect to a class of problems. Generally, heuristics are too complicated to permit a useful analytical relation between the parameter values and algorithm search cost. Instead, numerical optimization can find good parameter values for a sample of problem instances, in the hope such values will also work well on other instances from the same problem ensemble.

When optimizing algorithm parameters for a sample of instances on a quantum machine, each trial requires only polynomial time, provided we pick the number of steps j to grow only polynomially with n. On the other hand, at least for most parameter choices, the solution probability P_{soln} is exponentially small, thus requiring exponentially many trials to estimate P_{soln} on the sample instances because each trial gives only a single state. Hence a direct attempt to find parameter values minimizing the median search cost would require exponentially many trials on a quantum computer. One way to address this difficulty is to identify how good parameter choices scale with n and then extrapolate values based on optimization with smaller n. Another approach uses the shift in amplitudes towards low-cost states, shown later in section 5: instead of maximizing P_{soln}, we could minimize the expected cost of the state produced by a trial, that is, $\langle c \rangle = \sum_c c P_{\text{conf}}^{(j)}(c)$, a quantity easily estimated with a modest number of trials.

Currently, however, we must simulate the quantum algorithm on conventional machines, so each trial requires exponential search cost and memory but has the benefit of giving P_{soln} directly from evaluating a single trial.

Another approach to finding good phase functions, discussed in section 6.2, uses an approximate analytical theory of the algorithm performance. The theory allows rapid, though approximate, performance evaluation for large problem sizes. Numerical optimization then finds phases giving high performance according to this approximation.

Algorithms using problem structure usually take the phase functions to have the form:

$$\rho(f, c) = \rho(f) c \, \Delta \tag{6}$$
$$\tau(f, b) = \tau(f) b \, \Delta$$

with Δ a parameter used to characterize how these functions scale with the number of steps j. We also call the single-parameter functions $\rho(f)$ and $\tau(f)$ the cost and mixing phase functions. The mixing matrix becomes [250]:

$$u_d^{(h)} = \left(e^{-i\tau\Delta/2} \cos\left(\frac{\tau\Delta}{2}\right) \right)^n \left(i \tan\left(\frac{\tau\Delta}{2}\right) \right)^d \tag{7}$$

which depends on the step fraction f through $\tau(f)$.

Completing the algorithm requires explicit forms for the phase functions, $\rho(f)$ and $\tau(f)$, and values for Δ and the number of steps j. Ideally, these quantities would minimize the expected search cost for the particular problem instance. For hard problems, such optimal choices are not known *a priori*. We therefore focus instead on functional forms giving good performance on average for random k-SAT, so depending only on the ensemble parameters n, k and m. While the values could vary from one trial to the next, by analogy with the procedure described in sec. 3.1 for unstructured search when the number of solutions is not known, for simplicity we use the same values for each trial. The expected cost of finding a solution is then $j/P_{\text{soln}}(j)$, the number of steps per trial multiplied by the expected number of times trials must be repeated to give a solution.

3.4   HAMILTONIAN FORMULATION

An alternate formulation of the algorithm steps involves the Hamiltonians producing the unitary operators used with eq. (1). For the phases given by eq. (6), in matrix form the step is

$$e^{-i\tau(f)H_0\Delta}e^{-i\rho(f)H_c\Delta}\psi, \tag{8}$$

with H_0 and H_c defined as follows. The mixing Hamiltonian has $(H_0)_{r,s}$ equal to $n/2$ when $r = s$, $-1/2$ when states r and s differ by exactly one bit, and 0 otherwise. The problem cost Hamiltonian H_c is diagonal with values equal to the state costs: $(H_c)_{r,r} = c(r)$.

The algorithm's initial state, with all amplitudes the same, is the ground state of H_0, with eigenvalue 0. Since H_c encodes the costs of the search states, its ground state corresponds to having nonzero amplitudes only in solutions (or, if there are no solutions, in states with the minimum number of conflicts).

When Δ is small, eq. (8) gives [503] $\psi^{(h)} \sim \exp(-iH(f)\Delta)\psi^{(h-1)}$ where $H(f) \equiv \tau(f)H_0 + \rho(f)H_c$. Defining the state vector $\psi(f) \equiv \psi^{(h)}$ and $T = j\Delta$, eq. (1) becomes

$$\frac{d\psi(f)}{df} = -iTH(f)\psi(f). \tag{9}$$

Thus, for small Δ, the algorithm steps closely approximate the continuous evolution of this Schrödinger equation with the time-dependent Hamiltonian $H(f)$ for $0 \leq f \leq 1$.

A significant application of this correspondence is the adiabatic limit [151]: for T sufficiently large, $\psi(f)$ remains close to the ground state of $H(f)$ if it starts in the ground state of $H(0)$. Since the state with uniform amplitudes is the ground state of H_0, the initial condition is achieved if $H(0) \propto H_0$, i.e., $\rho(0) = 0$. If we also have $\tau(1) = 0$, then $H(1) = H_c$ so the final state $\psi(1)$ will be close to the ground state of H_c, namely the solution (or a linear combination of solutions in case of multiple solutions). Thus, for a fixed problem size n, taking $T \to \infty$

and $\Delta \rightarrow 0$ leads to the solution probability approaching one, $P_{\text{soln}} \rightarrow 1$. An important open question is how many steps are sufficient to achieve large values of P_{soln}: algorithms of this form cannot efficiently solve worst-case instances of SAT [508], but it remains to be seen how effective they are on average.

By comparison, the heuristic method using eq. (6) has $\Delta = 1/j$ so $T = 1$. Hence, the adiabatic limit does not apply and, in general, the solution probability P_{soln} is exponentially small in n. However, with suitable phase functions $\rho(f)$ and $\tau(f)$ the expected number of steps to find a solution is significantly lower than that of the adiabatic limit for random 3-SAT (see sec. 5.1), at least for small n. The approximation of section 6.2 suggests this favorable scaling continues for larger problems.

A third possibility is phase adjustments whose size is independent of the number of steps, with Δ held constant instead of approaching zero as j increases. We can take the constant to be $\Delta = 1$, since any other value amounts to rescaling ρ and τ. In this case eq. (1) does not closely approximate the continuous evolution of eq. (9). Nevertheless, when $\rho(0) = 0 = \tau(1)$, a discrete version of the adiabatic theorem applies: the state vector starts in an eigenstate of the initial mixing operator, and is then multiplied by a series of slowly changing unitary matrices. When the changes are sufficiently small, that is, with a large number of steps j, the state vector remains close to an eigenstate of the matrices. In particular, the final state will be close to an eigenstate of the final operator, corresponding to states with a particular cost. Unlike the continuous adiabatic method, this final eigenstate need not correspond to solutions [246]. Ensuring the final eigenstate does correspond to solutions requires that the phase functions, and hence Δ, not exceed a threshold value that depends on the problem instance. Above this threshold, the changing eigenvectors take the initial ground state to an eigenvector of H_c other than its ground state. Thus, applying the discrete adiabatic limit requires identifying a suitable threshold value for the problem ensemble and so requires some parameter tuning but, since any values below the threshold will give $P_{\text{soln}} \rightarrow 1$ as j increases, identifying suitable values need not be as accurate as for the heuristic method. On the other hand, exceeding this threshold slightly can sometimes be beneficial by giving high solution probabilities for intermediate numbers of steps j, even though $P_{\text{soln}} \rightarrow 0$ as $j \rightarrow \infty$ [246].

Table 1 summarizes the various approaches to incorporating problem structure in quantum algorithms. For good performance, it is not necessary that the solution probability P_{soln} be very close to one: somewhat smaller values give lower expected costs j/P_{soln}, an observation that applies to a variety of quantum [76, 151, 383] and classical methods [368].

TABLE 1   Summary of quantum search algorithms using problem structure. The single-step and heuristic methods require finding appropriate choices for the phase functions to give good performance. Here Δ characterizes the scaling of the phase functions with number of steps, and $T = j\Delta$ characterizes the total phase changes applied over the course of a trial.

algorithm	parameters	phase functions
single-step	$T = 1, \quad \Delta = 1$	with suitable ρ, τ
heuristic	$T = 1, \quad \Delta \to 0$	with suitable ρ, τ
adiabatic: continuous	$T \to \infty, \Delta \to 0$	$\rho(0) = 0 = \tau(1)$
adiabatic: discrete	$T \to \infty, \Delta = 1$	$\rho(0) = 0 = \tau(1)$,
		values for other f not too large

4   MAXIMALLY CONSTRAINED 1-SAT

For a simple illustration of quantum search behavior, consider 1-SAT problems with a single solution. In this case, the number of conflicts $c(s)$ in state s equals its distance to the solution. Thus, from eq. (1), the amplitudes for step h have the form $\psi_s \propto (Z_h)^{c(s)}$ with Z_h a complex number. Initially, all amplitudes are the same, so $Z_0 = 1$. Including the overall normalization, the solution probability corresponding to Z is

$$P_{\text{soln}} = \frac{1}{\sum_{c=0}^{n} \binom{n}{c} |Z|^{2c}} = (1 + |Z|^2)^{-n} \tag{10}$$

so $|Z| \ll 1/\sqrt{n}$ gives $P_{\text{soln}} \to 1$. How Z changes in one step is determined by eq (1), giving:

$$Z_{h+1} = \frac{-ie^{-i\rho\Delta} Z_h + v}{-i + e^{-i\rho\Delta} Z_h v} \tag{11}$$

with $v \equiv \tan(\tau\Delta/2)$, and the phase functions ρ and τ can depend on $f = h/j$.

If $\rho\Delta = \tau\Delta = \pi/2$ then $Z_1 = 0$, so all amplitude is in the solution after a single step, providing an example of the discussion of section 3.2. While this problem is simple enough to solve in a single step, it is also instructive to consider its behavior with other phase choices. The remainder of this section examines the limit $\Delta \to 0$ as the number of steps increases. In this case, eq. (9) gives

$$\frac{dZ}{df} = -iT \left(\rho Z - \frac{\tau}{2}(1 - Z^2) \right) \tag{12}$$

with $Z(0) = 1$.

We first consider $\Delta = 1/j$ so $T = 1$. For suitable choices of the phase functions, eq. (12) gives $Z(1) = 0$. One such choice is $\rho = \tau = \pi/\sqrt{2}$, independent of f, in which case

$$Z(f) = -1 + \frac{4 - i\sqrt{2}\sin(\pi f)}{3 - \cos(\pi f)}. \tag{13}$$

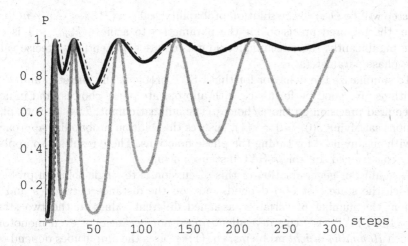

FIGURE 1 Solution probability, $P_{\text{soln}}(j)$, vs. number of steps j used in the trial, for a maximally constrained 1-SAT problem with $n = 100$ using $\Delta = 1/\sqrt{j}$, $\rho = \gamma f$, $\tau = \gamma(1 - f)$ with $\gamma = 0.828$ selected to give $P_{\text{soln}}(10) = 1$. Solid: exact (from eq. (11)), dashed: continuous approximation (eq. (12)). For comparison, the gray curve is for $n = 1000$.

The exact discrete map, eq. (11), for these parameters, gives $Z_j \sim \Theta(1/j)$ so $P_{\text{soln}} \sim \exp(-O(1)n/j^2)$. Hence, when the number of steps is $j \gg \sqrt{n}$, the solution probability approaches one, giving considerably lower search cost than the $\Theta(2^{n/2})$ value for unstructured search.

Second, consider the limit $T \to \infty$ using $1/j \ll \Delta \ll 1$. The adiabatic theorem applies if $\rho(0) = 0 = \tau(1)$. A simple choice is linear variation: $\rho = f$, $\tau = 1 - f$. In this limit eq. (12) gives $Z(f)$ close to the ground state of $H(f)$, which is proportional to $\Lambda(f)^c$ with

$$\Lambda(f) = \frac{\sqrt{1 - 2f + 2f^2} - f}{1 - f} \tag{14}$$

and $\Lambda(1) = 0$. Evaluating eq. (11) with $\Delta = 1/\sqrt{j}$ shows $Z(1)$ rapidly approaches the value from eq. (12) and this solution, in turn, approaches 0 as $1/T$. Thus the solution probability scales as $P_{\text{soln}} \sim \exp(-\Theta(1)n/T^2)$.

Figure 1 gives an example of this limit, showing $P_{\text{soln}} \to 1$ as the number of steps j increases. Furthermore, the probability oscillates, reaching values very close to 1 for relatively few steps, $T = \Theta(1)$. Exploiting these oscillations for rapid search requires identifying appropriate parameter values and ensuring that any implementation errors in these values are sufficiently small. If the parameter values vary by $O(\epsilon)$ from the ideal values giving $Z(1) = 0$, the value of Z at the final step will be $O(\epsilon)$ so the solution probability scales as $P_{\text{soln}} \sim \exp(-nO(\epsilon^2))$. Hence, the required precision for the parameters to achieve $P_{\text{soln}} \sim 1$ is $\epsilon \ll 1/\sqrt{n}$. Significantly, these techniques do not require exponentially precise values of the phase parameters.

To summarize the behavior for this 1-SAT problem, the one-step and $T = 1$ algorithms give good performance with appropriate phase choices, and indicate the required precision on those choices. The adiabatic limit, $T \to \infty$ with phase functions satisfying $\rho(0) = 0 = \tau(1)$, ensures the solution probability approaches one without any need for tuning the phase functions. These results also apply to highly constrained solvable k-SAT instances [249].

A significant generalization of this discussion is to single-solution problems in which the state cost $c(s)$ depends only on the distance between s and the solution: the number of variables assigned different values in the two states. Unlike the 1-SAT case, this dependence need not be linear or even monotonic. For such *Hamming-weight* problems, eq. (1) ensures the amplitudes depend only on the costs. This simplification allows studying the performance of quantum algorithms with a variety of cost structures [508], though classical algorithms can efficiently solve such problems by using the cost symmetry.

5 RANDOM K-SAT

On average, for k-SAT, the solution probability P_{soln} decreases exponentially with n for most phase function choices. Section 4 showed particular parameter choices leading to much better performance for 1-SAT with a single solution. As described in this section, the same options apply to hard random k-SAT problems, but, not surprisingly, do not perform as well.

5.1 SEARCH COST SCALING

For trials consisting of a single step or, more generally, a constant number of steps independent of n, the expected value of the solution probability P_{soln} always decreases exponentially when $m \propto n$ [251]: no choices are as good as those for the 1-SAT example of section 4. Nevertheless, selecting the best parameters for each value of the clause-to-variable ratio $\alpha = m/n$ exhibits the easy-hard-easy pattern as a function of α [251].

Better performance requires the number of steps in a trial, j, to increase with problem size. The approximate analysis of section 6.2 suggests choices for ρ and τ that appear to work well with $j \gg \sqrt{n}$. One example, with $\alpha = 4.25$ (close to the critical threshold), $j = n$ and $\Delta = 1/j$ is the phase functions [250]

$$\rho(f) = \pi(4.86376 - 4.18118(1 - f)) \tag{15}$$
$$\tau(f) = \pi(1.2 + 3.1(1 - f)).$$

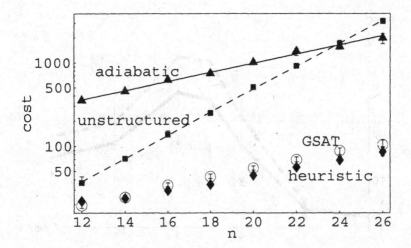

FIGURE 2 Log plot of median search cost vs. number of variables n for solvable random 3-SAT for the parameters of eq. (15) (diamond), unstructured search (box), GSAT with restarts after $2n$ steps (circle) and adiabatic search with $j = n^2$ (triangle). The error bars show the 95% confidence intervals [480, p. 124] of the medians estimated from the random 3-SAT instances with $m = 4.25n$ (when n is not divisible by 4, half the samples have $m = \lfloor 4.25n \rfloor$ and half have $m = \lceil 4.25n \rceil$). The same instances were solved with each method. We use 1000 instances for each n up to 20, and 500 for larger n. The lines are exponential fits to unstructured search (dashed) and the adiabatic method (solid).

The adiabatic limit gives solution probability $P_{\text{soln}} \to 1$ as $T \to \infty$. Empirically it appears to give good average performance for hard search problems [151] but the question of how rapidly T must grow to achieve this limit remains open. An example of the adiabatic limit is $\Delta = 1/\sqrt{j}$ and phase functions $\rho(f) = f$, $\tau(f) = 1 - f$. This gives good cost scaling with the number of steps $j = \Omega(n^2)$.

Figure 2 compares the median search costs of these algorithms as a function of the number of variables n. The figure also shows Grover's unstructured search [215] (without prior knowledge of number of solutions [76]) and the conventional heuristic GSAT [467].

The unstructured search cost grows as $\exp(0.32n)$. The exponential fit to the adiabatic method is $\exp(0.13n)$. The growth rate is about the same as that of GSAT. The phase functions of eq. (15) give the lowest median costs.

The discrete adiabatic method, using $\Delta = 1$ and number of steps $j = n$, gives costs similar to the heuristic. For both algorithms, using nonlinear variation of the phase functions with step fraction f gives significantly better performance

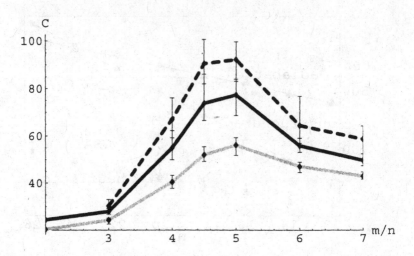

FIGURE 3 Median search costs for solvable random 3-SAT problems with $n = 28$ (dashed), $n = 24$ (black), and $n = 20$ (gray) as a function of clause-to-variable ratio m/n, using at least 100 instances for each point. Error bars show the 95% confidence intervals. The search uses the same algorithm parameters for each problem instance, namely $\Delta = 1$, number of steps $j = n$, and linear varation of the phase functions matching the requirements of the adiabatic theorem, namely $\rho(f) = f$, $\tau(f) = 1 - f$.

than for the linear functions shown here [246], a property also seen with the Hamming-weight problems [508].

As with the $j = O(1)$ case, the number of steps j increasing with n also shows the easy-hard-easy pattern, as illustrated in figure 3 for one choice of algorithm. In this example, the phase choices are the same for each value of α. The cost peak is quite wide for these small problems. This peak also appears with quantum algorithms including partial assignments [248], as arise in classical backtracking searches.

Figure 4 illustrates several properties of the algorithm using eq. (15) for one problem instance. At each step, probability concentrates in states with a fairly small range of costs. Each step shifts the peak in the probability distribution to assignments with fewer conflicts, until a large probability builds up in the solutions. This shift is also seen for other problem instances (with differing final probabilities) and when averaged over many instances.

The variation of amplitudes among states with the same cost is relatively large only in the last few steps of the algorithm and then primarily for higher-cost states for which the amplitudes are small. The shading in figure 4 shows this behavior, indicating the relative deviation of the amplitudes (ratio of standard deviation to mean) for states with the each cost, ranging from white for zero

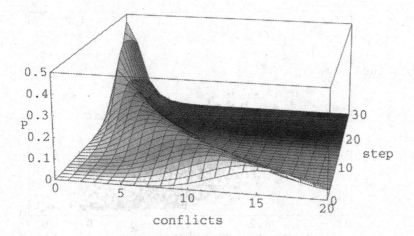

FIGURE 4 Solving a randomly generated 3-SAT problem with $n = 30$ and $m = 127$ clauses using eq. (15). For each step h, the figure shows the probability $P_{\text{conf}}^{(h)}(c)$ in assignments with each number of conflicts. Shading shows the relative deviations of the amplitudes, as described in the text. The small contributions for assignments with $c > 20$ are not included. This instance has 19 solutions.

deviation to black for relative deviations greater than 3. These observations motivate the approximate analysis of section 6.2.

5.2 PROBLEM STRUCTURE AND SEARCH COST

Each step of the algorithm, given in eq. (1), adjusts amplitudes based on the costs associated with the states and mixes them based on their Hamming distances. When cost and solution distance are perfectly correlated, as with the 1-SAT example of section 4, the quantum algorithm performs well. Thus we can expect a structural property of problem instances, namely the correlation between distance between states and their cost difference, to characterize search difficulty: high correlations should correspond to lower costs, on average.

As one example, figure 5 shows the relation between the search cost for the heuristic quantum algorithm (using eq. (15)) and the correlation between cost and distance to the nearest solution for all assignments. GSAT shows a similar relationship.

The adiabatic method provides another structural property relevant for search cost: the minimum energy gap g. The energy gap for step fraction f is the difference between the ground-state energy of $H(f)$ and the energy of the $(S + 1)$th state, where S is the number of solutions. The minimum gap is the smallest gap over the range $0 \leq f \leq 1$. The adiabatic limit requires $T \gg 1/g^2$. With multiple solutions, the distribution and sizes of gaps between successive

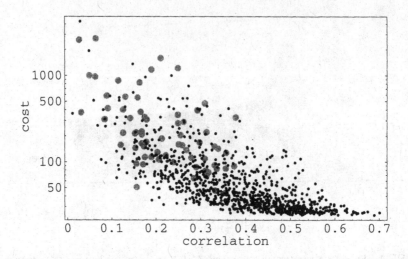

FIGURE 5 Expected search cost for the heuristic quantum method vs. correlation between state cost and distance to nearest solution for 1000 3-SAT instances with $n = 20$ variables and clause-to-variable ratio $\alpha = 4.25$. The large gray points are instances with a single solution.

states up to the $(S+1)$th can also affect performance. Except for simple, highly symmetric problems such as the 1-SAT example of section 4, the scaling behavior of the gap is not known.

Of broader interest for understanding phase transitions in combinatorial search, the minimum energy gap provides an algorithm-independent character- ization of search problems. Although directly related to the performance of the adiabatic method [151], the minimum gap g is also relevant for other algorithms. For instance, figure 6 shows high costs for the heuristic (using $T = 1$ and based on eq. (15)) generally correspond to small gaps. GSAT gives a similar plot. These observations suggest the minimum gap is a global property characterizing hard instances for a variety of algorithms. It thus may offer useful insights as an alternative to other such properties, such as the *backbone*, consisting of those variables with the same assigned values in all solutions [477]. Both the minimum gap and backbone are computationally expensive to determine, so they are most significant as theoretical constructs relating problem structure to search cost.

5.3   LONGER RANGE INTERACTIONS FOR AMPLITUDE MIXING

The algorithms discussed so far in this chapter incorporate information about the specific instance to solve only in the cost phase function ρ. As a further application of problem structure to designing quantum algorithms, this section

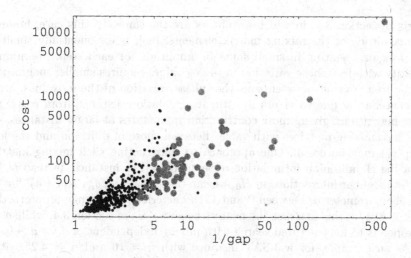

FIGURE 6 Expected search cost for the heuristic quantum method vs. $1/g$ for 500 3-SAT instances with $n = 16$ variables and clause-to-variable ratio $\alpha = 4.25$. The large gray points are instances with a single solution.

briefly examines one approach to adjusting the mixing phase function τ to match the mixing operator to the problem instance.

The mixing matrix in eq. (7) has the phase i^d associated with the mixing matrix elements u_d for all choices of τ. Thus, increasing τ increases the magnitude of the contribution from more distant states, but does not change the variation in phase of the u_d values.

Section 3.2 showed that good performance is possible if distances to the solution can be estimated well. On the other hand, errors in such estimates will make contributions from various distances tend to cancel, leading to less shift of amplitude towards lower-cost states during each step. For example, the phase factor i^d in u_d means an error in estimating distance to the solution by $2, 6, 10, \ldots$ changes the sign of the contribution to the new amplitude in eq. (1). Unfortunately, for problems such as 3-SAT, state costs are not perfectly correlated with solution distance (see fig. 5), so such phase errors in the mixing are inevitable for these algorithms.

One approach to this difficulty is using many steps, each with a small value of Δ, as described in section 3.3. This can be effective, but means that the mixing matrix for each step is close to the identity and that contributions to the amplitude of a state are concentrated among nearby states, which can lead to low performance in cases where solutions tend to be surrounded by high cost states. An alternative is using a mixing matrix in which the phases associated with successive u_d values vary more slowly than i^d. An extreme example is the diffusion matrix of section 3.1, in which all phases are the same. In this case, however, the magnitude of the mixing matrix elements, $|u_d|$, is exponentially small for $d > 0$, again resulting in small shifts in amplitude for each step. Maintaining unitarity with less phase variation in the u_d values requires smaller magnitudes, leading to a tradeoff between smoother phase variation of the u_d values, giving less cancellation due to errors in estimating solution distance from costs, and larger magnitudes giving more contribution from states at larger distances.

Thus, mixing matrices with values between those of diffusion and the form of eq. (7) may be useful. One approach to constructing such mixing matrices is via the Hamiltonian formulation of section 3.4. For instance, instead of just nearest-neighbor interactions in H_0, we can take $(H_0)_{r,s} \propto \delta_{r,s} - (1+a)^{-n} a^{d(r,s)}$ with the parameter a, between 0 and 1, characterizing the range of interaction. If $0 < a \ll 1/n$, this matches the nearest-neighbor H_0 of section 3.4, while $a = 1$ corresponds to unstructured search with mixing independent of d for $d > 0$.

As an example, for a 3-SAT instance with $n = 16$ and $\alpha = 4.25$ with a particularly small minimum gap (about 0.002, compared to the median minimum gap of about 0.5 for such problems), allowing the mixing to depend on a longer-range Hamiltonian with $j = 16$ steps increased the solution probability P_{soln} from 0.0006 to 0.12. Improvement is also seen with other instances with especially small gaps.

While difficult to provide definitive conclusions from these small instances, this additional flexibility in matching the mixing matrix to characteristics of the problem may improve performance. In particular, studies of how state costs vary through the search space to give local minima or plateaus [168, 254, 258] could suggest appropriate choices for the interaction range. Such information may help evaluate other types of quantum algorithms that rely on properties of the cost function throughout the space, such as those using partial assignments [86, 149, 248].

6　APPROXIMATE SCALING BEHAVIOR

As seen in the previous section, simulations for small problems show proper phase choices can give costs comparable to a good conventional heuristic. Unfortunately, these small problem sizes do not adequately address scaling of the cost behavior, particularly whether the quantum algorithms can perform significantly better than classical methods for hard random SAT problems, on average.

Approximate analytical techniques provide a complementary approach. The average properties of random k-SAT successfully help us to understand and improve search methods, both classical [91, 138, 195, 247] and quantum [250]. Quantum algorithms operate with the entire search space at each step, so performance depends on averaged properties of the search states. For simple ensembles, such as random k-SAT, such averages are readily computable.

This section discusses how the properties of random k-SAT provide a qualitative understanding of the phase transition behavior seen with the quantum methods and estimate the algorithm's behavior for large n.

6.1 PROBLEM STRUCTURE

The algorithms in this chapter adjust amplitudes based on the state costs and Hamming distances between them. Thus the selection of appropriate phase functions, and the resulting algorithm performance, depends on the relationship between these quantities for typical problem instances. This section describes this relationship for random k-SAT, and summarizes how its dependence on the number of clauses qualitatively explains the peak in search cost seen with the quantum algorithms.

For random k-SAT with m clauses, the probability an assignment has cost C is a binomial distribution: $P(C) = \binom{m}{C} p^C (1-p)^{m-C}$ where $p = 2^{-k}$ is the probability a single clause conflicts with a given assignment. The expected number of states with cost C is $v(C) = 2^n P(C)$. As one application, if the amplitudes after step h satisfy $\psi_s \propto Z^{c(s)}$ for some constant Z, then the probability of obtaining a state with c conflicts $P_{\text{conf}}^{(h)}(c)$ is

$$P_{\text{conf}}^{(h)}(c) = \frac{P(c)|Z|^{2c}}{(1 - p(1 - |Z|^2))^m} . \tag{16}$$

In particular, $P_{\text{soln}}(h) = P_{\text{conf}}^{(h)}(0)$ is the probability of obtaining a solution.

To relate distances and costs, the probability that two states separated by distance d have costs C and c, respectively, is given by a sum of multinomials depending on the number of clauses conflicting with both states [250]. The corresponding conditional probability $P(c|C, d)$ is peaked for c values close to C when the two states have the same assignments for most variables, that is, when $d \ll n$. This arises from the local nature of the constraints in k-SAT: two states that differ in assignments to only a few variables are very likely to violate many of the same clauses and hence have similar costs. Quantitatively, when n is large, the average cost c for a state at distance d from another state with cost C is

$$\langle c \rangle = \frac{m}{2^k - 1} \left(1 - \chi + (1 - \delta)(2^k \chi - 1) \right) \tag{17}$$

with $\delta \equiv d/n$, the fraction of variables with different values in the two states, and $\chi \equiv C/m$, the fraction of conflicting clauses in the first state. The variance of the distribution for c has a similar expression, proportional to $m(1 - (1 - \delta)^k)$.

Figure 7 is an example of how cost varies with distance from a state with given cost, and gives a qualitative understanding of the underlying cause of the easy-hard-easy behavior for the quantum algorithms. Specifically, since the relative deviation of c/m decreases as $1/\sqrt{m}$, the figure shows the distribution of costs c from the conditional probability $P(c|C, d)$ is narrow for either small

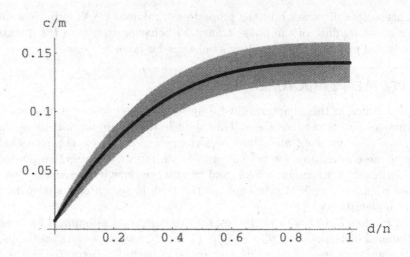

FIGURE 7   Expected fraction of conflicting clauses, $\langle c \rangle / m$, vs. the fraction of variables with different assignments, d/n, for the conditional distribution $P(c|C, d)$ when $n = 100$, $k = 3$, $m = 4n$ and $C = 3$. The gray region shows the extent of the deviation of the distribution: one standard deviation above and below the mean, multiplied by $\sqrt{m}$.

distances d or large numbers of clauses m. The smaller variance gives a stronger correlation between costs and distance, leading to an increased ability to pick phase adjustments to move amplitudes to desired states (as seen in sec. 4).

Underconstrained problems have many solutions, so distance to the nearest solution is typically fairly small. Hence amplitudes need only be shifted a small distance, so steps can mainly mix amplitudes of nearby states (i.e., use relatively small values of the mixing phase function τ, as is also suitable for the single-step method with underconstrained problems [251]). These small distances have high correlations between cost and distance, allowing fairly precise shifts of amplitude towards the lower-cost states and, hence, low overall search costs.

Conversely, overconstrained but solvable problems tend to have just a few solutions and long distances to them from most states. In this case, the increasing number of clauses results in small variance even for large distances, due to the $1/\sqrt{m}$ decrease in relative deviation of the conditional probability distribution $P(c|C, d)$. Thus, we can expect a good ability to shift amplitudes to lower-cost states for overconstrained problems.

Between these extremes we can expect larger search costs because, as shown in figure 7, the deviation grows rapidly with distance when the fraction of variables with different assignments, δ, is small, but more slowly as δ increases. Thus, when the clause-to-variable ratio α is small, we can expect the growth in variance due to increasing distance to outweigh the decrease due to the $1/\sqrt{\alpha}$ factor. As

α increases, $\delta \to 1/2$ when only one solution remains and hence the variance increases slowly due to changes in distance, and the $1/\sqrt{\alpha}$ factor dominates to give an overall decrease. These observations give a qualitative understanding of the cost peak for typical problems with an intermediate number of clauses: relatively long distances to solutions combined with higher variance in the relation between cost and distance. These factors reduce the ability to shift amplitudes reliably towards lower-cost states, giving higher overall search costs.

Using the expression for the relative deviation and estimating typical distances from the expected number of solutions, $\langle S \rangle = 2^n (1 - 2^{-k})^m$, matches this qualitative description with a peak in relative deviations for 3-SAT around a clause-to-variable ratio of $\alpha = 3$. Thus, while by no means a quantitatively accurate identification of the cost peak, the behavior of the conditional probability a state has cost c, given it is at distance d from another state with cost C, $P(c|C,d)$, provides a qualitative understanding of why the easy-hard-easy pattern arises in these quantum search algorithms. This also illustrates the usefulness of structurally simple ensembles such as random k-SAT: since each clause is selected independently at random, the state cost distributions are analytically simple to describe.

6.2 MEAN-FIELD APPROXIMATION

Evaluations of algorithm behavior, such as that of figure 4, show that amplitudes for states with the same number of conflicts are generally quite similar. This observation motivates an analysis based on the behavior of the average amplitude for states with each cost [250]. The resulting approximation corresponds to a *mean-field* approach in statistical physics.

Consider the average amplitude $A_C^{(h)} = \left\langle \psi_s^{(h)} \right\rangle$, with the average taken first over all states s with cost $c(s) = C$ in a problem instance, and then over all random k-SAT instances with given numbers of variables n and clauses m. Simulations show that the probability concentrates in a small range of cost values, as illustrated schematically in figure 8. Let us assume amplitudes for states with the same cost are the same, at least for states whose cost is near the dominant cost value at each step, that is, the peak in $P_{\text{conf}}^{(h)}$ of figure 4. Then eq. (1) becomes

$$A_C^{(h)} \approx \sum_{d,c} u_d^{(h)} e^{-i\rho(h/j,c)} v_d(G,c) A_c^{(h-1)} \qquad (18)$$

where $v_d(C,c) = \binom{n}{d} P(d|C,d)$ is the expected number of states with c conflicts at distance d from a state with C conflicts. The dominant costs are close to the average cost associated with the amplitudes, $\langle C \rangle = \sum_C C\, v(C) |A_C|^2$. We thus expand $A_c \approx A_C Z^{c-C}$ around the average cost, with Z a complex number depending on the step fraction $f = h/j$. This expansion is the same form as the exact expression for the amplitudes given in section 4 for the simple 1-SAT problem.

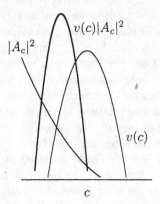

FIGURE 8 Schematic behavior of average amplitudes, on a logarithmic scale, as a function of number of conflicts c. The average number of states with c conflicts, $v(c)$, is sharply peaked around the average number of conflicts $m/2^k$. When the magnitude of the amplitudes decreases rapidly with c, as shown here, the probability in states with c conflicts is also sharply peaked, but at a somewhat lower value, corresponding to the shift towards lower-cost states seen in figure 4. Quantitatively, the values decrease exponentially with n, so the logarithms, shown here, are proportional to n and the relative width of each peak is $O(1/\sqrt{n})$.

For the case of $\Delta = 1/j$ and $j \gg 1$, using this expansion in eq. (9) gives [250]

$$\frac{dZ}{df} = i\left(-\rho Z + \frac{\tau}{2}kF\frac{1 - p(1 - Z)}{1 - p}(1 - Z)\right) \tag{19}$$

where $p = 2^{-k}$ is the probability that a clause conflicts with a given assignment, $\chi = |Z|^2 p/1 - p(1 - |Z|^2)$ is the expected fraction of conflicting clauses, $\langle C \rangle /m$, with this approximation for the amplitudes, and

$$F = \exp\left(-k\alpha(1 - Z)\left(\frac{p(1 - \chi)}{1 - p} - \frac{\chi}{Z}\right)\right).$$

Initially all amplitudes are equal so $Z(0) = 1$.

For $k = 1$ this reduces to the 1-SAT example, eq. (12) with $T = 1$, except for a factor of F due to the random choice of clauses of the ensemble, compared with the situation of section 4, where each clause must involve a distinct variable since the choices are required to give one solution.

With suitable choices for ρ and τ, such as those in eq. (15) for $k = 3, \alpha = 4.25$, eq. (19) gives $Z(1) = 0$ thereby predicting that most of the amplitude concentrates in states with the fewest conflicts, that is, solutions if the problem instance is solvable. More precisely, this predicts the solution probability $P_{\text{soln}}(j)$ is, at worst, only polynomially small with proper phase choices. This approximation relies on the decreasing size of the relative variance of the conditional probability $P(c|C, d)$ discussed in section 6.1 and, hence, the number of steps must satisfy $j \gg \sqrt{n}$. Combining these scaling behaviors, the mean-field approximation predicts that the average cost j/P_{soln} grows only polynomially with n for typical random k-SAT instances, with suitable choices of the phase functions depending on the clause-to-variable ratio α. Evaluating the error in this approximate result via simulation is difficult due to the requirements of many states with each cost and small relative deviation in problem structure among states with each cost value, that is, $\sqrt{n} \gg 1$.

The functions of eq. (15) do not have $\rho(0) = 0 = \tau(1)$. Choices for phase functions ρ and τ do exist that both satisfy these conditions for the adiabatic method and give $Z(1) = 0$ from eq. (19). However, empirically they require more steps than the choices of eq. (15).

7   DISCUSSION

This chapter has reviewed several approaches to quantum search. First, unstructured search is the quantum analog of generate-and-test. The probability of finding a solution is close to one after exponentially many steps for hard search problems. Second, the adiabatic method can also guarantee solution probability $P_{\text{soln}} \approx 1$ after sufficiently many steps, with the required number of steps related to an aspect of problem structure, the energy gap, not previously examined in the context of phase transitions. Third, for problems with a strong correlation between cost and distance to solution (such as 1-SAT or highly constrained k-SAT), appropriate phase choices allow solving the problem in $O(1)$ steps for any number of variables. Finally, the heuristic method gives good average performance for hard k-SAT problems based on empirical evaluation, but lacks an exact analysis of performance scaling. An approximate theory modeled on the behavior of the algorithm for 1-SAT suggests the possibility of polynomial scaling of average cost, but the accuracy of this prediction remains an open question. Moreover, even if the algorithm performs well on average, it has no guarantee for specific instances. At any rate, the approximation provides reasonably good choices for the phases, as seen in figure 2.

A number of extensions are possible. First, the amplitude shift of figure 4 means that even if a solution is not found after a trial, the result probably has low cost. Thus, like local classical search methods such as GSAT but unlike unstructured search, the heuristic and adiabatic methods apply directly to combinatorial optimization, that is, finding a minimal conflict state [174]. For example, the shift in amplitudes towards low-cost states is seen in satisfiability problems with no solutions, the traveling salesman problem [252] and graph coloring [149].

Second, the mean-field analysis also applies to other search problem ensembles, provided the probabilities relating problem properties can be determined. This is possible for a variety of commonly studied ensembles such as coloring random graphs. Ensembles of real-world problems lack analytically known probability distributions, but sampling representative instances allows estimating $P(c|C, d)$. Such estimates may even be useful for analytically simple ensembles, allowing some tuning of phase parameters for a particular problem instance.

Third, in common with amplitude amplification [76] and some classical methods [369], the growth of solution probability $P_{\text{soln}}(h)$ with step h during a trial, as seen in figure 4, means stopping a bit before the largest P_{soln} value reduces the expected search cost. More generally, a portfolio [205, 265] of trials with somewhat different parameter values can improve trade-offs between expected costs and their variation among different instances [383].

Fourth, the heuristic can readily incorporate other computationally-efficient properties of the search states as additional arguments to the phase function ρ. One such property, used by conventional heuristics such as GSAT, is how the number of conflicts in a state compares to those of its neighbors. Moreover, in an analogy with quadratically improving conventional heuristics with amplitude amplification [77], we could also evaluate a conventional heuristic, such as GSAT, for a fixed number of steps and use the cost of the resulting state to adjust phases (either instead of or in addition to the cost of the original state). In this case, we would be searching not for a solution state directly but rather for a "good" initial state, from which the conventional heuristic rapidly finds a solution. In fact, using just a few steps of GSAT with random SAT instances shows the same shift towards low-cost states as seen in figure 4, and the resulting solution probability, P_{soln}, is larger. However, for problem sizes amenable to simulation, P_{soln} of the original algorithm is sufficiently large that even if using a few steps of GSAT were able to make $P_{\text{soln}} = 1$, it would not reduce the overall trial cost due to the additional steps involved in evaluating GSAT. Nevertheless, this approach may be useful for larger problem sizes and illustrates the potential trade-off between the cost of the procedure evaluating search state properties and the resulting probability for a solution, which determines the expected number of trials.

An interesting open question is whether the heuristic can benefit from using different parameters and numbers of steps for each trial, as used for amplitude amplification when the number of solutions is not known. The simulations indicate a wide range of performance among different instances with the same numbers of variables and clauses, n and m, even if they have the same number of solutions. This approach would rely on the variation among problem instances not addressed by ensemble averages. Furthermore, the series of low-cost states returned by the unsuccessful trial may also be useful indications of problem structure as another example to apply dynamic adjustments based on algorithm behavior during search [304]. Finally, implementations of structured quantum searches [489] will allow a comparison of how the various algorithms respond to uniquely quantum mechanical sources of error, such as decoherence.

A quantum machine with even a modest number of bits could help address these issues by evaluating algorithm performance beyond the range of classical simulation. This will be particularly useful for more complicated heuristics, using additional problem properties, whose theoretical analysis is likely to be more difficult. Exploring their behavior will identify opportunities for quantum computers to use information available in combinatorial searches to improve performance significantly.

ACKNOWLEDGMENTS

I have benefited from discussions with Wim van Dam. I thank Miles Deegan and the HP High Performance Computing Expertise Center for providing computational resources for the simulations.

CHAPTER 11

Scalability, Random Surfaces, and Synchronized Computing Networks

Zoltan Toroczkai
György Korniss
Mark A. Novotny
Hasan Guclu

1 INTRODUCTION

In most cases, it is impossible to describe and understand complex system dynamics via analytical methods. The density of problems that are rigorously solvable with analytic tools is vanishingly small in the set of all problems, and often the only way one can reliably obtain a system-level understanding of such problems is through direct simulation. This chapter broadens the discussion on the relationship between complexity and statistical physics by exploring how the computational scalability of parallelized simulation can be analyzed using a physical model of surface growth. Specifically, the systems considered here are made up of a large number of interacting individual elements with a finite number of attributes, or local state variables, each assuming a countable number (typically finite) of values. The dynamics of the local state variables are discrete events occurring in continuous time. Between two consecutive updates, the local variables stay unchanged. Another important assumption we make is that the interactions in the underlying system to be simulated have finite range.

Computational Complexity and Statistical Physics, edited by
Allon G. Percus, Gabriel Istrate, and Cristopher Moore, Oxford University Press.

Examples of such systems include: magnetic systems (spin states and spin flip dynamics); surface growth via molecular beam epitaxy (height of the surface, molecular deposition, and diffusion dynamics); epidemiology (health of an individual, the dynamics of infection and recovery); financial markets (wealth state, buy/sell dynamics); and wireless communications or queueing systems (number of jobs, job arrival dynamics).

Often—as in the case we study here—the dynamics of such systems are inherently stochastic and *asynchronous*. The simulation of such systems is nontrivial, and in most cases the complexity of the problem requires simulations on distributed architectures, defining the field of parallel discrete-event simulations (PDES) [186, 367, 416]. Conceptually, the computational task is divided among n processing elements (PEs), where each processor evolves the dynamics of the allocated piece. Due to the interactions among the individual elements of the simulated system (spins, atoms, packets, calls, etc.) the PEs must coordinate with a subset of other PEs during the simulation. For example, the state of a spin can only be updated if the state of the neighbors is known. However, some neighbors might belong to the computational domain of another PE, thus, message passing will be required in order to preserve causality. In the PDES schemes we analyze, update attempts are self-initiated [155] and are independent of the configuration of the underlying system [365, 366]. Although these properties simplify the analysis of the corresponding PDES schemes, they can be highly efficient [342] and are readily applicable to a large number of problems in science and engineering. Further, the performance and scalability of these PDES schemes become independent of the specific underlying system, that is, we learn the generic behavior of these complex computational schemes.

The update dynamics, together with the information sharing among PEs, make the parallel discrete event simulation process a complex dynamical system in itself. In fact, it perfectly fits the type of complex systems we are considering here: the individual elements are the PEs, and their states (local simulated time) evolve according to update events which are dependent on the states of the neighboring PEs.

With the number and size of parallel computers on the rise, the problem of designing efficient parallel algorithms or update schemes becomes increasingly important. In passing, we can mention a few examples of large parallel computers: the 9472-node ASCII Red at Sandia, the 12288-node QCDSP Teraflop Machine at Brookhaven, and the 8192-node IBM ASCII White with 12.3 Teraflops. The 65536-node IBM Blue Gene/L with 360 Teraflops is due for delivery at Livermore as this volume goes to press. And the largest supercomputer ever built is by Nature itself: the brain, which does an immense parallel computing task to sustain the individual. In particular the human brain has 10^{11} PEs (neurons) each with an average of 10^4 synaptic connections, creating a bundle on the order of 10^{15} "wires" jammed into a volume of approximately 1400 cm^3.

The fact that the dynamics of the simulation scheme form a complex system, with properties hard to deduce using classical methods of algorithmic analysis, makes the design of efficient parallel update schemes a challenging problem. In this chapter we present a new approach to analyzing efficiency and scalability for the class of massively parallel conservative PDES schemes [87] by mapping the parallel computational process itself onto a non-equilibrium surface growth model [343]. This allows us to formulate questions of efficiency and scalability in terms of certain topological properties of this non-equilibrium surface. Then, using methods from statistical mechanics, developed some time ago to study the dynamics of such surfaces in a completely different context, we solve the scalability problem of the computational PDES scheme [343, 346]. Similar connections between computational schemes and complex systems behavior have recently been made [457, 478] for rollback-based PDES algorithms [280] and self-organized criticality [24].

The chapter is organized as follows. In the following section we present the problem of scalability in conservative PDES schemes. In section 3 we discuss the scalability of the computational phase and the failure of scalability of the measurement phase of the basic conservative scheme, for regular topologies. We then show how a simple modification of the communication topology (from a regular lattice to a small-world structure) leads to a fully scalable PDES scheme. In section 4 we study the scalability problem on scale-free network topologies, presenting numerical results for Barabási-Albert networks. Section 5 is devoted to conclusions.

2 SCALABILITY OF MASSIVELY PARALLEL DISCRETE-EVENT SIMULATIONS

Since one is interested in the *dynamics* of an underlying complex system, the parallel discrete-event simulation scheme must simulate the "physical time" variable of the complex system. When simulations are performed on a single-processor machine, a single (global) time stream is sufficient to *label* or time-stamp the updates of the local configurations, regardless of whether the dynamics of the underlying system are synchronous or asynchronous. When simulating asynchronous dynamics on distributed architectures, however, each PE generates its own physical, or virtual time, which is the physical time variable of the particular computational domain handled by that PE. Due to the varying complexity of the computation at different PEs, at a given wall-clock instant the simulated, virtual times of the PEs can differ—a phenomenon called *time horizon roughening*. We denote the simulated, or virtual time at PE i measured at wall-clock time t by $\tau_i(t)$. For noninteracting subsystems the wall-clock time t is directly proportional to the (discrete) number of parallel steps simultaneously performed on each PE, also called the number of Monte Carlo steps (MCS) in dynamic Monte Carlo simulations. Without altering the meaning, t will from now on be taken to denote the number of discrete steps performed in the parallel simulation. The set

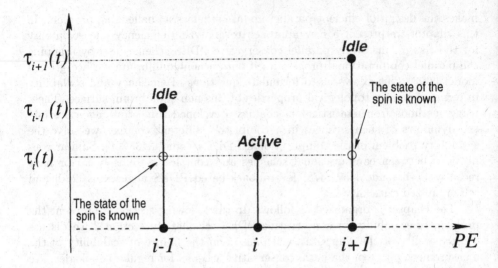

FIGURE 1 A simple diagram to illustrate the conservative PDES scheme for a one-dimensional system with nearest-neighbor interactions.

of virtual times $\{\tau_i(t)\}_{i=1}^n$ forms the *virtual time horizon* of the PDES scheme after t parallel updates.

In conservative PDES schemes [87], a PE can only perform its next update if it can obtain the correct information to evolve the local configuration (local state) of the underlying physical system it simulates, without violating causality. Otherwise, it idles. Specifically, when the underlying system has nearest-neighbor interactions, each PE must check with its "neighboring" PEs (mimicking the interaction topology of the underlying system) to see if those have progressed at least up to the point in virtual time where the PE itself has [365, 366]. Based on the fundamental notion of discrete-event systems that the value of a local state variable remains unchanged between two successive update attempts, the rule above guarantees the causality of the simulated dynamics [365, 366]. A simple illustration of this is given in figure 1. One may consider, for example, a magnetic system as the underlying physical system, where the spins are arranged on the sites of a one-dimensional lattice, and a single spin is handled by a single PE (for more realistic and efficient implementations see Korniss et al. [342], and Lubachevsky [365, 366]). In figure 1, showing the distribution of the virtual simulated times at a given wall-clock instant t, the only PE that can update from the set $\{i-1, i, i+1\}$ is in site i since the states of the neighboring spins at sites $i \pm 1$ are already known. However, PEs $i \pm 1$ cannot update their spin states at wall clock instant t, because the state of the neighboring spin i at *their* simulated times (at τ_{i-1} and τ_{i+1}) is not yet known. In other words PE i can only update at wall-clock instant t if $\tau_i(t) \leq \min\{\tau_{i-1}(t), \tau_{i+1}(t)\}$, that is, its virtual time is a *local minimum* among the virtual times of its neighboring PEs. It is easy to see that the same conclusion holds for arbitrary PE topologies. Let the topology for the communication among the processing elements be symbolized by a graph $G(V, E)$, where V is the vertex set of n nodes and E is the edge set of G. Given a node $i \in V(G)$, we denote by N_i the set of i's nearest neighbors on G. Then, node (PE) i can update its state in the conservative PDES scheme if and only if:

$$\tau_i^{(G)}(t) \leq \min_{j \in N_i}\{\tau_j^{(G)}(t)\} \ , i = 1, .., n \,. \tag{1}$$

In the following, the set of (active) nodes obeying condition (1) at time t will be denoted by $A(t)$.

Now we are in a position to formulate the scalability problem of PDES schemes for systems with asynchronous dynamics. For the PDES scheme to be fully scalable, the following two criteria must be met: (i) the virtual time horizon must progress on average at a nonzero rate, and (ii) the typical spread of the time horizon must be finite, as the number of PEs n goes to infinity. When the first criterion is ensured for large enough times t, the simulation is said to be *computationally scalable*. This simply means that when increasing the size of the computation to infinity, while keeping the average computational domain/load constant on a single PE, the simulation will progress at a nonzero rate. However, as we will show below, increasing the system size can cause the *spread* in the time horizon to diverge, severely hindering frequent data collection about the state of the simulated system. Specifically, when one needs to take a measurement of some physical property of the simulated system at (virtual or simulated) physical time τ, we have to *wait*, in wall-clock time, until all the virtual simulated times at all the PEs reach the value τ. Thus in order to collect system-wide measurements from the simulation, we incur a waiting time proportional to the *spread*, or width of the fluctuating time horizon. When condition (ii) is fulfilled for large enough times t, we say that the PDES scheme is *measurement scalable*. For PDES schemes for which the spread diverges with system size, however, the waiting time for the measurements will also diverge, and the scheme is *not* measurement scalable.

The scalability criteria above can be formalized in terms of the properties of the virtual time horizon, $\{\tau_i^{(G)}(t)\}_{i=1}^n$. The average of the time horizon after t parallel steps is:

$$\bar{\tau}^{(G)}(t) = \frac{1}{n} \sum_{j=1}^{n} \tau_j^{(G)}(t) \,. \tag{2}$$

At a given wall-clock time t the only PEs that can make progress, that is, are not idle, are those with virtual times obeying condition (1). Thus, the rate of progress of the time horizon average becomes:

$$\bar{\tau}^{(G)}(t+1) - \bar{\tau}^{(G)}(t) = \frac{1}{n} \sum_{l \in A(t)} \left[\tau_l^{(G)}(t+1) - \tau_l^{(G)}(t) \right] . \tag{3}$$

The difference in the square brackets on the right-hand side of eq. (3) is the *physical time* elapsed between two consecutive events in the physical domain simulated by the lth PE, and it is determined by the physical process responsible for the stochastic dynamics of the simulated complex system. If we replace the time intervals in square brackets in eq. (3) with their (clearly finite) average value Δ, we obtain that the average progress rate of the time horizon, or average *utilization* $\langle u^{(G)}(t) \rangle = \langle \bar{\tau}^{(G)}(t+1) - \bar{\tau}^{(G)}(t) \rangle$ is proportional to the *number* of non-idling, or active PEs. The average $\langle \cdots \rangle$ is taken over the stochastic event dynamics, assumed to be the same at all sites. For many cases, the Δ factor is independent of n due to the finite range of the interaction in the complex system, so the computational efficiency or average utilization of the simulation can simply be identified with the average density of the active PEs:

$$\langle u^{(G)}(t) \rangle = \frac{\langle |A^{(G)}(t)| \rangle}{n} , \tag{4}$$

where $|A^{(G)}(t)|$ denotes the number of elements of the set $A^{(G)}(t)$. Thus, the PDES scheme is computationally scalable if there exists a *constant* $c > 0$, such that:

$$\langle u^{(G)}(\infty) \rangle = \lim_{\substack{t \to \infty \\ n \to \infty}} \frac{\langle |A^{(G)}(t)| \rangle}{n} > c. \tag{5}$$

The measurement scalability of the PDES scheme is characterized by the spread of the virtual time horizon. Instead of dealing with the actual spread (difference between the maximum and minimum values) we shall consider the average "width" (or variance) of the time horizon defined as:

$$\langle [w^{(G)}]^2(t) \rangle = \frac{1}{n} \sum_{j=1}^{n} \left[\tau_j^{(G)}(t) - \bar{\tau}^{(G)}(t) \right]^2 . \tag{6}$$

A PDES scheme is measurement scalable if there exists a *constant* $M > 0$ such that:

$$\langle [w^{(G)}]^2(\infty) \rangle = \lim_{\substack{t \to \infty \\ n \to \infty}} \frac{1}{n} \sum_{j=1}^{n} \left[\tau_j^{(G)}(t) - \bar{\tau}^{(G)}(t) \right]^2 < M . \tag{7}$$

In reality, the number n of PEs or the simulation time t can never be taken to infinity, so for practical purposes, the scalability is deduced from the scaling behavior of the quantities for long times and for a large number of PEs. The setup presented above is perfectly suited to establishing a mapping between non-equilibrium surface growth models [29] and conservative PDES schemes. We discuss this mapping extensively in the next section.

3   SCALABILITY OF CONSERVATIVE PDES SCHEMES ON REGULAR AND SMALL-WORLD TOPOLOGIES

In many large complex systems the stochastic event dynamics can be characterized by a Poisson-distributed stream. To give one example, in an Ising magnet
with single spin-flip Glauber dynamics [342] the spin-flip attempts are Poisson-
distributed events. To give another example, in wireless cellular communications
the call arrivals obey Poisson statistics [209]. In the following, we restrict ourselves to such Poisson distributed stochastic processes for event dynamics. However, numerical simulations show that our conclusions for scalability hold for a
large class of other stochastic distributions as well. The evolution of the virtual
time horizon incorporating condition (1) for Poisson asynchrony is given by the
equation:

$$\tau_i^{(G)}(t+1) = \tau_i^{(G)}(t) + \eta_i(t) \prod_{j \in N_i} \theta(\tau_j^{(G)}(t) - \tau_i^{(G)}(t)). \tag{8}$$

Here, $\theta(x)$ is the Heaviside step function and $\eta_i(t)$ is the Poisson-distributed virtual time increment at PE i and time t. These increments are drawn at random,
independently of i and t, and independently of the existing time horizon.

3.1   THE BASIC CONSERVATIVE SCHEME ON REGULAR TOPOLOGIES

Next, we consider the *basic* conservative scheme, which is defined on regular,
cubic lattice communication topologies, in d dimensions, so that $n = L^d$. For
brevity we drop the superscript (G) in the notation for $\tau_i(t)$. In particular, we
first illustrate our analysis on the simplest regular topology, that of a regular
one-dimensional lattice with periodic boundary conditions, so that G is a ring.
Later, we discuss the general, d-dimensional case. The evolution equation on the
ring is simply:

$$\tau_i(t+1) = \tau_i(t) + \eta_i(t)\theta(\tau_{i-1}(t) - \tau_i(t))\theta(\tau_{i+1}(t) - \tau_i(t)). \tag{9}$$

with the boundary conditions $\tau_{n+1} = \tau_n = \tau_0$. The total number of active
sites/PEs is thus given by $|A(t)| = \sum_{i=1}^{n} \theta(\tau_{i-1}(t) - \tau_i(t))\theta(\tau_{i+1}(t) - \tau_i(t))$ so
the average utilization (4) becomes:

$$\langle u(L,t) \rangle = \frac{1}{n} \sum_{i=1}^{n} \langle \theta(\tau_{i-1}(t) - \tau_i(t))\theta(\tau_{i+1}(t) - \tau_i(t)) \rangle. \tag{10}$$

The average $\langle \cdots \rangle$ is performed over the random variables $\{\eta_i(t')\}_{\substack{i=1,\ldots,L \\ t'=1,\ldots t}}$, which
have an exponential distribution, $\mathbf{Pr}[x < \eta \le x + \delta x] = \int_x^{x+\delta x} dy\, e^{-y}$. In spite
of the simple appearance of the dynamics (9), and the exponential (or Poisson)
stochastic dynamics at nodes, calculating the average utilization (10) is very

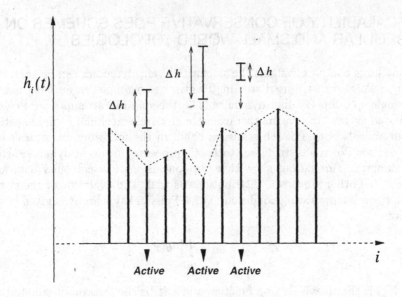

FIGURE 2 A simple surface growth model on a 1-d substrate corresponding to the basic conservative PDES scheme.

difficult. A rigorous proof even for the existence of the lower bound (5) using direct methods is still an open problem.

Here we present a different approach, first mapping the problem to a non-equilibrium surface grown via a molecular beam epitaxy model, where atoms or molecules are deposited from vapors or beams onto the surface. The analogies for the various quantities are as follows: the ith PE is the site i in the substrate; the number of parallel updates t is the number of deposited monolayers; $\tau_i(t)$ is the height $h_i(t)$ at site i and time t; and a virtual time increment of $\eta_i(t)$ at PE i in the tth step corresponds to a material "rod" of length $\eta_i(t)$ deposited onto the surface (see fig. 2). The length of the rod is a Poisson-distributed random variable. During the tth update, the rods are deposited only into *local minima* of the surface. The utilization of the PDES scheme corresponds to the *density* of local minima of the growing surface. Even though the lengths of the rods are independent random variables, the fact that they can only be deposited in local minima will generate lateral correlations into the surface fluctuations, and makes the problem hard to solve exactly. The rods are deposited onto the surface in a parallel update scheme: after all local minima are updated (deposited onto), the time t is incremented by unity. We call the surface growth analog of our basic conservative PDES scheme the massively parallel exponential update (MPEU) model.

Both the utilization (density of minima) and the width of the time horizon are quantities characterizing the fluctuations of the growing surface. The type of fluctuations can be classified into universality classes, each class having distinct statistical properties. Studying the PDES scheme as a surface growth model, we can describe its fluctuations and identify the surface growth universality class to which it belongs. In order to do this, we first introduce the *slope variables*, $\phi_i = \tau_i - \tau_{i-1}$. Provided $\tau_i(t)$ is a local minimum, depositing a rod of length η_i corresponds to taking an amount of η_i from ϕ_{i+1} and adding it to ϕ_i, since $\phi_i(t+1) = \tau_i(t) - \tau_{i-1}(t) + \eta_i(t)$ and $\phi_{i+1}(t+1) = \tau_{i+1}(t) - \tau_i(t) - \eta_i(t)$. Thus, in the *surface of slopes* $\{\phi_i\}_{i=1}^{L}$, the dynamics are those of *biased surface diffusion*, given by the equation:

$$\phi_i(t+1) - \phi_i(t) = \eta_i(t)\,\theta(-\phi_i(t))\,\theta(\phi_{i+1}(t)) - \eta_{i-1}(t)\,\theta(-\phi_{i-1}(t))\,\theta(\phi_i(t)) \quad (11)$$

with the constraint $\sum_{i=1}^{L} \phi_i = 0$ generated by the periodic boundary conditions in the τ variables. In terms of the local slope variables, the expression for the average density of minima or average utilization becomes: $\langle u(L,t) \rangle = \frac{1}{L}\sum_{i=1}^{L} \langle \theta(-\phi_i(t))\theta(\phi_{i+1}(t)) \rangle$. Translational invariance (no node is statistically special) implies $\langle u(L,t) \rangle = \langle \theta(-\phi_i(t))\theta(\phi_{i+1}(t)) \rangle$ for any $i = 1, \ldots, L$. From eq. (9) it follows that $\langle \tau_i(t+1) \rangle - \langle \tau_i(t) \rangle = \langle u(L,t) \rangle$. Therefore, *the average rate of propagation of the MPEU surface is identical to the average utilization of the PDES scheme*. It is also easy to see that in terms of the slope variables it is identical to the average current in the ring.

Next, we perform a naive coarse graining by using the representation $\theta(\phi) = \lim_{\kappa \to 0} \frac{1}{2}[1 + \tanh(\phi/\kappa)]$, and keeping only the terms up to first-order in ϕ/κ. This leads to:

$$\langle \phi_i(t+1) \rangle - \langle \phi_i(t) \rangle = \frac{1}{4\kappa}\langle \phi_{i+1}(t) - 2\phi_i(t) + \phi_{i-1}(t) \rangle - \frac{1}{4\kappa^2}\langle \phi_i(\phi_{i+1} - \phi_{i-1}) \rangle. \quad (12)$$

Strictly speaking, all of the $(\phi/\kappa)^j$, $j = 1, 2, \ldots$ terms are divergent. But by taking the proper continuum limit and introducing an appropriately scaled bias, one can show that the only relevant terms are those appearing in eq. (12). In the continuum limit, one thus obtains for the coarse-grained field:

$$\frac{\partial}{\partial t}\hat{\phi} = \frac{\partial^2}{\partial x^2}\hat{\phi} - \lambda \frac{\partial}{\partial x}\hat{\phi}^2 \quad (13)$$

where λ is a parameter related to the coarse-graining procedure. The nonlinear partial differential equation (13) is known as nonlinear biased diffusion, or the Burgers equation [83]. Returning to the coarse-grained equivalent of the height, or virtual times, $\hat{\tau}$, we obtain via $\hat{\phi} = \partial\hat{\tau}/\partial x$ the Kardar-Parisi-Zhang (KPZ) equation [296]:

$$\frac{\partial\hat{\tau}}{\partial t} = \frac{\partial^2\hat{\tau}}{\partial x^2} - \lambda\left(\frac{\partial\hat{\tau}}{\partial x}\right)^2. \quad (14)$$

To capture the fluctuations, one typically adds a delta-correlated noise term $\xi(x,t)$, to the right-hand side, conserved for eq. (13), that is, $\int \xi dx = 0$, and non-conserved for eq. (14). It is important to note that we obtained the KPZ equation as a result of a coarse-graining procedure. While this results in the loss of some of the microscopic details for the original growth model on the lattice, eq. (14) with noise added describes the long-wavelength behavior of the MPEU model. Thus, we claim that virtual time horizon for the basic conservative PDES scheme exhibits kinetic roughening and it belongs to the KPZ universality class. Identifying the universality class of a model is one of the main objectives of surface science, and is used extensively to classify fluctuation statistics. Our procedure above indicates that the long-wavelength statistics of the fluctuations of the time horizon for the basic conservative PDES scheme are in fact captured by the nonlinear KPZ equation.

In one dimension a steady state for the surface fluctuations is reached (in the long time limit) for any *finite* system size, and it is governed by the Edwards-Wilkinson (EW) Hamiltonian $H_{EW} \propto \int dx \left(\frac{\partial \hat{\tau}}{\partial x}\right)^2$ (see, e.g., Barabási and Stanley [29]). The corresponding surface is a simple random-walk surface, where the slopes are independent random variables in the steady state. This means that of the four local configurations of slopes around a point (down-up, down-down, up-up, up-down), only one contributes on average to a minimum (down-up), and since they are all equally likely, we conclude that $\langle u_{EW}(L \to \infty, t \to \infty)\rangle = 1/4 = 0.25$. (Zero slopes are statistically irrelevant, since the probability that two virtual times are exactly equal is zero, given that the updates are drawn from a continuous probability distribution.) Our numerical simulations for the MPEU model (see fig. 3(a)) indicate a value of $\langle u(L \to \infty, t \to \infty)\rangle = 0.24641 \pm (7 \times 10^{-6})$, a value close but not identical to that for the simple random walk surface. The reason for the obvious difference is that the coarse-grained version and the original microscopic model are not identical over the whole spectrum of wavelengths of the fluctuations. The coarse-graining procedure preserves the statistics of the long-wavelength modes, but it loses some information on the short-wavelength ones. In particular, the density of minima is heavily influenced by the short wavelengths (by how "fuzzy" the interface is). However, the density of minima cannot vanish in the thermodynamic limit (large n, large t): a zero density of local minima would imply that it is zero on all length scales, which would contradict the fact that it belongs to the EW universality class. The fact that the steady state of the MPEU model belongs to the EW universality class guarantees that the local slopes are *short-range* correlated (fig. 3(c)), and that the finite-size corrections for the density of local minima (average propagation rate of the surface) follow a universal scaling form [350]:

$$\langle u(L, \infty)\rangle \simeq \langle u(\infty, \infty)\rangle + \frac{\text{const.}}{L^{2(1-\alpha)}} . \tag{15}$$

Here α is the roughness exponent (equal to $1/2$ for the EW universality class), characterizing the macroscopic surface-height fluctuations, as described in detail

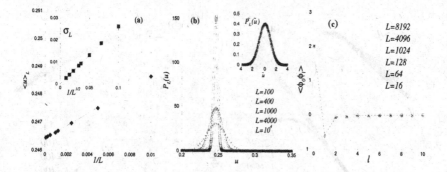

FIGURE 3   (a) Steady state average utilization as a function of the number of PEs L in a one-dimensional ring geometry; (b) The full distribution for the rescaled utilization in the steady state $\tilde{u} = (u(L) - \langle u(L)\rangle)/\sigma_L$, collapsed onto the normal distribution; (c) Slope-slope correlation function.

in the next paragraph. Figure 3 confirms this scaling behavior. Further, calculating the variance in the average utilization in the steady state as a function of system size, $\sigma_L^2 = \langle u^2(L,\infty)\rangle - \langle u(L,\infty)\rangle^2$, we obtain $\sigma_L \propto L^{-1/2}$. These findings suggest that the utilization is a self-averaging macroscopic quantity: its full distribution $P_L(u)$ for large L is a Gaussian (fig. 3(b)).

In the following we show numerical results supporting our claim that the MPEU model belongs to the KPZ universality class. One of the fundamental characteristic quantities strongly influenced by the long-wavelength modes is the average width of the height fluctuations, as given in eq. (6). As the surface grows due to deposition, after an initial transient the width will grow as a power law $\langle w^2(L,t)\rangle \sim t^{2\beta}$ along with the lateral surface correlations $\xi_{\|}(L,t) \sim t^{1/z}$, until the correlations reach the system size ($\xi_{\|} = L$) at a crossover time $t_\times$ [29]. After the crossover time $t_\times$ (for any finite system L) the surface fluctuations are governed by a steady-state distribution and the width scales as

$$\langle w^2(L,\infty)\rangle \sim L^{2\alpha} . \tag{16}$$

The exponent β is called the *growth exponent*, α is the called *roughness exponent*, and z is called the *dynamic exponent* in the surface growth literature [29]. It is easy to show that the three exponents are not all independent, and in fact $\alpha = z\beta$ [29]. Also, these scaling forms allow one to collapse all the different curves for the width onto a single function in the scaling regime, expressing the *dynamic scaling property* of the width: $\langle w^2(L,t)\rangle = L^{2\alpha}f(t/L^z)$ (f is easy to read off, after comparing it to the scaling behavior). For the KPZ interface, the exact values obtained analytically for the exponents are: $\beta = 1/3$, $\alpha = 1/2$ and $z = 3/2$. Figure 4 shows the scaling properties for the width of the MPEU model, measured numerically. For large system sizes ($L = 10^5$), the values

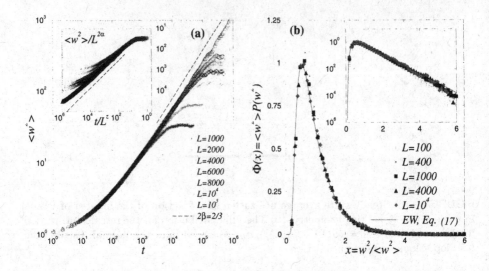

FIGURE 4   (a) The width of the time horizon fluctuations shows dynamical scaling and indicates KPZ universality; (b) The scaling function for the steady-state width distribution follows the scaling function for the EW (one-dimensional KPZ) universality class.

obtained numerically for the exponents, $\beta = 0.326 \pm 0.005$ and $\alpha = 0.49 \pm 0.01$, confirm the KPZ behavior including the dynamical scaling property (inset). Another confirmation for the EW universality class in the steady state comes from measuring the full width distribution $P(w^2)$. For systems belonging to the EW universality class and having the same type of boundary conditions imposed, the width distribution has a universal scaling form [162] $P(w^2) = \frac{1}{\langle w^2 \rangle} \Phi \left(\frac{w^2}{\langle w^2 \rangle} \right)$ with

$$\Phi(x) = \frac{\pi^2}{3} \sum_{n=1}^{\infty} (-1)^{n-1} n^2 e^{-\frac{\pi^2}{6} n^2 x} , \qquad (17)$$

for the case of periodic boundary conditions. Figure 4(b) is a confirmation that the MPEU indeed belongs to the steady state of the EW class, implying that the average utilization (density of local minima) approaches a non-zero, finite value in the thermodynamic limit (5) as reflected by eq. (15). Therefore, the basic conservative scheme is *computationally scalable*. For an in-depth and systematic analytical calculation of the density of minima (utilization) for a number of surface growth models (including the EW class) see Toroczkai et al. [501]. The measurement phase of the basic conservative scheme, however, is *not* scalable, as indicated by the power-law divergence of the width in the long-time large L limit (eq. (16)). For higher-dimensional topologies, using universality arguments, the conclusion remains the same: the basic conservative PDES is computationally scalable, but the measurement phase may not be, depending on what is known as the *upper critical dimension* [29] of the surface (see Korniss et al. [344, 345]).

3.2   THE CONSERVATIVE SCHEME ON SMALL WORLD NETWORKS

From the previous section it follows that the average width of the fluctuations scales in the steady state as $\langle w^2(L, t = \infty)\rangle \sim L^{2\alpha} = L$, and thus grows linearly with the system size. This means that the basic conservative PDES scheme is *not* measurement scalable. Standard methods to control the width of the virtual time horizon in a PDES scheme employ windowing techniques [186]. That is, the local simulated time at any PE cannot progress beyond an appropriately chosen and regularly updated "cap," measured from the global minimum of the time horizon [340]. Thus, a PDES scheme with a moving window relies on frequent global synchronizations or communications, which, depending on the architecture, can get costly for large number of PEs. Here we show how to modify the original conservative scheme such that the scheme is also measurement scalable *without* global "intervention" [346].

The divergence of the width of the surface fluctuations is closely related to the fact that the lateral surface correlations also grow with the system size. In particular, for the one-dimensional EW surface in the steady state, for large L (and fixed l)

$$\langle \hat{\tau}_i \hat{\tau}_{i+l}\rangle \propto \xi_{||}(L, \infty) - |l| \,, \tag{18}$$

where $\hat{\tau}_i$ are the coarse-grained height fluctuations measured from the mean and $\xi_{||}(L, \infty) \sim L$. Thus, $\langle w^2(L, \infty)\rangle = \langle \hat{\tau}_i^2\rangle \propto \xi_{||}(L, \infty) \sim L$. The "height-height" correlations can be characterized by introducing the structure factor for the heights:

$$S^{(\tau)}(k) = \frac{1}{L}\langle \tilde{\tau}_k \tilde{\tau}_{-k}\rangle \tag{19}$$

where $k = 2\pi\alpha/L$, $\alpha = 0, 1, 2, \ldots, L-1$ is the wave-vector, and $\tilde{\tau}_k = \sum_{j=0}^{L-1} e^{-ikj}$ $(\tau_j - \overline{\tau})$ is the discrete spatial Fourier transform of the fluctuations of the virtual time horizon. Then

$$\langle \hat{\tau}_i \hat{\tau}_{i+l}\rangle = \frac{1}{L}\sum_k e^{ikl} S^{(\tau)}(k) \tag{20}$$

and

$$\langle w^2(L, \infty)\rangle = \frac{1}{L}\sum_k S^{(\tau)}(k) \,. \tag{21}$$

Since the universality class for the time horizon evolution is EW, it follows that the expected behavior for the steady-state structure factor for small wave-numbers is

$$S^{(\tau)}(k) \propto \frac{1}{k^2} \tag{22}$$

(see, e.g., eq. (11) in Toroczkai et al. [501]). Indeed, this is also confirmed by

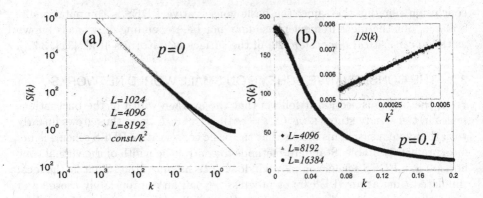

FIGURE 5 Steady-state structure factors for the virtual time horizon for the (a) basic conservative scheme on a regular one-dimensional lattice ($p = 0$) and (b) small-world scheme with $p = 0.1$.

our direct simulation results, shown in figure 5(a). This form of the structure factor implies that there are no length scales other than the lattice constant and the system size, and thus the correlation length and the width diverge in the thermodynamic limit, as can also be seen by evaluating eq. (21) directly.

To de-correlate the surface fluctuations, we modify the communication topology in the following way [346]: for every node i, at the onset of the simulation, we introduce one extra quenched (fixed for a given network realization) random communication link $r(i)$. Together with the existing regular topology, these extra communication links will form a small-world graph [326, 415, 519]. Note that in our specific construction of the small-world network, each node has exactly one random connection and $r(r(i)) = i$, so that there are exactly $L/2$ random links distributed. The updating on PE i will obey the following probabilistically chosen condition:

$$\tau_i \leq \begin{cases} \min\{\tau_{i-1}, \tau_{i+1}, \tau_{r(i)}\} & \text{with probability } p \\ \min\{\tau_{i-1}, \tau_{i+1}\} & \text{with probability } 1-p \end{cases} \tag{23}$$

The PE performs the update (generates the virtual time of the next update or deposits the rod at i in the MPEU surface) only if condition (23) is fulfilled. This means that for sites that would normally be updated within the basic conservative scheme, that is, $\tau_i \leq \min\{\tau_{i-1}, \tau_{i+1}\}$, the PE will make an extra check for the condition $\tau_i \leq \tau_{r(i)}$ with probability p. The parameter p allows us to tune the scalability properties of the corresponding PDES scheme on the quenched small-world network continuously from the pure basic conservative scheme ($p = 0$) to the "fully" small-world conservative scheme ($p = 1$). These occasional extra checks through the quenched random links are not necessary for the faithfulness of the simulation. Rather, they are used to *synchronize* the PEs in such a way that the fluctuations of the time horizon remain bounded in the limit of infinite system size. Most importantly, as the width is reduced from "infinity" (or some large number proportional to L for a finite number of PEs) to a finite, controlled value, the utilization still remains bounded away from zero.

To support this statement, we first use the same coarse-graining procedure used to derive the KPZ equations, as the continuum counterpart of the MPEU model. For the small-world topology we obtain

$$\frac{\partial \hat{\tau}}{\partial t} = -\gamma(p)\hat{\tau} + \frac{\partial^2 \hat{\tau}}{\partial x^2} - \lambda \left(\frac{\partial \hat{\tau}}{\partial x}\right)^2 + \xi(x,t) \qquad (24)$$

with $\gamma(p) = 0$ for $p = 0$, and $\gamma(p) > 0$ for $0 < p \leq 1$. This implies that the extra checking along the random links introduces a *strong* relaxation (first term on the right-hand side of eq. (24)) for the long-wavelength modes of the surface fluctuations, resulting in a finite width. A more transparent picture is gained if we look at the steady-state structure factor (19). Restricting our attention to the linear terms in eq. (24) we obtain

$$S^{(\tau)}(k) \propto \frac{1}{\gamma + k^2} \cdot \qquad (25)$$

In this approximation, the lateral correlation length $\xi_{||}$ scales as $1/\sqrt{\gamma}$, and remains *finite* (and independent of system size) in the thermodynamic limit for all $p > 0$, that is, for an arbitrary small probability of using the random links. Figure 5(b) shows the structure factor for the small-world network with $p = 0.1$, confirming the prediction of eq. (25) for small wave numbers. Consequently, the height-height correlations decay exponentially

$$\langle \hat{\tau}_i \hat{\tau}_{i+l} \rangle \propto \xi_{||} \, e^{-|l|/\xi_{||}} , \qquad (26)$$

and the width remains finite, $\langle w^2(L, \infty) \rangle \sim \xi_{||}$, where $\xi_{||}$ is independent of the system size for all $p > 0$. Further, for the structure factors of the local slopes (the Fourier transform of the slope-slope correlations) one obtains

$$S^{(\phi)}(k) = \frac{1}{L}\langle \tilde{\phi}_k \tilde{\phi}_{-k} \rangle = k^2 S^{(\tau)}(k) \propto \frac{k^2}{\gamma + k^2} = 1 - \frac{\gamma}{\gamma + k^2} \cdot \qquad (27)$$

Both terms above yield short-range correlations (delta function for the first term and exponential decay for the second one), thus the slopes remain short-range correlated, resulting in a non-zero density of local minima. Figure 6 shows two snapshots of the virtual time horizons for the basic conservative scheme $p = 0$, and the small-world scheme with $p = 0.1$. Figure 7(a) shows the scaling of the steady-state width with the system size for various p values and figure 7(b) shows the scaling of the average, steady-state utilization with the system size for the same set of p values. Notice that when increasing p (from $p = 0$ to $p = 0.01$),

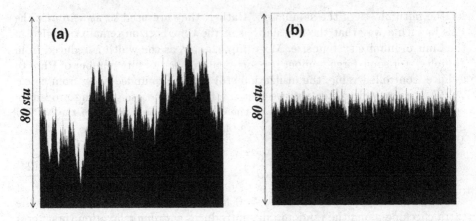

FIGURE 6   Steady-state virtual time horizon snapshots with $L = 10000$ after $t = 10^6$ parallel algorithmic steps (Monte-Carlo sweeps) for the (a) basic conservative scheme ($p = 0$) and (b) small-world scheme $p = 0.1$. Note that the vertical scales are the same in (a) and (b) (plotted in arbitrary simulated time units).

the width instantaneously drops from a linear divergence to a saturated value, while at the same time, the utilization hardly changes. In fact, an infinitesimally small p will make the width bounded, and only at an infinitesimal expense to the utilization. For example, for a hypothetical infinite system, taking $p = 0.01$, the width is reduced from infinity to about 40, while the utilization only from 0.2464 to about 0.246; for $p = 0.1$, the width is further reduced to about 5, while the utilization only to 0.242. By further increasing p, the width further reduces, and at $p = 1$ it is about 1.46, whereas the utilization decreases to 0.141, still clearly bounded away from zero in the thermodynamic limit.

4   SCALABILITY OF THE CONSERVATIVE PDES SCHEME ON SCALE-FREE NETWORK TOPOLOGIES

The internet is a spontaneously grown collection of connected computers. The number of webservers by February 2003 reached over 35 million [414]. The number of PCs in use (internet users) surpassed 660 million in 2002, and it is projected to surpass one billion by 2007 [105]. The idea for using it as a giant supercomputer is rather natural: many computers are in an idle state, running at best some kind of screen-saver software, and the "wasted" computational time is simply immense. Projects such as SETI@home [473] or the GRID consortium [198] are aiming to harness the power lost to screen-savers.

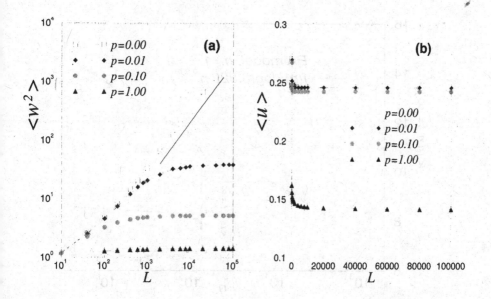

FIGURE 7   (a) The average steady-state width and (b) the utilization for various p values. In addition to ensemble averages over 10 realizations of the random links (solid symbols) a single realization is also shown (open symbols). The solid straight line has a slope of 1/2 and represents the asymptotic one-dimensional KPZ power-law divergence of the width for the basic conservative scheme ($p = 0$).

Most of the problems solved currently with distributed computation on the internet are "embarrassingly parallel" [318], in that the computed tasks have little or no connection to each other: for example, starting the same run with a number of different random seeds, and at the end collecting data to perform statistical averages. However, before more large-scale, complex problems can be solved in real time on the internet, a number of challenges have to be solved, such as the task allocation problem that is complex in itself [457].

Here we ask the following question. *Assuming* that task allocation is resolved and the PE communication topology on the internet is a scale-free network, what are the scalability properties of a PDES scheme on such networks? We present numerical results, for the PDES update scheme, as measured on the Barabási-Albert (BA) model [28, 30] of scale-free networks. The network is created through the stochastic process of preferential attachment: to the existing network of n nodes at time t, the $(n + 1)$th node with m links ("stubs") attaches at time $t + 1$, such that each stub attaches to a node with probability proportional to the existing degree (at t) of the node. We restrict ourselves here to the $m = 1$ case, where the network is a scale-free tree. We have repeated the simulations

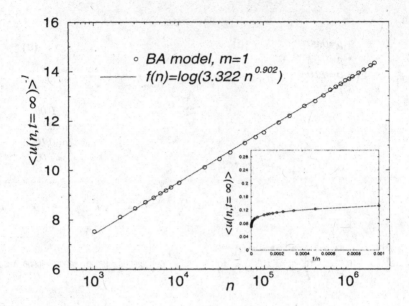

FIGURE 8 Steady-state utilization for the scale-free BA model.

with different m values (up to $m = 10$) and found no significant deviations from our conclusions below—numerical factors are different, but the generic behavior is the same. Once we reach a given number of nodes in the network, we stop the process and use the random network instance to run the MPEU model on top of it, using the evolution (eq. (8)) for the time horizon. While in case of regular topologies, the degree of a node is constant, such as, $P^{(L^d)}(k) = 2d\delta_{k,2d}$ for d-dimensional "square" lattices, for the BA network it is a power law in the asymptotic ($n \to \infty$) limit: $P^{BA}(k) \simeq 2m^2 k^{-3}$. The condition (1) for a site to be updated, namely that its virtual time is a local minimum, is a *local* property. Thus, we expect that the utilization itself will be correlated with local structural properties of the graph, such as the degree distribution.

To get a more detailed picture, we define two more quantities. The first is the *connectivity utilization*

$$u_k(n,t) = \frac{|A_k(t)|}{n}, \qquad (28)$$

which is the fraction of active nodes of degree k, and the second is the *relative connectivity utilization*

$$r_k(n,t) = \frac{|A_k(t)|}{n_k}, \qquad (29)$$

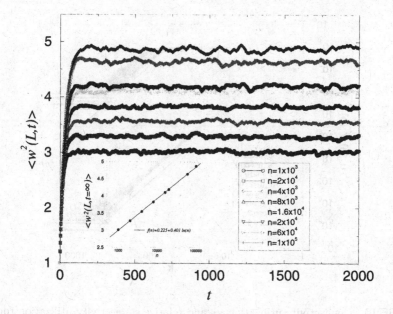

FIGURE 9 Behavior of the time horizon width for the scale-free BA network. The inset shows the scaling of the steady-state width as a function of the system size n. Notice the log-linear scale in the inset.

which is the fraction of nodes that are active and have degree k among the set of all nodes of degree k. From the definitions above, we find the following relations: $\sum_k u_k(n,t) = u(n,t)$ and $\sum_k r_k(n,t)n_k/n = \sum_k r_k(n,t)P^{BA}(k) = \sum_k u_k(n,t) = u(n,t) = \langle r_k(n,t)\rangle_{network}$ at all times, where $\langle \cdots \rangle_{networks}$ is an average over network realizations. Figure 8 shows the steady-state ($t \rightarrow \infty$, in the MPEU model on a fixed BA network of n nodes) values of the average utilization as a function of the network size n. The inset in figure 8 is analogous to figure 3(a), which showed the same quantity on a ring. Notice that strictly speaking, the PDES scheme is computationally *non-scalable*. However, an empirical fit suggests that $u^*(n) = \langle u(n, t = \infty)\rangle \simeq [\log(an^b)]^{-1}$ with $a \approx 3.322$ and $b = 0.902$, that is, the computation is only logarithmically (or marginally) non-scalable. For a system of $n = 10^3$ nodes we have found a steady-state utilization (for the worst case scenario) of $u^*(10^3) = 0.1328$ (13.3% efficiency), while for a system of $n = 10^6$ nodes, the utilization drops only to $u^*(10^6) = 0.073$ (7.3% efficiency), by less than half of its value! For practical purposes the PDES scheme can be considered computationally scalable, and we will call this type of non-scalability *logarithmic* (or marginal) non-scalability.

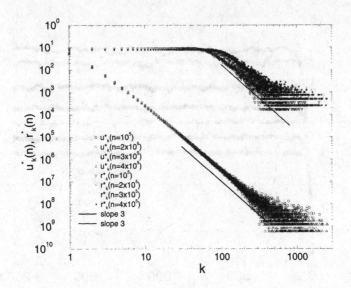

FIGURE 10 Connectivity utilization u_k and relative connectivity utilization (defined in the text) r_k as function of degree. Each data set is obtained after averaging over 200 independent runs.

Figure 9 shows the scaling of the width of the fluctuations for the time horizon as a function of time, and the scaling of its value in the steady state as a function of system size (inset). Notice that while the steady-state width diverges to infinity, it does so only logarithmically, $\langle w^2(n, t = \infty) \rangle \simeq \left[\log \left(cn^d \right) \right]$ with $c \approx 1.25$ and $d \approx 0.401$. Some specific values are $\langle w^2(10^3, t = \infty) \rangle \approx 3.01$, $\langle w^2(10^5, t = \infty) \rangle \approx 4.78$. This means that the measurement phase of the PDES scheme on a scale-free network is also non-scalable, but only logarithmically, and so for practical purposes the scheme can be considered scalable. Overall, the PDES update scheme has logarithmic (or marginal) non-scalability on scale-free networks. If one examines the connectivity utilization and relative connectivity utilization in the steady state, as shown in figure 10, one finds that to a good approximation $u_k^*(n) \sim k^{-3}$, and $r_k^*(n) = \text{const.} \simeq u^*(n)$ for $k \leq k_{\times}$ and $r_k^*(n) \sim k^{-3}$ for $k > k_{\times}$, with $k_{\times} \sim 1/u^*(n) = \log \left(an^b \right) \sim \log n$ being the crossover degree.

5 CONCLUSIONS

We have studied the fundamental scalability problem of conservative PDES schemes where events are self-initiated and have identical distributions on each

PE. First, we considered the scalability of the basic conservative scheme for systems with short-range interactions on regular lattices. By exploiting a mapping between the progress of the simulation and kinetic roughening in non-equilibrium surfaces, we found that while the average progress rate of the PEs $\langle u(\infty, \infty) \rangle$ is a finite *non-zero* value, the spread of the progress of the PEs about the mean $\langle w^2(\infty, \infty) \rangle$ *diverges*. This divergence makes the measurement phase of the algorithm non-scalable. In order to make the measurement part of the simulation scalable as well, we have introduced a small number of quenched random connections between PEs so that the resulting random links on top of the regular short-range connections form a *small-world* connection topology. Invoking the same conservative protocol used at an arbitrarily small (but strictly positive) rate through the random links is sufficient to achieve full scalability: the PEs progress at a non-zero, near-uniform rate *without* requiring global synchronization. This construction of a fully scalable algorithm for simulating large systems with asynchronous dynamics and short-range interactions is an example of the enormous "computational power and synchronizability" [519] that can be achieved by small-world couplings. The suppression of critical fluctuations of the virtual time horizon is also closely related to the emergence of mean-field-like phase transitions and phase ordering in *non-frustrated* interacting systems [34, 197, 256, 257, 316]. In particular, the fluctuations exhibited by the virtual time horizon with small-world synchronization should exhibit very similar characteristics to the fluctuations of the order parameter in the XY spin model on a small-world network [316].

Second, we have studied the scalability properties for a causally constrained PDES scheme hosted by a network of computers where the network is *scale-free* following a "preferential attachment" construction [28, 30]. Here the PEs simply have to satisfy the general criterion eq. (1) in order to advance their local time. Despite some nodes in the network having abnormally high degrees, as a result of the scale-free nature of the degree distribution, we find that the computational phase of the algorithm is only marginally non-scalable. The utilization exhibits slow logarithmic decay as a function of the number of PEs. At the same time, the width of the time horizon diverges logarithmically slowly, rendering the measurement phase of the simulations marginally non-scalable as well. An intriguing question to pursue is how the logarithmic divergence of the surface fluctuations observed here can be related to the collective behavior (in particular, the finite-size effects of the magnetic susceptibility) of Ising ferromagnets on scale-free networks [15, 51, 127, 359] with the same degree distribution.

ACKNOWLEDGEMENTS

Discussions with P. A. Rikvold, B. D. Lubachevsky, Z. Rácz, and G. Istrate are gratefully acknowledged. Zoltan Toroczkai was supported by the Department of Energy under contract W-7405-ENG-36. This research is supported in part by the National Science Foundation, through Grant No. DMR-0113049 and the Research Corporation through Grant No. RI0761.

CHAPTER 12

Combinatorics of Genotype-Phenotype Maps: An RNA Case Study

Christian M. Reidys

1 INTRODUCTION

The fundamental mechanisms of biological evolution have fascinated generations of researchers and remain popular to this day. The formulation of such a theory goes back to Darwin (1859), who in the *The Origin of Species* presented two fundamental principles: genetic variability caused by mutation, and natural selection. The first principle leads to diversity and the second one to the concept of survival of the fittest, where fitness is an inherited characteristic property of an individual and can basically be identified with its reproduction rate. Wright [530, 531] first recognized the importance of genetic drift in evolution in improving the evolutionary search capacity of the whole population. He viewed genetic drift merely as a process that could improve evolutionary search. About a decade later, Kimura proposed [317] that the majority of changes that are observed in evolution at the molecular level are the results of random drift of genotypes. The neutral theory of Kimura does not deny that selection plays a role, but claims that no appreciable fraction of observable molecular change can

Computational Complexity and Statistical Physics, edited by
Allon G. Percus, Gabriel Istrate, and Cristopher Moore, Oxford University Press.

be caused by selective forces: mutations are either a disadvantage or, at best, neutral in present day organisms. Only negative selection plays a major role in the neutral evolution, in that deleterious mutants die out due to their lower fitness.

Over the last few decades, there has been a shift of emphasis in the study of evolution. Instead of focusing on the differences in the selective value of mutants and on population genetics, interest has moved to evolution through natural selection as an abstract optimization problem. Given the tremendous opportunities that computer science and the physical sciences now have for contributing to the study of biological phenomena, it is fitting to study the evolutionary optimization problem in the present volume. In this chapter, we adopt the following framework: assuming that selection acts exclusively upon isolated phenotypes, we introduce the following compositum of mappings

$$\textbf{Genotypes} \longrightarrow \textbf{Phenotypes} \longrightarrow \textbf{Fitness} . \qquad (1)$$

We will refer to the first map as to the genotype-phenotype map and call the preimage of a given phenotype its neutral network. Clearly, the main ingredients here are the phenotypes and genotypes and their respective organization. In the following we will study various combinatorial properties of phenotypes and genotypes for RNA folding maps.

In the context of the RNA toy-world pioneered by Peter Schuster et al. [160, 443, 462, 463, 464], the phenotypes are secondary structures that allow for a mathematical modeling of their corresponding neutral networks as random graphs. Many significant properties of these neutral networks, such as connectivity, density, and path-connectivity are monotonic: they are maintained after adding any number of edges. One may then ask whether the montonic property in question displays a sharp threshold or phase transition—as, for instance, in the classical random graph ensemble $\mathcal{G}_{n,p}$ where *every* monotonic property satisfies a 0-1 law. The application of random graph theory to biology and particularly computational biology is not new. Bollobás and Rasmussen have used directed random graphs [63] to study the evolution of autocatalytic networks. Lynch has analyzed phase transitions [371, 372] in Kauffmann's random Boolean networks used for the modeling of gene regulatory networks. Finally, Frieze et al. have studied optimal sequencing [181, 184] and the ordering of clone libraries [136] using methods and theory of random graphs.

The mapping of RNA sequences to their secondary structures plays an important role in the understanding of evolutionary optimization, as the generic properties of this mapping dictate to a large extent the dynamics of the optimization process itself. Populations of sequences subject to selective pressures, such as virus populations pressured by immune systems, constantly search for new fitter structures and try to realize them. During this search, however, the current "best" phenotype must necessarily be preserved while new mutants simultaneously emerge. In most cases the search process is essentially a "white noise computation," such as point mutations in single stranded RNA, where there is no rational design according to which the mutations occur. Accordingly, the generic structure of folding maps must allow for effective random search. In the case of folding maps from RNA sequences to their secondary structures, we will show that the key feature for enabling effective search by point mutations is a specific type of redundancy. The sequences folding into one specific secondary structure form networks with giant components. As a consequence, some fraction of random point mutations will have virtually no effect on the phenotype; that is, the RNA folds into the same secondary structure and, complimentarily, some fraction of point mutants will fold into new structures. Additionally, we will show that the combinatorics of secondary structures itself guarantees that any two structures can have neutral nets that are close in sequence space. Kimura's neutral theory fits smoothly into the genotype-phenotype map framework, since it reflects the relation between genotypes and phenotypes. Our main goal consists of providing insight into folding maps exhibiting the type of redundancy above and how generic such maps are, as well as investigating additional properties of these maps that are of key relevance to evolutionary optimization.

In the following section we introduce some basic facts about RNA sequences and RNA secondary structures. In section 3 we introduce the notion of compatible sequences with respect to a secondary structure and prove that for any two secondary structures there exists some RNA sequence that is compatible with both of them. This result guarantees the closeness of the corresponding neutral nets. In section 4 we address the actual modeling of preimages. Our approach consists of employing a certain random graph model for the preimages of a secondary structure. As we are interested in the question of how generic certain properties of these preimages are, a random graph model and its 0-1 laws are of particular relevance. We will state and discuss a suite of generic connectivity and path-connectivity results.

2 DEFINITIONS

2.1 RNA

In the following we will consider single-stranded RNA molecules. In viruses and cells RNA acts as a messenger (mRNA), carrying the genetic information from the DNA to the translation apparatus. As transfer RNA (tRNA), it plays the role of an adapter for the synthesis of proteins. Finally, as ribosomal RNA (rRNA), it is an integral part of the ribosome and exhibits catalytic activities in natural polypeptide synthesis [84, 85, 526]. RNA is thus able to serve two purposes: (i) storage of genetic information based on a one-dimensional template that can be read and copied on request, and (ii) catalytic properties as ribozymes that require three-dimensional structures in order to gain efficiency and specificity in processing specific substrates. As demonstrated by Spiegelman, *in vitro* evolution experiments can be applied to selection of RNA molecules that are capable of fast replication [399]. Indeed, replication rates are optimized in serial transfer experiments [144, 284, 452]. In case one wants to optimize properties other than replication, intervention is required making use of special techniques that interfere with natural selection. A well-known example is represented by the SELEX method—standing for *systematic evolution of ligands by exponential enrichment*—which allows the creation of molecules with optimal binding constants [507]. The SELEX procedure is a protocol that isolates high-affinity nucleic acid ligands for a target, such as a protein, from a pool of variant sequences. Multiple rounds of replication and selection exponentially enrich the population of species that exhibits the highest affinity, that is, that fulfills the required task. This procedure thus permits simultaneous screening of highly diverse pools of nucleic acid molecules for different functionalities (for a review, see Ellington [143] and Klug and Famulok [331]). Results from those experiments clearly demonstrate the essential property of RNA molecules: genotype, meaning the RNA sequence, and phenotype, associated with the structure, are combined in one molecule.

Here we will consider RNA sequences of constant length, represented by n-tuples, $(x_1, \ldots, x_n)$, with $x_i \in \mathcal{A}$, $\mathcal{A}$ being a finite alphabet formed by the nucleotides. The basic mutational mechanism is made up of random point mutations that occur with independent probability. This motivates calling two sequences *adjacent* if they differ by exactly one nucleotide. The sequence space with this adjacency relation is referred to as $\mathcal{Q}_\alpha^n$ (the generalized n-cube), where $\alpha = |\mathcal{A}|$. In $\mathcal{Q}_\alpha^n$ each sequence has $(\alpha - 1)n$ neighbors and the maximal (Hamming) distance between two sequences is n.

2.2 SECONDARY STRUCTURES

A secondary structure is a graph whose vertices are the nucleotides of its underlying sequence, and whose edges are base pairs formed among them. For biophysical reasons, one nucleotide can only establish exactly one Watson-Crick bond with another nucleotide. As we will see below, the fact that the edges of a secondary structure are Watson-Crick base pairs implies a number of additional graph properties. Following Waterman [518] we will consider RNA secondary structures over n vertices $\{1, \ldots, n\}$, which we denote by s_n.

A secondary structure is a vertex-labeled graph with an adjacency matrix $A(s_n) = (a_{i,k})_{1 \le i, k \le n}$ such that

- $a_{i,i+1} = 1$ for $1 \le i \le n - 1$;
- for each i there is at most a single $k \ne i - 1, i + 1$ such that $a_{i,k} = 1$; and
- if $a_{i,j} = a_{k,l} = 1$ and $i < k < j$ then $i < l < j$.

We call an edge $\{i, k\}$, $|i - k| \ne 1$ a *base pair*. A vertex i connected only to $i - 1$ and $i + 1$ is called *unpaired*.

The enumeration of secondary structures has been studied in detail in a series of excellent papers by Waterman et al. [263, 456]. A particular result from asymptotic combinatorics on secondary structures—with certain restrictions, such as minimum helix length—is that their number asymptotically becomes $O(a^n)$ with $a < 2$ [241]. This result immediately implies that there are structures having preimages of exponential size. The RNA model allows, moreover, several generic choices for the fitness assignment, for example using the thermodynamic stability and the degradation constant of the corresponding secondary structure.

3   SECONDARY STRUCTURES AND COMPATIBLE SEQUENCES

In this section we introduce the notion of compatible sequences with respect to a secondary structure. While Waterman et al. have extensively studied the combinatorics of secondary structures, their compatible sequences play a central role in the understanding of the mapping between RNA sequences and their structures. Theorem 3.1 below is central for evolutionary optimization as it guarantees the existence of at least one sequence that is compatible with any two given secondary structures. This fundamental property of secondary structures has been used, for example, in the *Science* publication "One Sequence Two Ribozymes," [461] in which the authors construct a sequence that can assume either of two ribozyme folds and catalyze the two respective reactions.

Let us now introduce compatible sequences. We call a sequence (x_i) compatible with respect to a secondary structure, s_n, if and only if for all $a_{i,k}$ with $a_{i,k} = 1$ and $k \neq i - 1, i + 1$, the nucleotides x_i and x_k can in principle form a Watson-Crick base pair.

In terms of combinatorics, the uniqueness property of the Watson-Crick base pairs of an RNA secondary structure corresponds to an involution (an operator of period 2), viewing the base pairs as transpositions within the symmetric group S_n [442, 443]. Now, any two involutions form a dihedral group that, in our situation, acts upon the nucleotides regularly and whose orbits are either even-length cycles or lines, as illustrated in figure 1.

Theorem 3.1 (Reidys et al. [443]). *Let s_n^1, s_n^2 be two secondary structures with the sets of compatible sequences $C(s_n^1)$, $C(s_n^2)$. Then*

$$C(s_n^1) \cap C(s_n^2) \neq \varnothing . \tag{2}$$

Accordingly, for any two secondary structures there exists at least one sequence that could, in principle, realize both. We will call such a sequence bicompatible with respect to the pair of structures. From this we can conclude that their corresponding neutral networks come relatively close in sequence space.

At this point we may speculate that populations performing evolutionary search by point mutations are capable of switching between any two networks. This speculation turns out to be not entirely correct but has, however, led to some

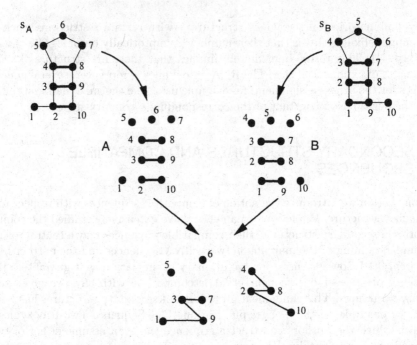

FIGURE 1   The key idea in theorem 3.1: any two involutions yield either non-closed paths or loops of even length. The two secondary structures s_A and s_B are decomposed into their paired and unpaired regions (see also fig. 2) yielding two graphs of identical order, A and B. Finally, the edge sets of these graphs are joined, resulting in the graph shown at the bottom. One will always find a sequence that is compatible with this graph. We may then conclude—for instance, by taking a segment composed of pairs of complementary nucleotides—that for any two secondary structures there exists at least one sequence compatible with both structures.

understanding of the transition phenomenon [165, 465, 520]. In fact, the group action above suggests the definition of a distance measure between secondary structures [442] from which the probability of a transition can be computed. Structural similarity thus plays an important role in the transition phenomenon.

4   NEUTRAL NETWORKS

4.1   MODELLING NEUTRAL NETWORKS VIA RANDOM GRAPHS

In the following we will model the preimage of a given structure as a random graph. The main motivation is that folding maps will always vary as a function of their underlying biophysical parameters. Hence "generic properties" of classes of maps are of particular interest. We will restrict ourselves to RNA secondary structures as phenotypes but, in principle, an analogous construction can be obtained for random structures, a more general class of phenotypes. The random graph model is constructed in the following two steps:

1. *Creating two new cubes.* One first determines the set of sequences $C(s_n)$ that are compatible with the given structure s_n. Each compatible sequence is decomposed into an unpaired and a paired segment, consisting of all unpaired and paired nucleotides respectively, as shown in figure 2. While the unpaired segment (of length n_u) is simply again a sequence of a sequence space of reduced dimensionality, the paired segment (of length n_p) is interpreted as a sequence over the alphabet of base pairs. For example, a paired segment in the case of the biophysical $\{A, U, G, C\}$-alphabet would have $\{(A-U), (U-A), (G-C), (C-G), (G-U), (U-G)\}$ as its new alphabet, that is, an alphabet of size 6. Accordingly, the set $C[s_n]$ of compatible sequences can be written as

$$C[s_n] = \mathcal{Q}_\alpha^{n_u(s_n)} \times \mathcal{Q}_\beta^{n_p(s_n)} . \tag{3}$$

2. *Randomization.* We now proceed by selecting the unpaired and paired segments with independent probabilities λ_u and λ_p. Accordingly, each compatible sequence is selected with probability $\lambda = \lambda_u \lambda_p$. Interestingly, it is not difficult to determine λ_u and λ_p for RNA folding maps by introducing the corresponding mutations systematically and then folding the mutants.

From the biophysical point of view, there is a significant difference between λ_u and λ_p: in the case of a $\{G, C\}$-alphabet a point mutant is produced with probability p and a base pair mutation occurs with probability p^2. From the combinatorial perspective, however, up to an isomorphism there is none. The selection processes of the unpaired and paired segments both take place in generalized n-cubes, and accordingly, we may formulate our results only for generalized n-cubes. It is worth pointing out that the random graph model above does not aim *a priori* to construct particular neutral networks, but to identify generic properties of the probability space formed by all neutral networks. In this sense, the present model follows an approach that is very natural in statistical physics.

In the following, we will denote a probability measure by μ_n where n refers to some index of the corresponding probability space Ω_n (here: a random graph). A random variable is a mapping $X : \Omega \to \mathbb{Z}$. Let P_n be some property or event in Ω_n. We then write that P_n holds *asymptotically almost surely* (a.a.s.) if $\lim_{n \to \infty} \mu_n \{P_n\} = 1$.

The random graph model. Let $\mathcal{Q}_\alpha^n$ be a generalized n-cube over an alphabet of length α. Let Γ_n be a subgraph of $\mathcal{Q}_\alpha^n$ and $\mu_n \{\Gamma_n\} = \lambda_n^{|\Gamma_n|} (1 - \lambda_n)^{\alpha^n - |\Gamma_n|}$. Then we call $\mathcal{Q}_{\alpha, \lambda_n}^n$ the random induced subgraph model.

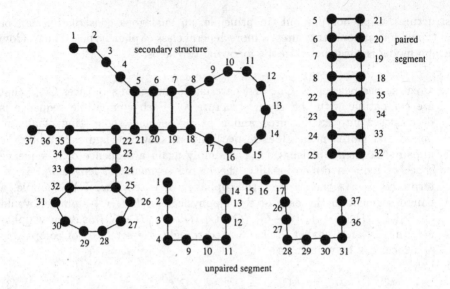

FIGURE 2   Decomposing a compatible sequence into the unpaired and paired segments. Note that the resulting alphabet over which the paired segment is considered is, in general, different from the alphabet of the unpaired nucleotides. In the case of Watson-Crick base pairing rules we obtain for the biophysical $\{A, U, G, C\}$-alphabet of size 4 the alphabet $\{(A - U), (U - A), (G - C), (C - G), (G - U), (U - G)\}$ of size 6.

4.2   GIANT COMPONENTS AND CONNECTIVITY

In the following, we will analyze generic properties of $\mathcal{Q}^n_{\alpha, \lambda_n}$, all of which deal with connectivity. The idea will be to let the picking probability, λ, gradually increase. Let us start our investigations with the probability $\lambda = O(\log n/n)$.

It is beyond the scope of this chapter to present full proofs of our results. Instead we will discuss and outline the proofs, and present the main ideas. For details, the reader is referred to the papers by Reidys and Stadler [441] and Reidys [440]. The proof of theorem 4.1 in [441] is inspired by the paper of Ajtai, Komlós, and Szemerédi [11] but differs significantly in the estimation of the respective vertex boundaries. While Ajtai et al. rely on Harper's isoperimetric inequality [225] for estimating the edge boundary, a completely new method has to be employed in order to estimate the vertex boundary. The proof of theorem 4.2 in Reidys and Stadler [441] is completely different from the proof of the classical result for $\mathcal{Q}^2_{2,p}$, which localizes the connectivity threshold at $p = 1/2$. Our proof is entirely constructive and additionally allows for the development of algorithms connecting two sequences on a neutral network above the threshold.

Theorem 4.1. *Let $C_n^{(1)}$ be the largest component of a Q_α^n-subgraph Γ_n. Then there exists a constant $c > 0$ such that for $\lambda_n \geq \frac{c \log n}{n}$,*

$$|C_n^{(1)}| \sim |\Gamma_n| \qquad \text{a.a.s.}$$

The existence of this giant component is proven indirectly in two steps.

1. One shows that a.a.s. any vertex of a random graph is contained in a component of at least size n^h, for some natural number h. That is, given a picking probability $\lambda_n = O(\log n/n)$, the random graph is composed almost entirely of connected components of polynomial size, where the degree of the polynomial is arbitrarily high. The idea to prove this involves estimating the vertex-boundary of subsets of n-cube vertices and computing the mean of all such components. Step 1 can be proven exclusively using the fact that n-cubes are Cayley graphs over Z_α^n, Z_α being the cyclic group of order α.
2. From step 1 we know that the random graph is composed almost exclusively of connected components of at least polynomial size, and potentially many of these. Clearly, if there exist at least two such components, then there must be a bipartition of the set of all components such that no edge connects the two parts. We show that the probability of such a bipartition existing, formed by two sets of vertices of the same order, tends to 0. Thus, the size of the second-largest component can be at most subpolynomial in n.

It is important to note that our argument proves the existence of a giant component indirectly. The proof of theorem 4.1 gives no clue as to how to construct a path between two vertices, and, moreover, as to how long such a path might be. The explicit construction of (short) paths between vertices of neutral networks would, therefore, be of particular interest and would lead to a deeper understanding of how likely such a path would be realized in an evolutionary search. We address this question by studying paths and distances in generalized n-cubes in theorem 4.3 below.

Let us next analyze connectivity of generalized n-cubes. We now assume a constant probability $\lambda > 0$.

Theorem 4.2. *In the random graph $Q_{\alpha,\lambda}^n$ the probability $\lambda^* = 1 - \sqrt[\alpha-1]{\alpha^{-1}}$ is the threshold value for connectivity. That is, a.a.s. no random graph is connected for $\lambda < \lambda^*$ and a.a.s. every random graph is connected for $\lambda > \lambda^*$.*

Let P, Q be arbitrary vertices of the random graph. As will be seen for theorem 4.3, we can reduce the case to P, Q having finite Hamming distance. For $\lambda > \lambda^*$, one then shows that any vertex has an arbitrary finite number of neighbors in the random graph. Using these neighboring vertices one proceeds analogously to the proof of theorem 4.3 below. To prove that λ^* is a threshold value, we show that there exist isolated vertices when $\lambda < \lambda^*$. This can be proven by considering the random variable counting the isolated vertices, Y. It is clear that Y has mean $\mu = \lambda \, \alpha^n \, (1 - \lambda)^{(\alpha - 1)n}$ and for finite μ one can show that Y becomes Poisson-distributed in the limit of large n. From this we can conclude that a.a.s. for $\lambda < \lambda^*$ and arbitrary natural number ℓ, there are at least ℓ isolated vertices in the random graph.

4.3 DISTANCES

As we have seen, theorem 4.1 does not provide insight into the path structure of the giant component. Theorem 4.2 on the other hand is proven constructively, but only works for constant λ. In this section we present a framework that allows us to bound the length of shortest paths between two $\mathcal{Q}_\alpha^n$ vertices for the probability $\lambda_n \geq n^{-a}$ with $0 \leq a < 1/2$.

In the following, we write Γ instead of Γ_n. Intuitively, our result guarantees the a.a.s. existence of very short paths between any two Γ-vertices. Technically this fact is a little delicate to express in probabilistic language, since it is impossible to have terms like a.a.s. as a predicate of a property in a probability space. Our strategy will be to use conditional probabilities in the statement of the result. The main question is how the distance $d_\Gamma(P, Q)$ between two vertices P, Q in a random graph Γ relates to the distance $d_{\mathcal{Q}_\alpha^n}(P, Q)$ between P, Q in $\mathcal{Q}_\alpha^n$, which is known to be very small. Let us denote the least integer greater or equal to c by $\lceil c \rceil$.

Theorem 4.3. *Let* $0 \leq a < 1/2$, $k = \lceil \frac{1+3a}{1-2a} \rceil$ *and* $p_n = n^{-a}$. *Then for any two vertices* $P, Q \in \mathcal{Q}_\alpha^n$ *we have* $d_\Gamma(P, Q) < \lceil 2k + 3 \rceil d_{\mathcal{Q}_\alpha^n}(P, Q)$ *a.a.s. conditional on* $P, Q \in \Gamma$, *and for any constant* $p > 0$ *we have* $d_\Gamma(P, Q) \leq 7 d_{\mathcal{Q}_\alpha^n}(P, Q)\} = 1$.

Essentially, theorem 4.3 means that for probabilities larger than $1/\sqrt{n}$ and in the limit of large sequence length, the distance between almost all pairs of vertices is, up to a constant factor, equal to their distance in the n-cube itself. One consequence of this result is that the distances between sequences on a neutral network are surprisingly small. The diffusion process, performed by the error-prone relication of haploid RNA sequences in the course of their evolutionary optimization, enables visiting every region of the neutral network.

The main idea for the proof of theorem 4.3 is as follows: from two different $\mathcal{Q}_\alpha^n$-vertices (sequences) one tries to branch *simultaneously*, that is, by performing successively identical point-mutations on the sequences in positions where P and Q do not differ, into some kth sphere centered at P and Q, respectively. The trick with respect to the simulaneous mutations consists of being able to guarantee that the resulting pairs of sequences have the same distance as P and Q. We then have to show that there are sufficiently "many" of these pairs in the kth sphere and that the collection of their associated paths connecting them is vertex disjoint. This is illustrated in figure 3.

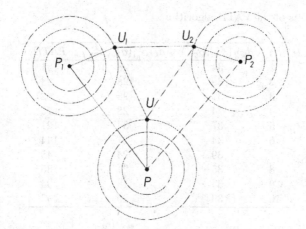

FIGURE 3 The branching process with initial vertex P in an $(n - m)$-cube can be considered by lifting as a simultaneous branching process in an n-cube initialized at the two vertices P_1, P_2. By concatenation, any point U of the branching process yields two points U_1, U_2 in the n-cube. The key feature of this construction is that any two points reached by the simultaneous branching have *constant* distance m, and for increasingly high Hamming distances, more and more such pairs are constructed.

4.4 AN ALGORITHM

Let s_n be a secondary structure. The proof of theorem 4.3 motivates the algorithm *PATH*, which tries to determine short paths on the neutral network of s_n. The algorithm works in $C[s_n] \cong \mathcal{Q}_\alpha^{n_u(s_n)} \times \mathcal{Q}_\beta^{n_p(s_n)}$. The *input* of *PATH* consists of (i) a secondary structure, s_n, and (ii) two $\mathcal{Q}_\alpha^n$-vertices P, Q that map into s_n. Its *output* is the length of a $C[s_n]$-path connecting P and Q on the neutral network of s_n if the algorithm finds one of length $\leq 11\, d_{C[s_n]}(P, Q)$, and "$-$" otherwise. *PATH* can be sketched as follows:

1. Write P, Q in the form (cf. eq. (3))

$$(\xi_1, \ldots, \xi_{n_u}, (\eta_1, \eta_1'), \ldots, (\eta_{n_p}, \eta_{n_p}')),$$

   where $\xi_l < \xi_{l+1}$, $l \in \mathbb{N}_{n_u-1}$, $\eta_i < \eta_i'$, $i \in \mathbb{N}_{n_p}$ and $\eta_j < \eta_{j+1}$, $j \in \mathbb{N}_{n_p-1}$
2. Construct a $C[s_n]$-path, $\pi(P, Q) = (P, V_1, \ldots, V_\ell, Q)$, from P to Q successively replacing ξ_i^P by ξ_i^Q and then $(\eta_i^P, \eta_i'^P)$ by $(\eta_i^Q, \eta_i'^Q)$ according to the ordering given in step 1
3. For $1 \leq i \leq \ell$, try to find a vertex $G_i \in \mathcal{B}_3(V_i)$ which maps to s_n and store the family $(P, G_1, \ldots, G_\ell, Q)$
4. Try to connect the pairs of vertices (P, G_1), (G_ℓ, Q) and (G_i, G_{i+1}), $1 \leq i \leq \ell - 1$. This is done using (a) the branching process shown in figure 3

TABLE 1 Results of the PATH algorithm.

(i)	$d_{\mathcal{Q}^n_\alpha}(\sigma^*_{76}, \sigma^{(i)}_{76})$	$d_{\mathcal{C}[s^*_{76}]}(\sigma^*_{76}, \sigma^{(i)}_{76})$	$PATH$
1	34	28	131
2	39	28	98
3	37	25	62
4	74	28	73
5	37	34	121
6	34	27	144
7	39	24	43
8	35	21	35
9	34	18	94
10	39	29	43

with concentric sphere $k \leq 3$, and (b) families of independent paths that are employed in the proof of theorem 4.3.

We finally present some results of $PATH$ [171]. As the underlying map from sequence to structure, we use the bio-physical folding algorithm RNAfold [242]. As the input secondary structure we select the tRNA

$$s^*_{76} = (((((((((.((((.(\ldots\ldots.)))).((((((\ldots\ldots)))))\ldots\ldots.((((\ldots\ldots))))))))))))))\ldots.$$

where "(" and ")" represent paired bases and "." represents an unpaired base. We take the natural RNA-sequence σ^*_{76}:

$$GCGGAUUU\,AGCUC\,AG\ddagger\ddagger GGG\,AG\,AGC\ddagger CC\,AG\,ACU\,G\,AA\ddagger$$
$$AUCUGG\,AG\ddagger U\,CCU\,GUG\ddagger\ddagger CG\,AUCC\,AC\,AG\,AAUU\,CGC\,ACC\,A,$$

where "$\ddagger$" denotes a *special base* that is kept fixed. Using the algorithm RNAinverse [242] we determine a set of sequences, $\sigma^{(i)}_{76}$, $i = 1, \ldots, 10$ of the neutral network of s^*_{76}. We then use s^*_{76}, σ^*_{76} and $\sigma^{(i)}_{76}$ as input for $PATH$. Preliminary data indicate that the success rate of $PATH$ for $n = 76$ is approximately 50%. We present some data in table 1.

5 CONCLUSION

We have investigated folding maps from RNA sequences to their secondary structures. We have first shown that the combinatorics of the structures themselves allows for sequences that are compatible with two given structures. This intersection result for the corresponding sets of compatible sequences indicates that

secondary structures as phenotypes can be searched effectively by point mutations. We have then modeled the preimage (neutral network) of a secondary
structure via random subgraphs of (generalized) n-cubes, proving that there exists a 0-1 law for connectivity and establishing probabilities above which giant
components in these neutral networks exist.

Random graph theory not only provides insight into the structure of neutral
nets of RNA secondary structures but also contributes on a conceptual level to
the understanding of evolutionary optimization. We have studied short paths
in neutral networks, which are of key importance for the dynamics of the optimization process. We have shown that the shortest path between two sequences
on a neutral network is longer only by a constant factor than the shortest path
between these sequences in the n-cube itself. Finally, we have presented an algorithm that computes the length of these paths in the neutral network of the
tRNA structure.

In this context, it is of interest to note that Grüner et al. have performed
[217, 218] an exhaustive folding of GC sequences of lengths 30, according to
a minimum free energy folding algorithm, into their corresponding secondary
structures. This study allows the comparison of the probabilistic results on the
structure of neutral networks with those of biophysical folding maps. One particular finding is that the existence of certain structural motifs (at this sequence
length) can cause a multi-partition of the corresponding neutral network into distinct components, since the preservation of these motifs induces a certain bias in
the sequences of the corresponding neutral network. As the probabilistic model
is based on a uniform picking probability, the findings above were anticipated.
However, at this point it is not obvious whether or not this phenomenon will
persist for significantly longer sequences, as the dimensionality of the n-cube
increases.

CHAPTER 13

Towards a Predictive Computational Complexity Theory for Periodically Specified Problems: A Survey

Harry B. Hunt III
Madhav V. Marathe
Daniel J. Rosenkrantz
Richard E. Stearns

1 INTRODUCTION

The preceding chapters in this volume have documented the substantial recent progress towards understanding the complexity of *randomly* specified combinatorial problems. This improved understanding has been obtained by combining concepts and ideas from theoretical computer science and discrete mathematics with those developed in statistical mechanics. Techniques such as the cavity method and the replica method, primarily developed by the statistical mechanics community to understand physical phenomena, have yielded important insights into the intrinsic difficulty of solving combinatorial problems when instances are chosen randomly. These insights have ultimately led to the development of efficient algorithms for some of the problems.

A potential weakness of these results is their reliance on random instances. Although the typical probability distributions used on the set of instances make the mathematical results tractable, such instances do not, in general, capture the realistic instances that arise in practice. This is because practical applications of

Computational Complexity and Statistical Physics, edited by
Allon G. Percus, Gabriel Istrate, and Cristopher Moore, Oxford University Press.

graph theory and combinatorial optimization in CAD systems, mechanical engineering, VLSI design, transportation networks, and software engineering involve processing large but regular objects constructed in a systematic manner from smaller and more manageable components. Consequently, the resulting graphs or logical formulas have a regular structure, and are defined systematically in terms of smaller graphs or formulas. It is not unusual for computer scientists and physicists interested in worst-case complexity to study problem instances with regular structure, such as lattice-like or tree-like instances. Motivated by this, we discuss periodic specifications as a method for specifying regular instances. Extensions of the basic formalism that give rise to *locally random but globally structured* instances are also discussed. These instances provide one method of producing random instances that might capture the structured aspect of practical instances. The specifications also yield methods for constructing *hard* instances of satisfiability and various graph theoretic problems, important for testing the computational efficiency of algorithms that solve such problems.

Periodic specifications are a mechanism for succinctly specifying combinatorial objects with highly regular repetitive substructure. In the past, researchers have also used the term *dynamic* to refer to such objects specified using periodic specifications (see, for example, Orlin [419], Cohen and Megiddo [103], Kosaraju and Sullivan [347], and Hoppe and Tardos [260]). However, since "dynamic" has also been used by researchers to mean other things, we have elected to use periodic specifications in the rest of the chapter to avoid ambiguity. The kinds of objects considered here include graphs, logical formulas; and systems of equations/constraints. These specifications arise naturally in engineering and VLSI designs, as well as in scheduling and routing models for airline industry. They have been studied for over 40 years, since the work of Ford and Fulkerson on dynamic network flows [163, 164] and extensively thereafter [103, 104, 245, 260, 274, 276, 347, 379, 380, 419, 420, 421]. In this chapter, we survey a number of results on the complexity and efficient approximability of problems, for periodically specified objects. We also propose several new extensions of the basic formalism that may be of interest to researchers studying phase transition phenomena for combinatorial problems.

Generally speaking, periodic specifications are extensions of the standard specifications used to represent combinatorial objects. An example of a standard specification for satisfiability problems on Boolean formulas is the conjunctive normal form, where the formula is represented as a set of clauses, with each clause being a set of literals. For problems in graph theory, a standard specification of the graph is the adjacency list representation or the adjacency matrix representation of the edges in the graph. Periodic specifications can represent succinctly—and in a space-efficient way—certain kinds of objects with highly regular structure. For example, consider a graph G_n consisting of a simple path with n vertices. At best, the standard specification represents G_n by each of its vertices and edges separately, and is thus of size $\Theta(n)$. In contrast, G_n can be specified succinctly by a *one-dimensional periodic finite graph specification* with

$O(\log n)$ symbols, by replicating a single edge (u, v) n times, and specifying that for $1 \le i \le (n-1)$, the ith copy of v is connected to the $(i+1)$th copy of u. Thus, the periodic specification of G_n results in exponential savings in space as compared to the standard specification of G_n. The simple example shows that, for all $n \ge 1$, periodic specifications of size $\Theta(n)$ can represent certain objects of size $2^{\Omega(n)}$, any of whose standard specifications are also of size $2^{\Omega(n)}$. Typically, the complexity of solving a problem is measured in terms of the size of the specifications of the problem's instances. This suggests that complexity of problems can be different depending on whether the instances are specified periodically, or are specified standardly. That is indeed true. For example, assuming NP$\neq$PSPACE, the 3-coloring problem for graphs, is NP-complete when graphs are specified by standard specifications such as adjacency matrices or adjacency lists [191]. On the other hand, it is PSPACE-complete when graphs are specified by the *one-dimensional infinite periodic specifications* of Orlin [419]. In contrast, however, the 2-coloring problem for graphs is solvable in polynomial time, *even* when instances are specified by one-dimensional infinite periodic specifications. Such results lead us to investigate the complexity and efficient approximability of solving graph theoretic, combinatorial, and algebraic problems, when instances are periodically specified.

In this chapter, for demonstration purposes, we focus mainly on periodically specified constraint satisfaction problems. Previously, constraint satisfaction problems with instances specified using standard specifications have been used to model a number of problems in such areas as automated reasoning, computer-aided design [219], computer-aided manufacturing [220], machine vision [220], database, robotics, integrated circuit design [219, 220], computer architecture, and computer network design. See Gu et al. [220] for a recent survey. In addition, constraint satisfaction problems have served as a rich collection of base problems, for proving NP-hardness, #P-hardness, APX-hardness, and a number of similar properties for numerous combinatorial problems (see Garey and Johnson [191], Schaefer [455], and Papadimitriou [423]). Here we outline how, analogously, periodically specified constraint satisfaction problems are useful in modeling problems arising in practical applications and serve as base problems for proving both easiness and hardness results for periodically specified combinatorial, logical, and algebraic problems. The results outlined here enable the development of a *predictive* complexity theory for periodically specified problems (section 7).

There are two main reasons why a discussion of periodically specified problems is of interest in the context of the relationship between computational complexity and statistical physics. First, periodically specified problems are an algebraic generalization of tiling problems (see section 8) and thus provide a natural parametric class of lattice-like structured problem instances. Lattice-like structured problems have been a topic of active research by physicists and computer scientists in the context of designing "hard" instances for heuristics solving satisfiability and graph problems [6]. Second, as we discuss in section 9, it is possible to define random periodically specified graphs and formulas. Such instances are locally random but globally structured, and provide parametrized classes of random finite and infinite satisfiability and graph problems. Random graph and satisfiability problems and questions related to their phase transitions have been an active topic of recent research, as seen throughout this volume. Random periodically specified satisfiability and graph problems are introduced here in the hope that their study will provide interesting insights into the phase transitions associated with combinatorial problems.

The rest of the chapter is organized as follows. Section 2 consists of examples illustrating how periodic specifications can *naturally* model a number of realistic problems. Section 3 outlines the basic definitions of periodically specified graphs and formulas, as well as simple variants of the main formalism. Section 4 consists of several broader extensions of the basic formalism of periodic specifications and the objects they specify. We also illustrate several situations where these extensions are likely to occur. Section 5 contains a brief description of the techniques developed for obtaining both easiness and hardness results for periodically specified problems. We also argue how these techniques form the basis for developing a *predictive complexity theory* for periodically specified problems: informally, we illustrate that many reductions between standardly specified problem instances can be translated mechanically into efficient reductions for the corresponding periodically specified problems. Section 7 outlines the complexity theoretic implications of the general results for periodically specified problems. Section 8 argues that periodically specified constraint satisfaction problems can be used as alternatives to tiling problems for proving bounds on complexity. Finally, section 9 presents concluding remarks and directions for future work.

2   MOTIVATION

Formally, a *one-dimensional finite periodic graph specification* $\Gamma(G(V,E),M)$ consists of (1) a finite directed labeled graph $G(V,E)$ called the *static graph* of the specification, each of whose edges is labeled by a non-negative integer, together with (2) a non-negative integer M. The finite directed graph $G^M(V^M, E^M)$ *specified* by $\Gamma(G(V,E),M)$ is defined as follows. V^M consists of $M+1$ distinct copies of each vertex $v \in V$, denoted $v_0, \ldots, v_M$, respectively. E^M consists of $M - l + 1$ distinct copies of each edge $(u,v) \in E$ labeled with $l \leq M$, namely (u_r, v_{r+l}) for all $0 \leq r \leq M - l$. M is called the range or the span of the specification. A *k-dimensional periodic graph specification* is defined analogously for $k \geq 2$, except now all edges are labeled by k-tuples of non-negative integers, M is a k-tuple of non-negative integers. Examples of a *one-* and a *two-dimensional periodic graph specification* and the graphs they specify are given in figures 1 and 2. These concepts can be extended to define 1-, 2- and k-dimensional periodic graph specifications that are *infinite* in some of their k-dimensions. They can also be extended quite naturally to define one-, two- and k-dimensional periodically specified formulas and systems of equations/constraints. See section 3 for formal definitions.

Periodically specified graphs and logical formulas occur naturally when modeling practical problems in VLSI design, transportation science, and program optimization. We discuss four examples that illustrate the range of applications. Many others can be found [104, 244, 260, 302, 358, 377, 379, 380, 419, 420, 421, 423, 437].

- *Routing.* The *tramp steamer problem* is discussed by Orlin [420]. Consider a steamer that visits n distinct ports. Traveling from port u to port v takes t_{uv} days and earns a profit of p_{uv} dollars, and both the transit time and the profit are independent of the starting time for the trip. The objective is to determine an infinite-horizon tour that maximizes the average daily profit. The static graph has n nodes, one for each port, and for each pair u, v of distinct nodes there is an arc with transit time t_{uv} and unit cost $-p_{uv}$. The upper and lower bounds on arc flows are 1 and 0, respectively, and the throughput is restricted to 1, representing the steamer. Formulating the problem with this static network, Dantzig, Blattner, and Rao [115] observed that each basic solution to the tramp steamer problem is a flow around a *circuit*, which is a simple directed cycle. Each circuit induces an infinite-horizon tour. Ports are traveled in the order that they appear on the circuit and the average daily cost is the ratio of the cost of traveling the circuit to the transit time. Thus, an optimal circuit has the minimum cost-to-time ratio and induces an optimal tour.

- *Network scheduling and dynamic network flows.* Applications of dynamic network flow problems arise when one wishes to model transit time on edges. The following example is from Hoppe and Tardos [260]. We are given a directed graph $G(V, E)$ with sources, sinks, non-negative edge capacities c_{uv} and transit times t_{uv} for each edge $(u, v) \in E$. Time is assumed discrete here. In a feasible dynamic flow, at most c_{uv} units of flow can enter edge (u, v) at each integer time step. The flow leaving u along edge (u, v) at time θ reaches the other endpoint v at time $\theta + t_{uv}$. For example, an edge with capacity 2 and transit time 3 can accept 2 units of flow at any given time step, for a total of up to 6 units of flow on the edge at any time. The *quickest transshipment* problem is defined by a dynamic network with a set of sources and sinks; each source has a specified supply of flow and each sink has a specified demand, with the standard assumption that total supply equals total demand. The problem is to find a way to schedule the flow so that each source and sink sends and receives the specified amount of flow in a *minimum amount of time.*

The problem of finding a feasible and quickest dynamic flow in dynamic networks reduces to finding "usual static flow" in time-expanded graphs, following the periodic specification above. Formally, for a given time horizon T, we construct a time expanded network $G(T) = (V(T), E(T))$ as follows: Each vertex

$v \in V$ has $T + 1$ copies in $V(T)$, denoted by $v(i)$, $0 \leq i \leq T$. Each edge $(u, v) \in E$ has $T - t_{uv} + 1$ copies in $E(T)$, each with capacity c_{uv} and denoted by $(u(\theta), v(\theta + t_{uv}))$ with the provision that such edges exist if both end points are within the time horizon bounds, that is, $0 \leq \theta \leq T - t_{uv}$. In addition, we add holdover edges $(u(\theta), v(\theta + 1))$ with infinite capacity, representing flow that remains at a given node over a time step. An *infinite-horizon dynamic flow* is a static flow in the infinite time-expanded dynamic network. Note that the time-expanded dynamic network is essentially a periodically specified graph with additional holdover edges, and the infinite version corresponds similarly to an infinite periodically specified graph. The single-source and single-sink version of the problem was originally defined by Ford and Fulkerson [164]; the work of Hoppe and Tardos extends it to the multi-source and multi-sink case. Note that as specified, since the numbers are given in binary, we have infinite graphs in which the end points of an edge can be exponentially far apart in time. Recently, Fleischer [161] gave a faster algorithm for the quickest transshipment problem when the transit times are zero. Periodic specifications that allow us to specify such "long edges" are called *wide specifications.*

The quickest transshipment problem and its variants have a number of applications. One such application is to find the quickest way to evacuate a building in emergencies. Another application arises in network scheduling problems where there is a transit cost for moving jobs from one processor to another: the goal is to minimize the make-span of the schedule. See Hoppe and Tardos [260] for a detailed discussion of these applications. A related problem that can be cast in much the same terms is to find the *quickest path in a temporal network*; that is, the fastest way to reach a destination from a source when travel times on edges change over time.

- *Phase space properties of discrete dynamical systems.* One-dimensional cellular automata consist of a sets of vertices placed on a one-dimensional grid. Each vertex has an associated Boolean function that depends on the Boolean values associated with adjacent vertices. The system evolves synchronously: at each time step, the automata corresponding to nodes synchronously update their state using the Boolean transition function that takes as input the values stored at the vertex and at its neighbors. A two-dimensional periodic specification can easily be seen to represent the dynamic changes in the configuration of finite one-dimensional cellular automata over time, where the second dimension represents time [528]. Using this representation, the configuration reachability problem for a finite one-dimensional cellular automata is simply the circuit value problem for periodically specified circuits. Thus, periodic specifications provide a succinct method for representing the phase spaces of cellular automata and finite discrete dynamical systems.

- *Parallel programming.* The following problem was introduced in Iwano and Steiglitz [276]; more efficient algorithms were given by Kosaraju and Sullivan [347], and by Cohen and Megiddo [104]. An essentially similar problem was first considered by Karp, Miller, and Winograd [302]. The problem arises in the implementation of regular iterative algorithms on systolic arrays. We are given n functions $F_1, F_2, \ldots, F_n$, on the k-dimensional integer lattice defined recursively as follows:

$$F_u(z) = \psi_u(F_1(z - w_{u1}), \ldots, F_n(z - w_{un})).$$

Here, the w_{uv}'s are integer vectors. In order for the functions to be well-defined it is necessary and sufficient that no cycle have a total vector weight that is non-negative. This fundamental problem also arises while implementing simulations on parallel computers. A closely related problem arises in the context of the design of memory-efficient simulations of iterative programs consisting of **for** loops [243, 244, 358]. We can model the problem as follows. We have a static graph consisting of n vertices, one corresponding to each function. Each directed edge (u, v) has an integer weight w_{uv} on it, denoting the dependency of $F_u(z)$ on $F_v(z - w_{uv})$. The expanded graph is constructed by placing a copy of each vertex in the static graph at the lattice point in $\mathbb{N}^k$. The vertex corresponding to F_u at lattice point z is connected by a directed edge to the vertex corresponding to F_v at lattice point $(z - w_{uv})$. The problem is to find if the *expanded infinite graph* has a directed cycle. Note that we seek to find a cycle in the expanded graph as opposed to the static graph. Cohen and Megiddo give a strongly polynomial algorithm for detecting cycles in such expanded infinite graphs. Note also that w_{uv} is given in binary: this makes the problem substantially harder, since as in the dynamic network flow problem, the vertices of a cycle can now span time periods that are exponentially apart.

3   PRELIMINARY DEFINITIONS

Basic definitions are used in algebra, graph theory, computational complexity, dynamical systems, and approximation algorithms [23, 54, 373, 423, 447, 525, 528]. We have already defined *one-dimensional finite periodic graph specifications* and the *finite graphs* they specify. Here we discuss related concepts that yield variant periodic specifications.

3.1   PERIODICALLY SPECIFIED GRAPHS

Definition 1. *Let the static graph* $G(V, E)$ *be a finite undirected graph such that each edge* (u, v) *has an associated non-negative integer weight* t_{uv}. *The two-way infinite graph* $G^{\mathbb{Z}}(V^{\mathbb{Z}}, E^{\mathbb{Z}})$ *is defined as follows.* $V^{\mathbb{Z}}$ *and* $E^{\mathbb{Z}}$ *are multiple copies of the vertex and edge set:*

1. $V^{\mathbb{Z}} = \{v(i) \mid v \in V \text{ and } i \in \mathbb{Z}\}$

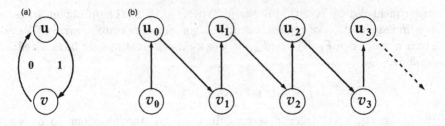

FIGURE 1   (a) The static graph with one-dimensional integer vectors associated with each edge. (b) Part of the one-way infinite graph it represents.

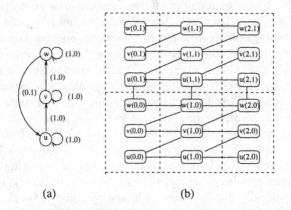

(a)   (b)

FIGURE 2   (a) The static graph with two-dimensional integer vectors associated with each edge. (b) The graph $G^{2,1}$ specified by $\Gamma(G, 10, 01)$.

2. $E^{\mathbb{Z}} = \{(u(i), v(i + t_{uv})) \mid (u, v) \in E$, t_{uv} *is the weight associated with the edge* (u, v) *and* $i \in \mathbb{Z}\}$

A one-dimensional two-way infinite periodic specification, or $(\mathbb{Z})$W-*specification, is given by* $\Gamma(G(V, E))$ *and specifies the graph* $G^{\mathbb{Z}}(V^{\mathbb{Z}}, E^{\mathbb{Z}})$.

Γ *is said to be a* narrow *specification, or* $(\mathbb{Z})$N-*specification, if* $\forall (u, v) \in E$, $t_{uv} \in \{0, 1\}$. *This implies that* $\forall (u(p), v(q)) \in E^{\mathbb{Z}}$, $|p - q| \leq 1$. ∎

Note that if we replace $\mathbb{Z}$ by $\mathbb{N}$ in definition 1, we obtain *one-way* infinite periodic specifications and the graphs they define. It may be useful to imagine a narrow periodically specified graph $G^{\mathbb{Z}}$ as being obtained by placing a copy of the vertex set V at each lattice point on the x-axis (or the timeline), and joining vertices placed on neighboring lattice (time) points in the manner specified by the edges in E.

Definition 2. *Let $G(V, E)$ denote a static graph. Let $G^{\mathbb{Z}}(V^{\mathbb{Z}}, E^{\mathbb{Z}})$ denote the two-way infinite $(\mathbb{Z})$N-specified graph as in definition 1. Let $M \geq 0$ be an integer specified using binary numerals. Let $G^M(V^M, E^M)$ be the subgraph of $G^{\mathbb{Z}}(V^{\mathbb{Z}}, E^{\mathbb{Z}})$ induced by the vertices $V^M = \{v(i)|v \in V \text{ and } 0 \leq i \leq M\}$. A one-dimensional finite periodic specification, or (B)N-specification, is given by $\Gamma(G(V, E), M)$ and specifies the graph G^M.* ∎

It is important to observe that M is specified using binary notation. If we use unary notation, then we denote such specifications as (U)N-specifications. The size of the (B)N-specification $\Gamma(G(V, E), M)$ is given by $\text{size}(\Gamma) = |V| + |E| + \text{bits}(M)$, where $\text{bits}(M)$ is the number of bits in the numeral M. (Note that in the rest of the chapter we use M to denote both integer and binary representation; its intended meaning will be clear from context.) An example of a periodic specification and the associated graph appears in figure 1.

It is easy to extend the definition above to define two-dimensional periodic specifications and associated graphs. As before we have a static graph, but now for each edge (u, v) we have a two-dimensional label (l, b). The *two-dimensional four-way infinite periodically specified graph* $G^{\mathbb{Z},\mathbb{Z}}(V^{\mathbb{Z},\mathbb{Z}}, E^{\mathbb{Z},\mathbb{Z}})$ is defined as follows: $V^{\mathbb{Z},\mathbb{Z}} = \{v(i, j) \mid v \in V \text{ and } i, j \in \mathbb{Z}\}$, and $E^{\mathbb{Z},\mathbb{Z}} = \{(u(i, j), v(i + l, j + b) \mid (u, v) \in E(i, j) \text{ and } i, j \in \mathbb{Z}\}$. For narrow periodic specifications this is called the $(\mathbb{Z},\mathbb{Z})$N-specification. For a non-negative integer vector (M, N), the *two-dimensional finite periodically specified graph* $G^{M,N}(V^{M,N}, E^{M,N})$ is the subgraph of $G^{\mathbb{Z},\mathbb{Z}}(V^{\mathbb{Z},\mathbb{Z}}, E^{\mathbb{Z},\mathbb{Z}})$ induced by the vertices $V^{M,N} = \{v(i, j) \mid v \in V \text{ and } 0 \leq i \leq M, 0 \leq j \leq N\}$. As mentioned previously, the method of representation used to specify M and N yields various kinds of specifications: when they are binary numerals, this results in the (B,B)N-specification. An example of such a periodic specification and its associated graph appears in figure 2. It is easy to extend the definitions above to obtain $G^{N,\mathbb{Z}}(V^{N,\mathbb{Z}}, E^{N,\mathbb{Z}})$, $G^{\mathbb{Z},N}(V^{\mathbb{Z},N}, E^{\mathbb{Z},N})$, and $G^{N,N}(V^{N,N}, E^{N,N})$ as well. Similarly, one can define variants where one of the dimensions is finite while the other dimension is infinite. For example $G^{M,\mathbb{Z}}(V^{M,\mathbb{Z}}, E^{M,\mathbb{Z}})$, is a graph in which the x-dimension has a span represented by M while the graph is infinite in both directions along the y-axis.

Periodically specified logical formulas can be defined in a similar manner. An example of periodically specified Boolean formulas is as follows.

Example 1. *Let $U = \{x(t), x(t+1), y(t), y(t+1), z(t), z(t+1)\}$ be a set of static variables. Let C be a set of static clauses given by $[x(t) \vee \overline{y}(t) \vee z(t)] \wedge [\overline{x}(t + 1) \vee z(t)] \wedge [z(t + 1) \vee y(t)]$. Let $F = (U, C, 3)$ be a (B)N-specification. Then F specifies the 3-CNF formula $F^3(U^3, C^3)$ given by*

$$\left([x(0) \vee \overline{y}(0) \vee z(0)] \wedge [\overline{x}(1) \vee z(0)] \wedge [z(1) \vee y(0)] \right) \bigwedge$$

$$\left([x(1) \vee \overline{y}(1) \vee z(1)] \wedge [\overline{x}(2) \vee z(1)] \wedge [z(2) \vee y(1)] \right) \bigwedge$$

$$\left(\, [x(2) \vee \overline{y}(2) \vee z(2)] \wedge [\overline{x}(3) \vee z(2)] \wedge [z(3) \vee y(2)] \, \right) \bigwedge$$
$$[x(3) \vee \overline{y}(3) \vee z(3)] \, . \qquad \blacksquare$$

It is easy to see that the basic formalism is quite rich: one can define many different combinatorial objects, including graphs, logical formulas, systems of equations, and inequalities.

3.2   TYPES OF PERIODIC SPECIFICATIONS

Different kinds of periodic specifications can be obtained either by changing the basic definition of the specifications or the algorithm used to construct the expanded object. We discuss this briefly.

- *Dimension.* The number of dimensions in which the expansion is carried out can be varied, e.g., 1-, 2-, or k-dimensions.
- *Finite vs. infinite object.* For finite objects, we can specify the bounds either in *unary* (U) or *binary* (B) notation for specifying the range M and N. For infinite objects we have two options: one-way infinite objects, represented by natural numbers $\mathbb{N}$, or two-way infinite objects, represented by integers $\mathbb{Z}$. Note that one dimension can be finite while another is infinite. If an object is infinite, it can be infinite in any of its dimensions.
- *Narrow vs. wide specification.* Most generally, in the case of narrow specifications, the weights on the edges of the static graph (or the difference between the indices of the static variables) are specified in unary. For wide specifications, they are specified in binary. We denote narrow specifications with the letter N, and wide specifications with the letter W. When we have more than one dimension, edges for certain dimensions may be specified in unary and for others in binary. In the case of narrow specifications, intuitively, vertices having an edge are not too far apart (in terms of the distance in index space). For wide specifications, two vertices that are exponentially far apart can have an edge between them.
- *Boundary conditions.* We can allow initial or final boundary conditions— explicit assignments to the variables at the beginning or at the end. In case we have more than 1-dimension, we could allow boundary conditions for a subset of the dimensions. We use the suffix (BC) to denote a specification with boundary conditions. As described here, the concept only applies to Boolean formulas: an extension to graphs is possible, but more problem-dependent.

In the interest of simplifying notation, we specify the dimension and the finite vs. infinite nature implicitly. Note that in the computational complexity literature, these are often specified explicitly with a prefix such as 1-, 2-, to indicate dimension followed by an F or I to indicate finite or infinite. A letter P often appears as well to indicate that the specification is periodic. Thus, for example, the $(\mathbb{Z},\mathbb{Z})$N-specification is more fully written out as 2-I$(\mathbb{Z},\mathbb{Z})$PN, and the (B)N-specification as 1-F(B)PN.

Technically, some of the variants discussed above refer to specifications while others refer to the algorithms used to construct the expanded object. In other words, we are talking about both the specification (syntax) and the specified object (semantics). This distinction is important, although for the most part, it can be understood from the context. We omit the formal definitions of these extensions.

Let Π be a problem whose instances are specified using one of the standard specifications in the literature. For example, instances of CNF satisfiability problems are specified by CNF formulas and by sets of clauses, each clause being a set of literals. Let α be one of the periodic specifications. Then we use α-Π to denote the problem Π when instances are specified using periodic specification α. For example, one-dimensional finite narrow periodic 3-SAT (denoted by (B)N-3-SAT, or more fully as 1-F(B)PN-3-SAT) is the problem of determining if a one-dimensional finite narrow periodically specified 3-CNF formula is satisfiable.

4   EXTENSION OF THE BASIC FORMALISM

The extensions outlined in section 3.2 are straightforward. We now discuss four other extensions that are somewhat less straightforward. The first extension concerns periodically specified constraint satisfaction problems, the second describes a different algorithm for constructing the expanded graph, the third concerns how to define satisfiable formulas, and the fourth describes how to define quantified formulas using periodic specifications. Note that these extensions change one or more of the basic elements used to define periodic specifications and the associated graphs, formulas or system of equations.

4.1   PERIODICALLY SPECIFIED CONSTRAINT SATISFACTION · PROBLEMS

Let D be an arbitrary nonempty set (not necessarily finite); C a finite set of constant symbols denoting elements of D; and S an arbitrary finite set of finite-arity relations on D. An S-clause is a relation in S applied to variables on elements in D. An S-formula is a finite nonempty conjunction of S-clauses. We denote the problem of determining the satisfiability of finite conjunctions of S-clauses by SAT(S). The corresponding problems including (B)N-SAT(S) and (B,B)N-SAT(S) are defined analogously. We give a simple example to illustrate the one-dimensional case.

Example 2. *Let* $D = \{0,1\}$*, i.e., we have a Boolean domain. Let* $S = \{\mathrm{XOR}(\alpha,\beta),$ $\mathrm{XNOR}(\alpha,\gamma)\}$*, be the set of relations on* D*, where* $\mathrm{XNOR}(\alpha,\beta) \equiv \mathrm{NOTXOR}(\alpha,\beta)$*.*

Let $V = \{w, x, y, z\}$ be the set of variables and the S-clauses be given by $P = \mathrm{XOR}(x, y)$, $\mathrm{XOR}(\overline{w}, y)$ and $\mathrm{XNOR}(y, z)$. Let $F(V, P)$ be an instance of the $\mathrm{SAT}(S)$ problem given by

$$F(V, P) = \mathrm{XOR}(x, y) \wedge \mathrm{XOR}(\overline{w}, y) \wedge \mathrm{XNOR}(y, \overline{z})\,.$$

Then F is TRUE with $x = 0, y = 1, z = 0, w = 1$. Now let $U = \{w, x, y, z\}$ be a set of static variables and $H = (U, C, 2)$ an instance of (B)N-SAT(S) with the set S above, where C is given as

$$C = \mathrm{XOR}(x(0), y(1)) \wedge \mathrm{XOR}(\overline{w}(0), y(0)) \wedge \mathrm{XNOR}(y(0), \overline{z}(1))\,.$$

Then $H^2(U^2, C^2)$ is the expanded $\mathrm{SAT}(S)$ formula given by

$$\left[\, \mathrm{XOR}(x(0), y(1)) \wedge \mathrm{XOR}(\overline{w}(0), y(0)) \wedge \mathrm{XNOR}(y(0), \overline{z}(1)) \,\right] \bigwedge$$

$$\left[\, \mathrm{XOR}(x(1), y(2)) \wedge \mathrm{XOR}(\overline{w}(1), y(1)) \wedge \mathrm{XNOR}(y(1), \overline{z}(2)) \,\right] \bigwedge$$

$$\mathrm{XOR}(\overline{w}(2), y(2))\,. \quad \blacksquare$$

4.2 RULES FOR CONSTRUCTING EXPANDED GRAPHS

Our original definition of time expanded graphs used certain specific semantics for interpreting the meaning of edge weights. There are other ways to construct expanded networks. We illustrate this via an example in epidemiology.

The *contact graph* is constructed as follows. Let V be a set representing a population; consider a complete graph G on V. Each edge $e \in E$ of G consist of a list of time intervals $L_1^e, L_2^e, \ldots$. Each $L_i^e = [a_i^e, b_i^e]$, where a_i and b_i are integers and we assume that $a_i^e > b_{i-1}^e$. The semantics of the lists are simple: they give the time ranges when the two people were in contact. These graphs can model certain time-varying phenomena. Let us first consider a simple version of this [305].

Definition 3. *A temporal network is an undirected graph $G(V, E)$ in which each edge has a time label $\lambda(e)$ representing the time when the two end nodes of the edge come in contact (or communicate). In general, each edge can have multiple labels capturing the fact that the nodes can come in contact more than once. A path P in G is time respecting if the labels on the edges of the path are non-decreasing.* $\blacksquare$

A time-expanded temporal network is constructed as follows. Assume for the present purposes that we have only one label per edge, with $\lambda_{min}(e)$ being the minimum value and $\lambda_{max}(e)$ the maximum value. A copy of the vertices in G are placed at each discrete time step t between $\lambda_{min}(e)$ and λ_{max}. A copy of the node $v(t)$ is joined to $v(t + 1)$ by a directed edge. An edge (u, v) in G with label

λ is replaced by directed edges between the copies of the vertices at time λ, that is, between $(v(\lambda), w(\lambda))$ and $(w(\lambda), v(\lambda))$. Finding a critical path from $v(t)$ to $w(t + x)$ is merely finding a path in this time-expanded network between these two nodes. This representation can also be extended easily to the case in which edges have multiple labels. A more interesting situation arises when a person is infected by a communicable disease at time t and then becomes non-contagious at some other time $t + x$. This can also be represented quite easily using the formalism discussed above.

Note the difference between how an edge is added to time-expanded temporal networks and to expanded periodic graphs. The representation above is used commonly for routing in networks with time-dependent edge delay functions. The basic idea is quite general: it allows us to define rules for specifying how to add edges in the temporal networks on the basis of the static network. Also note that the procedure for constructing time-expanded temporal networks can be combined with the procedure for constructing expanded periodic networks.

4.3 SEMANTICS OF SATISFACTION

Recall that a periodically specified CNF formula was said to be satisfiable if and only if all the clauses in the expanded formula can be made TRUE. In other words, $F^M(U^M, C^M)$ is said to be satisfiable iff $\forall i, 0 \le i \le M$, $C(i)$ is satisfiable. This suggests a generalization allowing us to write a quantified formula consisting of i and basic integer inequalities. For instance, we could say that $F^M(U^M, C^M)$ is satisfiable iff

$$\forall i, \ L \le i \le U, \ C(i) \ \text{is satisfiable}.$$

Such an extension lets us, in a natural way, relate periodic satisfiability problems to satisfiability problems for temporal logics [19, 476] and, in general, to reasoning about any temporal phenomenon, such as questions in epidemic modeling and ad hoc wireless networks.

4.4 PERIODICALLY SPECIFIED QUANTIFIED FORMULAS

As a final extension, let us consider periodic Boolean formulas where not all the variables necessarily are existentially quantified. Quantified formulas have been well studied in the literature. We consider periodically specified quantified formulas. We have a static formula as before, but now each variable template used in the static clause is either existentially or universally quantified. The semantics we use in our expanded formula are that all copies of the variable have the same associated quantifier. For example, let F be a static formula given by

$$F = \forall x \ \exists y \ \forall z \ ([x(0) \vee y(1)] \wedge [y(0) \vee \overline{z}(0)]) .$$

Then the expanded formula $F^{\mathbb{N}}(U^{\mathbb{N}}, C^{\mathbb{N}})$ is given by:

$$\forall x(0), \forall x(1) \ldots \exists y(0), \exists y(1) \ldots \forall z(0), \forall z(1) \ldots \bigwedge_{t \in \mathbb{N}} ([x(t) \vee y(t+1)] \wedge [y(t) \vee \overline{z}(t)]) .$$

5 TECHNIQUES: HARDNESS AND EASINESS RESULTS

We now discuss how to combine four concepts: (i) local transformations (possibly augmented with fixed size *enforcers*), (ii) relational representability, (iii) *simultaneous* reductions based on local transformation, and (iv) *lifting* of simultaneous reductions based on local transformation to characterize relative complexities, or efficient approximability of various periodically specified problems. In conjunction with certain translation results (algorithms to transform periodic specifications of logical formulas into other succinct specifications of the same logical formula, up to a renaming), discussed in Hunt et al. [266] and Marathe et al. [378, 377], we also get as a corollary unified complexity results for problems specified using other succinct specifications. We discuss each of these techniques in some detail below. The ideas form a first step towards building a predictive complexity theory of periodically specified problems. We explain this further in subsequent sections. The theory is similar in spirit to very general results presented by Bálcazar, Lozano, and Toran [26] on the complexity of problems when they are encoded using circuits. However, the circuit model is a complexity theoretic model. The approach does not naturally model real-life problems, and the corresponding results do not hold for problems specified using periodic specifications. See Marathe et al. [378, 379, 380] for additional discussion on this topic.

5.1 LOCAL TRANSFORMATIONS

Reductions by local transformation have been used extensively in the literature (see Garey and Johnson [191]). The first step in formalizing this concept is to separate the concept of replacement from that of reduction. Transformation using local replacement constructs target instances from source instances by replacing each object (clause/variable in a formula) by a collection of objects (conjunction of clauses) in the target instance. A schematic diagram of this is shown in figure 3.

For the purposes of this chapter, it suffices to observe that the *local* transformations from a problem $\mathrm{SAT}(S)$ to a problem $\mathrm{SAT}(T)$ used here are of the following two kinds:

1. *Simple-local (SL) transformations.* Let $F = C_1 \wedge \ldots \wedge C_n$ where the C_i are S-clauses. Then, the T-formula $F' = \mathcal{R}(F)$ equals $C_1' \wedge \ldots \wedge C_n'$, such that the following holds:
   (a) Each C_i' is a fixed conjunction of T-clauses depending only upon the relation C_i.
   (b) The variables of C_i' are the variables of C_i plus *new* variables *local* to the clauses of C_i'.
2. *Simple-local-enforcer (SLE) transformations.* Let F be defined as in 1 immediately above. Then, the T-formula $F' = \mathcal{R}(F)$ equals $C_0' \wedge C_1' \wedge \ldots \wedge C_n'$, such that the following holds:

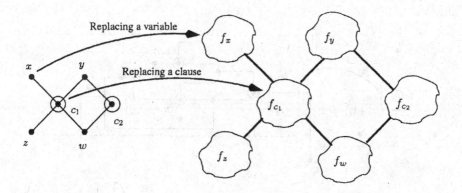

FIGURE 3   A schematic diagram illustrating the concept of local transformation.

(a) C_0' is a fixed T-formula, called the *enforcer*, the variables of which are called the *enforcer* variables.

(b) The clauses C_i' satisfy 1a and 1b, *except* that their variables can include *enforcer* variables.

For a simple demonstration of a local transformation, we first define the following problems. 1IN3-SAT is the problem of determining whether a 3-CNF formula has a satisfying assignment where exactly *one* literal in each clause is satisfied. 1EX3-SAT is the same problem but on a CNF formula in which each clause contains exactly 3 literals. We now give an example of an SL-transformation of 3-SAT to 1EX3-SAT.

Example 3.   *Consider a transformation of an instance F of 3-SAT to an instance F' of 1EX3-SAT. Each clause $C_j = (z_p \vee z_q \vee z_r)$ of F is transformed into a set of clauses C_j' of F' given by: $C_j' = \mathrm{EO}(z_p, u^j, v^j) \wedge \mathrm{EO}(\overline{z_q}, u^j, w^j) \wedge \mathrm{EO}(v^j, w^j, t^j) \wedge \mathrm{EO}(\overline{z_r}, v^j, x^j)$. $\mathrm{EO}(x, y, z)$ is a logical relation that takes the value of TRUE when exactly one of x, y, z is TRUE, and takes the value of FALSE otherwise. Here u^j, v^j, w^j, t^j and x^j are new variables local to C_j'. It is easy to see that this is an SL-transformation. The transformation is shown in figure 4.* ∎

Note that the definitions of SL- and SLE- transformations are fully syntactic in nature, since they do *not* require such transformations to be reductions. Essentially all the reductions discussed here are by SL- or SLE-transformations. Local replacements have a number of desirable properties. First, it is straightforward to show the following:

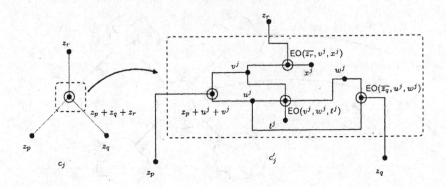

FIGURE 4   Figure showing how to replace a single clause in an instance of 3-SAT by a set of clauses to obtain an instance of 1IN3-SAT.

Proposition 5.1. *SL- and SLE-transformations are ultra-efficient, in the sense that they are reductions on multiple tape deterministic Turing machines that are simultaneously $O(n \log n)$ time-, linear size-, and $O(\log n)$ intermediate space-bounded.* ■

Second, they simultaneously preserve a number of both semantic and structural properties of instances. By structure of instances we mean the graph-theoretic structure as well as the structure of its specification. Third, they are extremely efficient in terms of the resources used, and preserve power and polynomial indices [487, 488].

5.2   RELATIONAL REPRESENTABILITY

For local replacement based transformations to be useful as reductions, we need the notion of relational representability. Let S and T be sets of relations/algebraic constraints on a common domain D. Relational representability formalizes the intuitive concept that the relations in S are *expressible* (or, extending the terminology from Schaefer [455], *representable*) by finite conjunctions of the relations in T. This is formalized in Definition 4 below:

Definition 4.

1. *We denote by $Rep(S)$ the set of all finite-arity relations on a non-empty set D logically equivalent to finite existentially quantified conjunctions of relations/algebraic constraints in S applied to variables.*
2. *We say that a set of relations S is representable by a set of relations/algebraic constraints T if and only if $S \subset Rep(T)$.*

Example 4. *Let* $S = \{XOR(\alpha, \beta)\}$ *and* $T = \{XNOR(\alpha, \beta)\}$. *Clearly,* $XOR(x, y) \equiv$ $XNOR(x, z) \wedge XNOR(y, \overline{z})$. *Thus each S-clause can be represented by a conjunction of T-clauses. The same holds for S-formulas. For example,*

$$XOR(x, y) \wedge XOR(x, w) \equiv XNOR(x, z) \wedge XNOR(y, \overline{z}) \wedge XNOR(x, z_1)$$
$$\wedge XNOR(w, \overline{z_1}). \quad \blacksquare$$

Variants of the concepts of definition 4 on the *relative representability* of ordered-pairs (S, T) of sets of *relations*, henceforth denoted collectively by *relational representability*, are well known, especially in mathematical logic. Previously in complexity theory, *relational representability* as used here and the individual constraint satisfaction problems studied have usually been restricted to finite sets S of finite-arity relations on *finite* sets D, generally the set $\{0, 1\}$. In contrast, the results discussed here apply to *both* finite and infinite domains and sets of relations/constraints.

Transformations that preserve a semantic property of interest are called reductions. To show this usually requires one to use some form of relational representability. For example, reductions that preserve decision complexity are simply called "reductions" in the literature. Reductions that preserve the number of solutions are known as parsimonious reductions. Several approximation-preserving reductions have been studied in the literature, including L-reductions that preserve polynomial time approximation schemes and A-reductions that preserve the approximation ratio. Given the concepts above, it is intuitively clear how to go about constructing reductions based on local transformations. The concept of relational representability must be modified when we wish to construct local transformations that also preserve other semantic properties. For example, number-preserving relational representability is a special form of relational representability that also preserves the number of solutions. The variants of relational representability used to construct A-reductions and L-reductions are a bit more subtle and do not necessarily have to be a decision-preserving transformation. Approximation-preserving versions of relational representability are also called *implementations* in Creignou [111] and Creignou et al. [113]. We will have more to say about this in the next section.

5.3 SIMULTANEOUS REDUCTIONS BASED ON LOCAL TRANSFORMATIONS

In general, it is easy to see that a local transformation preserving one type of semantic property does not necessarily preserve another type of semantic property. In fact, under standard complexity theoretic assumptions, designing a (local) transformation that simultaneously preserves more than one semantic property is not always possible. For example, consider the two widely studied problems 3-SAT and 2-SAT. The problem 3-SAT is NP-hard and 2-SAT is polynomial time solvable [191]. On the other hand, both MAX-3-SAT and

MAX-2-SAT are APX-hard, that is, unless P = NP they cannot have a poly-nomial time approximation scheme. Therefore, a natural question to ask in this context is: when can we design single transformations that are *simultaneously* decision-preserving, number-preserving, and approximation-preserving? We have found that for a large class of algebraic problems, it is indeed possible to devise such transformations: we call them *simultaneous reductions* [113, 267, 268, 310]. Moreover, most of these are based on local transformations. For example, a (parsimonious + A + L)-reduction is a reduction that is simultaneously a parsi-monious reduction, an A-reduction and an L-reduction. Simultaneous reductions have the advantage that they simultaneously preserve a variety of semantics and the structure of instances. By structure of instances we usually mean the variable-clause interaction graph structure and the structure of the specification used to specify the problem. The existence for a wide class of natural algebraic problems of simultaneous reductions based on local transformations is a bit surprising.

For example, we can show that there is a local transformation from the problem 3-SAT to the problem 1IN3-SAT that is simultaneously a (decision + parsimonious + A + L)-reduction. Consequently, using the known results on the complexity of 3-SAT and its variants, we *simultaneously* obtain the follow-ing: 1IN3-SAT is NP-hard, #-1IN3-SAT is #P-complete, MAX-1IN3-SAT is APX-complete and MAX-DONES-1IN3-SAT is MAX-Π_1-complete (MAX-DONES-1IN3-SAT is the problem of finding a satisfying 1IN3-SAT assignment maximizing the total number of variables set to TRUE). These constitute results on the complexity of 1IN3-SAT for standardly specified instances. As discussed in the next section, the transformation can be translated to obtain the rela-tive hardness of the periodically specified 1IN3-SAT problem and its variants, showing notably that one-dimensional finite narrow periodic 1IN3-SAT ((B)N-1IN3-SAT) is PSPACE-hard.

Simultaneous reductions based on local transformations induce natural equiv-alence classes of combinatorial problems. Obtaining general techniques showing when two problems are in the same equivalence class is an interesting direction for future research.

5.4   PUTTING IT ALL TOGETHER: LIFTING OF SIMULTANEOUS REDUCTIONS BASED ON LOCAL TRANSFORMATIONS

How does one prove complexity bounds for periodically specified problems? Our approach consists of two natural steps and builds on the concept of simultaneous reductions based on local replacement, as laid out in preceding sections.

First, by direct reductions from Turing machines, we characterize the com-plexity of a number of basic CNF satisfiability problems when specified peri-odically. The proof technique used is fairly generic; results characterizing the complexity of these satisfiability problems when the underlying periodic spec-ifications change can thus be obtained directly. For example, we prove that two-dimensional finite narrow periodic 3-SAT with explicit boundary conditions for both dimensions ((B,B)N(BC)-3-SAT) is NEXPTIME-complete. This proof together with a few simple observations shows that one-dimensional finite narrow periodic 3-SAT ((B)N-3-SAT) is PSPACE-complete, and that two-dimensional narrow periodic 3-SAT with one dimension finite and specified in binary and the other dimension two-way infinite in one direction ((B,$\mathbb{Z}$)N-3-SAT) is EXPSPACE-complete. A summary of our results for 3-SAT as well as for the two problems 3-HornSAT and CLIQUE appears later in table 2.

Second, we show that efficient reductions involving *local replacement* (possibly augmented with fixed-size enforcers) [191] including the problems 3-SAT, 1IN3-SAT, NAE3-SAT, to a problem Π can be extended to obtain efficient reductions from the problems α-3-SAT, α-1IN3-SAT, α-NAE3-SAT, to the problem α-Π. These problems include most of the basic problems in Karp [299], Garey and Johnson [191], as well as several basic P-complete problems [283]. An important property of our reductions is that they preserve the underlying specifications. *We note that the same reduction works when the specification α is changed,* thus avoiding the need for devising a new reduction for each result.

The idea, in fact, applies to simultaneous reductions based on local replacement. In other words, given a simultaneous reduction $\mathcal{R}$ from Π_1 to Π_2 that is based on local transformations, there is an efficient algorithm that takes as input $\mathcal{R}$ and an instance of the problems α-Π_1 and α-Π_2 (recall that α denotes a periodic specification), and constructs a transformation $\mathcal{R}'_\alpha$ such that $\mathcal{R}'_\alpha$ is an efficient reduction between α-Π_1 and α-Π_2. We call this *lifting* the reduction: transforming the static formulas into another static graph so as to obtain the needed correspondence between the expanded formulas. Lifting can be thought of as a compiler. It takes a known local transformation between two standardly specified problems and constructs a new transformation between their periodic counterparts in such a way that the semantics of transformation are preserved. Our idea, then, is to lift the known reduction from 3-SAT to problem Π when the instance is specified using standard specifications, and thus obtain a suitable reduction from α-3-SAT to the problem α-Π. In algebraic terms, the process can be seen in the form of the following *commutative diagram*:

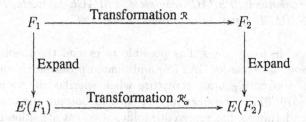

An example of this general technique of lifting is given below.

Theorem 5.1. *Let S and T be finite sets of finite-arity relations on a (possibly infinite) nonempty set D. Let α be one of the following periodic specifications:* (B)N, (ℤ)N, (B,B)N, (B)W, ...} *Then, if $S \subset Rep(T)$, the problem SAT(S) is reducible by a local replacement $\mathcal{R}$ to the problem SAT(T). Moreover, the transformation $\mathcal{R}$ can be extended to transformations $\mathcal{R}'_\alpha$ of the problems α-SAT(S) to α-SAT(T) such that the following hold:*

1. *Both transformations $\mathcal{R}$ and $\mathcal{R}'_\alpha$ are $O(n \log n)$ time-, $O(\log n)$ space-, and $O(n)$ size-bounded.*
2. *Both transformations $\mathcal{R}$ and $\mathcal{R}'_\alpha$ are decidable in parallel logarithmic time using only $O(n)$ processors.*
3. *Both transformations $\mathcal{R}$ and $\mathcal{R}'_\alpha$ preserve bandwidth, treewidth, and pathwidth of instances.*
4. *If the transformation $\mathcal{R}$ is a parsimonious reduction, then so is the transformation $\mathcal{R}'_\alpha$.*
5. *If the transformation $\mathcal{R}$ is an A- or L- reduction, then so is the transformation $\mathcal{R}'_\alpha$.*
6. *If the transformation $\mathcal{R}$ is a metric reduction, then so is the transformation $\mathcal{R}'_\alpha$.*
7. *If the transformation $\mathcal{R}$ preserves strong planarity, then so does the transformation $\mathcal{R}'_\alpha$.*

∎

As one example of theorem 5.1, we can show the following very general result. The theorem shows that as long as SAT(S), is NP-complete when instances are specified using standard specifications, it becomes hard for the appropriate complexity class when instances are specified using periodic specifications.

Theorem 5.2. *Let D be a finite nonempty set. Let S be a finite set of finite-arity relations on D such that $Rep(S)$ = Boolean Relations. Then, the problems* (B)N-SAT(S), (ℤ)N-SAT(S), (B,B)N-SAT(S), (B)W-SAT(S) *and* (ℕ,ℕ)N-SAT(S) *are, respectively, PSPACE-complete, PSPACE-complete, NEXPTIME-complete, EXPSPACE-complete and undecidable.* ∎

As an aside, in many cases it is possible to extend the theorem above to obtain *dichotomy theorems* for SAT(S) and related problems of equational satisfiability over a given algebraic structure when specified by variant periodic specifications [376, 377]. The discovery of such dichotomy theorems, for standardly specified formulas, has received significant recent attention in the literature [113, 310, 455].

Finally, we note that using the complexity theoretic results for periodic constraint satisfaction problems, we can characterize the complexity of a number of combinatorial problems when specified using periodic specifications.

6 TECHNIQUES: EASINESS RESULTS

Next, we discuss various methods by which easiness results can be obtained for periodically specified problems. We first consider methods by which exact polynomial time algorithms can be obtained for such problems. We then discuss how this idea can be extended to obtain approximation algorithms for "hard" problems.

6.1 POLYNOMIAL TIME ALGORITHMS

Almost all the polynomial time algorithms for periodically specified problems use the idea of periodic certificates. The basic idea is that if the given problem has a solution, then it has a periodic solution with a small period. Once this is established, the problem reduces to finding a solution on a polynomially sized instance that is a function of the periodicity of the solution. We explain the idea in the case of a specific kind of satisfiability problem.

A logical relation R is said to be *weakly negative* if $R(x_1, x_2, \ldots)$ is logically equivalent to some CNF formula having at most one unnegated variable in each conjunct. A weakly negative formula F is one in which each conjunct is weakly negative. The problem of deciding if a weakly negative formula is satisfiable is called *Horn satisfiability* or HornSAT: this problem has been discussed in chapter 9. The problem 3-HornSAT is the restriction of HornSAT to clauses containing no more than three literals. The problem $(\mathbb{Z})$N-HornSAT is the problem of deciding whether a one-dimensional two-way infinite periodically specified weakly negative formula is satisfiable. Here is a simple algorithm for solving the problem.

The algorithm works on the static formula F representing $F^{\mathbb{Z}}$ and is based on the following two observations. The first observation is that if there is a clause with only one literal, all copies of the corresponding variable must have the same value. For instance, if there is a clause consisting of the single literal $\overline{x_i(t+1)}$, then all copies of variable x_i have to be set to FALSE. The second observation is that after simplifying the set of clauses as much as possible on the basis of the first observation, every remaining clause has either no literals or more than one literal. Weak negativity implies that each clause with more than one literal contains at least one negative literal, so setting all remaining variables to FALSE will satisfy all such clauses. Since each simplification of the set of clauses based on the first observation assigns a value to a variable in the static formula that has not previously been assigned a value, the algorithm will terminate in polynomial time. When the algorithm terminates, either we are left with no clauses (in which case the formula is satisfiable by the discussion above) or we obtain a contradiction.

Note that if the expanded formula $F^{\mathbb{Z}}$ for the given instance F of $(\mathbb{Z})$N-HornSAT is satisfiable, then there exists a satisfying assignment assigning the same value to all copies of a given variable in the static formula.

Other researchers [103, 104, 243, 244, 260, 275, 276, 302, 338, 420, 421] have given efficient algorithms for solving problems such as determining strongly connected components, testing for existence of cycles, finding minimum cost paths between a pair of vertices, bipartiteness, planarity, quickest transshipment, and minimum cost spanning forests for periodically specified graphs. One particular problem that has received a lot of attention is that of detecting cycles in periodic graphs [104, 276, 302, 347]. Currently, the best known algorithm for this problem is by Cohen and Megiddo [104] and works even in the case of wide specifications. In Marathe et al. [377, 378], we gave polynomial time algorithms for various satisfiability problems when instances are specified using variant one-dimensional narrow periodic specifications. Interestingly, most of the results above are for one-dimensional periodically specified graphs. Höfting and Wanke [243, 244] and Wanke [516] have considered (finite/infinite) periodically specified (toroidal) graphs when the dimension in which the graph is replicated is given as part of the instance. In general, their results show that very simple problems become computationally intractable. On the other hand, certain problems are still solvable in polynomial time. As an example [243], polynomial time algorithms are obtained for solving path problems when the static graph is strongly connected or has a constant number of strongly connected components. The results rely crucially on the polynomial time solvability of solving linear Diophantine equations (integer linear equations with integral solutions). Finding polynomial time algorithms for two-dimensional narrow and one-dimensional wide periodically specified problems is an interesting direction for future research.

We finally discuss the notion of real-time certificates proposed by Orlin [419]. Here, instead of finding the solution for the complete expanded instance, we seek to find the solution for the first i periods in time that is polynomial in i and the instance representation. Unfortunately, as shown in Orlin [419], this does not make the problem any easier: all the problems for which he shows PSPACE-hardness results continue to be PSPACE-hard even when we wish to find a real-time solution.

6.2   APPROXIMATION ALGORITHMS

As we have seen, problems tend to become harder when specified succinctly using periodic specifications. Given the hardness results for solving the problems exactly when periodically specified, we investigate the existence of polynomial time approximation algorithms for these problems. We present a uniform approach for developing efficient approximation algorithms, as well as schemes for a number of optimization problems when specified using one- or two-dimensional finite, narrow, periodically specified problems. For the rest of the section, let α be one of the periodic specifications:

- (B)N: one-dimensional finite narrow periodic specifications,

- (B)N(BC): one-dimensional finite narrow periodic specifications with explicit boundary conditions,
- (B,B)N: two-dimensional finite narrow periodic specifications, and
- (B,B)N(BC): two-dimensional finite narrow periodic specifications with explicit boundary conditions in both dimensions.

It is important to understand what is meant by a *polynomial time approximation algorithm* for a problem Π, when Π's instances are specified by one of the periodic specifications α. We illustrate this by an example:

Example 5. *Consider the maximum independent set problem, when graphs are specified by (B,B)N-specifications. We provide efficient algorithms for the following versions of the approximate maximum independent set problem:*

1. Approximation problem: *compute the size of a near-maximum independent set in G.*
2. Query problem: *given any vertex v of G, determine whether v belongs to the approximate independent set so computed.*
3. Construction problem: *output a (B,B)N-specification of the set of vertices in the approximate independent set.*
4. Output problem: *output the approximate independent set computed.*

We require that algorithms for versions 1, 2, and 3 above run in time poly-nomial in the size of the (B,B)N-specification rather than in the size of the graph obtained by expanding the specification. The algorithm for version 4 should run in time polynomial (ideally linear) in the size of the expanded graph but use space which is polynomial (ideally linear) in the size of the periodic specification. ∎

The requirements above are a natural extension of the requirements imposed on approximation algorithms when instances are specified using standard specifications. This can be seen as follows for graph problems, and a similar argument holds for satisfiability problems. When instances are specified using standard specifications, the number of vertices is polynomial in the size of the description. Given this, any polynomial time algorithm to determine if a vertex v of G is in the approximate maximum independent set can be modified easily into a polynomial time algorithm that lists all the vertices of G in the approximate maximum independent set. For an optimization problem or a query problem, our algorithms use space and time that are low-level polynomials in the size (η) of the periodic specification and thus $O(\text{poly}\log\eta)$ in the size of the graph. Moreover, when we need to output, for example, the subset of vertices, or the subset of edges, corresponding to a vertex cover, or maximal matching, in the expanded graph, our algorithms take essentially the same time but substantially less (often exponentially less) space than algorithms that work directly on the expanded graph. It is important to design algorithms that work directly on the periodic specifications by exploiting the regular structure of the underlying graphs, because graphs resulting from expansions of given periodic descriptions are frequently too large to fit into the main memory of a computer. Hence, standard algorithms designed for flat graphs are impractical for periodically specified graphs.

We outline how to obtain approximation algorithms satisfying the performance requirements above for a number of problems, including problems Π in table 1, when instances are given by (B)N-, (B)N(BC)-, (B,B)N-, (B,B)N(BC)-specifications.

The basic technique consists of two main steps. First, by an extension of ideas in Baker [25] we show that for each fixed finite set S there is a polynomial time approximation algorithm as well as a scheme for planar instances (corresponding bipartite graphs are planar), for the problems MAX-SAT(S) specified periodically using one of the specifications mentioned earlier in the section. Next, we show that a number of important classes of problems, when specified periodically, can be reduced in an approximation-preserving way to appropriate problems MAX-SAT(S) specified using the same type of periodic specifications. We call these *structure preserving* L-reductions. This step uses the concept of simultaneous reductions outlined earlier.

Let us elaborate a bit more on the first step. The idea behind our approximation algorithms involves the conversion of solutions obtained from a *local* algorithm on small sub-grids to a solution of the *global* problem. The method of partial expansion involves the application of a divide and conquer algorithm iteratively by considering different subsets of the given graph, solving each subset with a local algorithm, constructing a global solution and finally choosing the best solution among these iterations as the solution to Π. The method can be seen as an extension of the shifting strategy devised by Baker [25] for finding efficient approximation algorithms to several combinatorial problems.

We illustrate the idea by discussing our polynomial time approximation scheme (PTAS) for the *maximum independent set* problem. Recall that an approximation algorithm for an optimization problem Π provides a *performance guarantee* of ρ if, for every instance I of Π, the value returned by the approximation algorithm is within a factor ρ of the optimal value for I. A PTAS for problem Π is a family of algorithms $\mathcal{F}$ such that for any fixed $\epsilon > 0$ there is a polynomial time algorithm $A \in \mathcal{F}$ that for all $I \in \Pi$ returns a solution within a factor $(1 + \epsilon)$ of the optimal value for I.

Consider a (B,B)N-specification of a graph G, and an integer $l > 1$. To begin with, for each i, $0 \le i \le l$, partition the graph G into k disjoint sets $G_1, \ldots, G_k$ by removing vertices with horizontal coordinates congruent to $i \bmod (l+1)$. For each subgraph G_p, $1 \le p \le k$, we find an independent set of size at least $l/l + 1$ times the optimal value of the independent set in G_p. The independent set for this partition is simply the union of independent sets for each of G_p. By an averaging argument, it follows that the partition yielding the largest solution value contains at least $(l/l + 1)^2 OPT(G)$ nodes, where $OPT(G)$ denotes the value of the maximum independent set in G. (For simplicity, we use a symbol to denote a set as well as its cardinality: the intended meaning will be clear from context.)

It is important to note that the size of the graph we are dealing with is, in general, exponential in the size of the specification. Hence, a naive application of the idea above will lead to algorithms that take an exponential amount of time. However, the regular structure of the graph allows us to solve the problems considered here in time polynomial in the size of the specification. The key observation is that for each iteration, although the total number of subproblems to be considered is exponential, they can be divided into a small number of equivalence classes. Moreover, it is easy to compute in polynomial time the number of elements in each equivalence class. Combining these two observations immediately yields the desired results.

Theorem 6.1. *For each fixed $l \geq 1$, and for each of the problems Π listed in table 1, the problem α-Π, has a polynomial time approximation algorithm with performance guarantee $(l + 1/l)^2 \cdot FBEST_\Pi$ and running time $O(RT_\Pi(l^2|G|))$. Here, $FBEST_\Pi$ denotes the best-known performance guarantee of an algorithm for the problem Π for non-succinctly specified instances, $RT_\Pi(n)$ denotes the running time of the algorithm with input size n guarantees the performance of $FBEST_\Pi$ for the problem Π and $|G|$ denotes the size of the specification.* ■

In fact, we can show that the theorem holds for most problems α-Π such that Π is in syntactic MAX-SNP.

As an example, using recent results in Goemans and Williamson [199], we find that for all $\epsilon > 0$, the problems (B,B)N-, (B,B)N(BC)-, (B,B)N- and (B)N-MAX-2-SAT have a PTAS that outputs solutions within a factor of $(1 + \epsilon)1.137$ of an optimal solution. As a corollary of theorem 6.1, using recent non-approximability results [21] we get the following:

Theorem 6.2. *For all the problems Π listed in table 1, the problems α-Π have a PTAS if and only if $P = NP$.* ■

A second result following from the proof of theorem 6.1 is as follows.

Theorem 6.3. *For all the problems Π listed in table 1, the problems α-Π have a PTAS when restricted to planar instances.* ■

We can show that many of these problems remain NEXPTIME-complete, even when restricted to planar instances.

The general approximation algorithms and schemes for the problems MAX-SAT(S) are an attempt to answer the fundamental question: which "hard" periodically specified optimization problems have efficient approximations? In this

TABLE 1 Performance guarantee results for optimization problems corresponding to problems specified using (B,B)N-specifications. All these problems can be shown to be **NEXPTIME**-hard using the method outlined in this chapter. Similar results hold for problems specified using (B,B)N(BC)-specifications. The symbol b denotes the degree bound and the symbol p denotes the maximum arity of a relation in S. Approximation results for the standard case for arbitrary and planar instances can be found in Ausiello et al. [23].

Problem	(B,B)N Specifications		Standard Specifications	
	Planar	Arbitrary	Planar	Arbitrary
MAX-3-SAT	$(\frac{l+1}{l})^3$	$(\frac{l+1}{l})^2 \cdot 4/3$	$(\frac{l+1}{l})$	$4/3$
MAX-SAT(S)	$(\frac{l+1}{l})^3$	$(\frac{l+1}{l})^2 \cdot 2^p$	$(\frac{l+1}{l})$	2^p
MIN-VERTEX-COVER	$(\frac{l+1}{l})^3$	$(\frac{l+1}{l})^2 \cdot 2$	$(\frac{l+1}{l})$	2
MAX-INDEPENDENT-SET	$(\frac{l+1}{l})^3$	$(\frac{l+1}{l})^2 \cdot b$	$(\frac{l+1}{l})$	b
MIN-DOMINATING-SET	$(\frac{l+1}{l})^3$	$(\frac{l+1}{l})^2 \cdot b$	$(\frac{l+1}{l})$	$\log b$
MAX-EDGE-DOMINATING-SET	$(\frac{l}{l-1})^3$		$(\frac{l}{l-1})$	2
MAX-PARTITION-INTO-TRIANGLES	$(\frac{l+1}{l})^3$	$(\frac{l+1}{l}) \cdot 3$	$(\frac{l+1}{l})$	3
MAX-H-MATCHING	$(\frac{l+1}{l})^3$	$(\frac{l+1}{l}) \cdot (|V_H|/2 + \epsilon)$	$(\frac{l+1}{l})$	$(|V_H|/2 + \epsilon)$
MAX-CUT	$(\frac{l+1}{l})^2$	$(\frac{l+1}{l}) \cdot 1.137$	polynomial	1.137

direction, the general theory developed here and discussed above provides a sufficient condition:

> Periodically specified graph and other optimization problems have an ϵ-approximation algorithm (or PTAS) when the semantics of the problem can be described by a SAT(S) formula in such a way that the formula interaction graph inherits the structure of the graph.

In recent years there has been significant interest [23, 113, 269, 309] in providing syntactic characterizations of optimization problems in an attempt to provide a uniform framework for solving such problems. Our results provide a syntactic (algebraic) class of problems, namely, MAX-SAT(S) whose closure under L-reductions and other appropriate approximation-preserving reductions define one such characterization for problems that have an ϵ-approximation (or PTAS). The algebraic model (characterization) is general enough to express the optimization version of: (i) the generalized satisfiability problems of Schaefer [455]; (ii) feasibility of systems of linear equations over a variety of algebraic structures; (iii) a class of nonlinear optimization problems; and (iv) several well-known graph theoretic problems. We refer the reader to Hunt et al. [269] and Khanna and Motwani [309] for more details.

The approximation algorithms have three desirable features: they are conceptually simple, they apply to large classes of problems Π and they apply to problems specified using any of the periodic specifications considered here. The polynomial time approximation algorithms for *natural* (as opposed to specifically constructed) NEXPTIME-hard problems yield a large class of problems for which there is a proven exponential—and possibly doubly exponential—gap between the time complexities of finding exact and approximate solutions. Previous non-approximability results have show that many optimization problems are NP-hard or PSPACE-hard to approximate beyond a certain factor. While those hardness results point out that it is *unlikely* in general to find "good" polynomial time approximation algorithms, the possibility is not ruled out. The results presented here, on the other hand, show a *provable* gap between approximation and decision versions of the problem. To see this, note that the decision problems are NEXPTIME-complete, and hence require *at least* 2^{cn} steps—and *possibly* $2^{2^{cn}}$ steps—to solve, assuming NEXPTIME $\neq$ DEXPTIME.

The study of approximation algorithms for NP-hard optimization problems has received a great deal of attention [23]. In contrast, efficient approximability of PSPACE-, NEXPTIME-hard problems has been considered only very recently. We refer the reader to Condon [106], Feigenbaum [154], and Marathe [376] for survey articles and Agarwal and Condon [10], Condon [107, 108], Hunt et al. [266], and Marathe et al. [379, 380] for related results. The NEXPTIME-hardness results for periodically specified problems show that the very regular structure of problem instances does *not* suffice to make problems easy. But the efficient approximation algorithms and schemes developed here show the following:

> The very regular structures of problem instances specified by the periodic specifications does make many of the basic problems approximable.

Interestingly, approximating many of the optimization problems considered here when instances are specified using *small circuit specifications* [26, 426] can be shown to be NEXPTIME-hard by extensions of the arguments in Arora et al. [21]. Thus, our results highlight an important difference between problems specified using multiple-dimension finite periodic specifications and small circuit specifications.

7 COMPLEXITY THEORETIC INSIGHTS

We briefly discuss certain complexity theoretic implications of the results summarized in the preceding sections. Additional discussion can be found in Hunt et al. [267, 268] and Marathe et al. [376, 380].

7.1 PREDICTIVE COMPLEXITY THEORY

Simultaneous local reductions and their lifts allow us to relate the computational complexities of variant combinatorial problems in a very strong sense. For example, we can show the following.

Theorem 7.1. *The problem* 3-SAT *is* (decision + parsimonious + A + L)*-reducible by local replacement to* EM-SAT, *the restriction of* 1EX3-SAT *to formulas having no negated literals. This together with the known results about* 3-SAT *and its periodically specified variants directly implies that:*

1. EM-SAT *is NP-complete*
2. MAX-EM-SAT *is APX-complete*
3. #-EM-SAT *is #P-complete*
4. *It is NP-hard to approximate* MAX-DONES-EM-SAT *beyond a factor* n^ϵ
5. (B)N-EM-SAT *is PSPACE-complete,* (B)W-EM-SAT *is EXPSPACE-complete and* (B,B)N-EM-SAT *is NEXPTIME-complete*
6. (B)N-MAX-EM-SAT *does not have a PTAS unless* $P = NP$
7. *It is PSPACE-hard to approximate* (B)N-MAX-DONES-EM-SAT *beyond a factor* n^ϵ

∎

This is a step towards developing a predictive complexity theory for periodically specified problems. The predictive aspect implies that the relationship between problems specified using standard specifications and the way it was derived is sufficient to deduce the relationship between the corresponding periodically specified problems. In other words, (i) a single transformation can serve to simultaneously relate the complexity of several variants of a standard problem (e.g., decision, counting, optimization), and (ii) sufficient conditions on the reductions between standardly specified problems can be used to predict the relationships between the corresponding periodically specified problems. The general technique presented here *simultaneously* applies to a large collection of problems α-SAT(S) when one varies (i) the periodic specification, (ii) the set S, and (iii) the objective function. Moreover, it applies for obtaining easiness as well as hardness results.

7.2 NATURAL MORPHISMS FOR COMPUTATIONAL COMPLEXITY

The results discussed in this chapter show in a number of cases—and strongly suggest in others—that *strongly-local reducibility* degrees for constraint satisfaction problems are preserved, when problems in P or in NP are generalized to periodically specified and to infinite recursive versions of these problems. In contrast, this is *not* true for *polynomial time reducibility* degrees. (Following Ladner [352], polynomial degrees are equivalence classes of languages or sets induced

TABLE 2   Table summarizing how the complexity of three basic problems changes
when the periodic specification is changed. Note that the complexity of 3-SAT and 3-
HornSAT$_C$ changes drastically while the complexity of CLIQUE remains unchanged.

Problems	Flat Specifications (boundary conditions) (U,U)N-specifications	2-D Periodic Specifications (boundary conditions) (N,N)N-specifications
3-SAT	NP-complete	undecidable
3-HornSAT$_C$	P	undecidable
CLIQUE	NP-complete	NP-complete

by polynomial time reducibility.) Indeed, Marathe et al. [380] have shown that
the problems 3-SAT and 3-HornSAT$_C$ (weakly negative SAT when clauses can
contain variables as well as constants) are *polynomial time inter-reducible*, when
specified by several different kinds of succinct specifications. For example, they
are both PSPACE-complete, both EXSPACE-complete, and even both undecidable, for certain specifications. In contrast, for other kinds of succinct specifications, such as two-dimensional periodic narrow finite specifications, 3-SAT is
NEXPTIME-hard but 3-HornSAT$_C$ is DEXPTIME-hard. Interestingly, for each
of these specifications the problem CLIQUE remains in NP. These results are
summarized in table 2.

Thus, the results show that the specific method of constructing infinitary
versions of standardly specified problems considered in Freedman [173] cannot
be used to resolve the P versus NP question. In fact, our results suggest that
many other infinitary extensions of combinatorial problems cannot be used to
resolve the P versus NP question either, since they also fail to distinguish versions of certain basic NP- and P-complete problems. In other words, equivalence
classes (degrees) induced by polynomial time or Turing reducibility are *not invariant*, across variant periodic specifications. In contrast, the results in Marathe
et al. [380] and Hunt et al. [267, 268] show that local transformation-based (simultaneous) reductions and the degrees induced by such reductions may be the
natural *morphisms* for complexity theory.

In addition, *none* of the very general structural extension properties discussed here for *strongly-local reductions*, hold for *simultaneous* linear time-, linear size-, and $O(\log n)$ space- bounded reductions. (To see this, just observe that
a suitably $\Theta(n^2)$-padded version of 3-SAT is *simultaneously* linear time-, linear
size-, and $O(\log n)$-space-bounded reducible to the problem CLIQUE, which
remains NP-complete for various periodic specifications.)

Additional evidence for this is provided by the fact [267] that the *strongly-
local reductions* as defined here are actually algebraic morphisms, or crypto-
morphisms as defined in Birkhoff [54]. Given the central importance of the proper
definition of morphisms in many areas of modern mathematics including topol-

ogy, algebra, and dynamical systems, it seems reasonable to conjecture that a successful relation of the P versus NP question to other areas of mathematics—as alluded to in Freedman [172, 173]—will require appropriate definitions of *morphism* possessing *good* preservation properties with respect to various classes of instance specifications. This may turn out to be an important implication of the results discussed here and in Hunt et al. [267, 268] and Marathe et al. [376, 380].

Interestingly, constructing infinitary versions of standardly specified problems has been used successfully for showing that certain optimization problems do not belong to the class syntactic MAX-SNP: a class defined syntactically using an existential second-order formula [240]. The elegant results of Hirst and Harel [240] use an idea very similar to the one proposed in Freedman [172] to achieve this.

8   PERIODIC SATISFIABILITY FOREVER

The title of this section is influenced by papers of Savelsberg and van Emde Boas [454], van Emde Boas [509], and Harel [224]. They, as well as other authors [48, 82, 206], have elegantly articulated the use of tiling or domino problems for obtaining lower bounds, especially for decision problems for various logical theories. Here, we present several advantages of using periodically specified satisfiability problems over the use of domino problems in proving *both* hardness and easiness results.

Domino problems were introduced by Wang [515] and Büchi [82] and have been studied extensively in the literature [224, 454, 509]. Usually a *domino system* is described as a finite set of *tiles* or *dominoes*, each tile being of a fixed shape (e.g., unit square) with a fixed orientation and colored edges. We have an unlimited supply of copies of every tile. Technically, if one arbitrarily shapes tiles, then one does not need colors and vice-versa. A *domino problem* asks whether it is possible to tile a prescribed subset of the Cartesian plane with elements of a given domino system, such that adjacent tiles have matching colors on their common edges. There may also be certain constraints on the tiles that are allowed at specific places, such as the origin.

1. Quoting Harel [224]: *"Since all domino problems owe their complexity to the correspondence with Turing machine computations and since this correspondence applies to non-deterministic models as well, domino problems can apparently not distinguish between deterministic and non-deterministic classes."* In contrast, the hardness results for periodically specified generalized CNF satisfiability problems include complete problems for the deterministic classes P, DSPACE(n), DEXPTIME, DEXPSPACE(n), etc. For example, when instances are specified periodically with explicit boundary conditions, the hardness results for 3-HornSAT$_C$ imply that exactly analogous hardness results hold for the monotone circuit value problem, when instances are periodically

specified with boundary conditions. The last result can be used to prove that a number of P-complete problems become DSPACE(n)-, DEXPTIME-, or DEXPSPACE-complete when periodically specified. One such problem is linear programming feasibility.

2. It is natural to consider periodically specified formulas with clauses containing variables defined at times $t, t + c_1, t + c_2$, etc, where c_1 and c_2 are integers specified using binary numerals. Following Orlin [419], we say that such periodic specifications are *wide*. As stated earlier, periodic specifications only containing clauses in which all the variables are defined at times $t, t + 1$ and $t - 1$ are called *narrow*. In contrast, domino problems are based on adjacency, and thus, are intrinsically narrow. (It is, of course, possible to define consistency relationships between tiles that are far apart, but in general, this is not natural.) The hardness results for wide, periodically specified satisfiability problems imply exactly analogous results for a number of problems specified using wide periodic specifications. Furthermore, these results show that there can be a significant difference between the complexities of the narrow and wide periodically specified versions of the same problem.

3. As mentioned earlier, efficient local simultaneous reductions to/from the problems 3-SAT, NAE3-SAT, 2-SAT can be extended to efficient approximation preserving reductions to/from 3-SAT, NAE3-SAT, 2-SAT, when instances are specified by various kinds of periodic specifications considered here. These reductions, taken together with the easiness/hardness results, imply analogous easiness/hardness results for a number of variant problems for periodically specified problems in graph theory and logic. Developing analogous theory using tiling problems, although plausible, appears to be much more cumbersome.

This is not to say that tiling problems are not useful starting points, nor does it imply that they are not interesting. The simplicity of tiling problems certainly makes them a natural starting point for proving lower bounds.

9 CONCLUSIONS AND FUTURE WORK

We have discussed instances of graph and satisfiability problems created in natural and simple ways: namely, by repeating a single graph or formula in a multidimensional grid and then connecting vertices placed at given grid points to vertices placed at neighboring grid points. The size of large objects created in this way can be exponential, or even infinite, in the *size* of the object being replicated. In spite of the simple repetitive nature of the constructed object, the difficulty in solving certain NP-complete problems blows up with the size of the object being specified. Thus, several of these problems are NEXPTIME-complete or even undecidable when complexity is measured in the size of the original (periodic) description. However, at the same time, the simple repetitive nature enables us to design efficient polynomial time approximation algorithms with good performance guarantees *even when* complexity is measured in the size of the periodic specification. The complexity of approximation algorithms remains polynomial in the size of the description. Thus we have a striking contrast: problems that are NEXPTIME-complete to solve exactly can be efficiently approximated in polynomial time.

The fact that there is an exponential gap between solving the problem exactly and approximately for such succinctly specified objects may prove to be useful in trying to tackle other questions in complexity theory. For example, the results obtained in this chapter raise the possibility of proving the recent non-approximability results *without* using the machinery of interactive proof systems [21, 23]. Obtaining formal proofs that this is not possible is an equally interesting direction for future research.

The simplicity of the graph or formula obtained in the proof of theorem 5.2 makes it a good candidate for being specified using other kinds of succinct and recursive descriptions. In particular, it can be specified using graph construction representation (GCR) specifications [189, 190] and by the recursive specifications of Beigel and Gasarch [42]. This shows that natural problems specified using the GCR model are NEXPTIME-hard to solve and problems specified by very simple recursive graphs are undecidable. The GCR model is generally acknowledged as a natural and useful way of describing large real-world objects such as circuits and VLSI designs.

We have outlined a collection of techniques that are a step towards developing a predictive complexity theory for periodically specified problems. These ideas are in fact much more general; we believe that they can be used to develop a predictive complexity theory of succinctly specified problems. We refer the reader to Hunt et al. [267, 268] and Marathe [376] for additional details. Further general results for characterizing the approximability of PSPACE-hard and NEXPTIME-hard periodically specified problems is an interesting direction for future research. As an example, it is an open question if a dichotomy theorem such as the one for MAX-SAT(S) exists for the one-dimensional, finite, wide, periodically specified MAX-SAT(S) problem. We conclude with a brief discussion of two topics currently under investigation.

9.1 PERIODICALLY SPECIFIED RANDOM CONSTRAINT SATISFACTION PROBLEMS

Recently there has been substantial interest in understanding the phenomenon of phase transitions in satisfiability problems. See Istrate [270], Kirkpatrick and Selman [319], and Monasson et al. [406] for more details on the subject. Here we propose a method to construct random instances of periodic constraint satisfaction problems. Similar models can be given for periodically specified random graphs and other combinatorial objects. The model presented here is proposed in the hope that it will provide a hierarchy of locally random but globally structured instances of constraint satisfaction problems whose worst-case complexity is, for example, NP-hard, PSPACE-hard, or NEXPTIME-hard.

Recall that a periodic specification of a constraint satisfaction problem consists of two different parts: the static formula and the rule used to expand it. A natural way to construct a random instance is to construct a random instance of the static formula. For this, we have the four literals $x_i(t), \overline{x_i(t)}$, $x_i(t+1), \overline{x_1(t+1)}$ for each variable x_i. Call this the set X. We can now construct a random static 3-CNF formula as is done for standardly specified instances: namely, for each clause the three literals appearing in it are chosen uniformly at random from X (we ignore the technicalities of whether these literals are chosen with or without replacement). Note that as constructed, the static formulas can yield finite as well as infinite formulas depending on the particular periodic specification. It would be of interest to investigate the similarities (and differences) between the phase transition behavior of such finite and infinite formulas. The formalism is very rich and can be extended to construct other constraint satisfaction problems. It can also be applied to periodic specifications of higher dimensional objects. Defining wide specifications is more subtle. In such a case, we might decide to choose the index of difference at random.

We reiterate that an interesting aspect of such formulas is that they are *locally random but globally structured*; that is, the set of clauses in two consecutive grid points is chosen randomly, but this random set is repeated by a simple replication rule. Furthermore, note that we do not have to study exponentially large instances: specifying the range in unary yields a polynomially sized formula. By observing the proof given by Cook [191], it is easy to see that even such formulas are NP-hard to decide in the worst case. Investigating the phase transition behavior of such formulas is an interesting direction for further research.

9.2   LATTICE-LIKE INSTANCES FOR SATISFIABILITY PROBLEMS

There has been substantial interest in generating structured and random instances of satisfiability to test the efficacy of the SAT solvers proposed in the literature [6, 303]. In this context, physicists have long studied and developed methods inspired by statistical mechanics for understanding physical phenomena on lattice-like structures. Recently, methods inspired by statistical mechanics have also been proposed to solve constraint satisfaction problems, as has been seen in chapter 4. Periodically specified random constraint satisfaction problems, graph problems, and feasibility of system of linear/nonlinear inequalities are in a parametrized class of *lattice structured* problems and thus might serve as test cases for physics-inspired methods for solving such problems. The value of the formalism lies in the fact that one can construct a large number of variant problems by specifying a simple set of parameters. In addition, work done on quasi-group completion and related Latin square completion methods can be viewed as special instances of periodic satisfiability problems. One way to see this is that a Latin square completion method often starts with a consistent tiling of the grid and then randomly punches holes or, stated alternatively, removes some of the tiles. It then presents this satisfiable instance to SAT solvers. In view of the discussion on tiling and its relationship to periodic satisfiability problems, it is clear that periodic satisfiability problems offer a rich parameterized class of such instances. More interestingly, the method yields hard instances of higher complexity classes such as PSPACE and NEXPTIME. Such instances might be significant, given the close correspondence of periodically specified satisfiability problems to problems in temporal logics.

To see the close similarity of periodically specified formulas to Latin square completion methods for specifying satisfiable formulas, consider an alternate definition of periodically specified formulas proposed by Freedman [172, 173]. As an example, the infinite version of 3-SAT in Freedman [172] is obtained as follows: take a finitely generated group $\mathcal{G}$ and a subgroup $\mathcal{H}$ of $\mathcal{G}$, of finite index. The elements of $\mathcal{G}$ are our alphabet and a literal is a symbol g or $\overline{g} \in \mathcal{G}$. An instance of 3-SAT is specified as a conjunction of clauses of the form $(g'_1 \vee g'_2 \vee g'_3)$, where g'_i, $1 \leq i \leq 3$ denotes either a negated or an unnegated literal. Thus a 3-CNF formula will be given as $F = \bigwedge_{j=1}^{m} (g'_{1,j} \vee g'_{2,j} \vee g'_{3,j})$. The infinite instance is now created as follows:

$$F^{\mathbb{Z}} = \bigwedge_{h \in \mathcal{H}} \bigwedge_{j=1}^{m} ((hg)'_{1,j} \vee (hg)'_{2,j} \vee (hg)_{3,j}).$$

Freedman considers the special cases $\mathcal{G} = \mathbb{Z}$ and $\mathcal{G} = \mathbb{Z} \oplus \mathbb{Z}$. In the latter case, he effectively considers four-way, infinite, wide formulas with periods (p_1, p_2) and thus $\mathcal{H} = \{(n_1 p_1, n_2 p_2) \mid n_1, n_2 \in \mathbb{Z}\}$. Assuming natural representations of integers, it is easy to see that these special cases are simply one-dimensional, two-way, infinite, wide periodic specifications and two-dimensional four-way infinite, wide, periodic wide specifications, denoted as $(\mathbb{Z})$W and $(\mathbb{Z},\mathbb{Z})$W-specifications respectively. The $(\mathbb{Z},\mathbb{Z})$N-specifications considered here can be easily seen to be special cases with $p_1 = p_2 = 1$. Using the close correspondence between tilings and periodic formulas, it is now possible to generate satisfiable formulas with periodic solutions as well as satisfiable formulas that do not have periodic solutions. See Freedman [172] and Grunbaum and Shephard [216] for more details.

ACKNOWLEDGMENTS

We thank the editors for inviting us to include our results in this volume. We thank the referees and the editors of the volume for a careful reading of the manuscript and suggesting changes that have substantially improved the readability of the text. We thank S. Arora, A. Condon, J. Feigenbaum. D. Harel, G. Istrate, T. Lengauer, J. Orlin, V. Radhakrishnan, S. Ravi, S. Shukla, M. Sudan, and E. Wanke for valuable discussions and comments. The work presented here was partly motivated by the questions raised in Orlin [419] and Freedman [173].

Bibliography

1. Achlioptas, D. "Lower Bounds for Random 3-SAT via Differential Equations." *Theor. Comp. Sci.* 265 (2001): 159–186.
2. Achlioptas, D. "Setting Two Variables at a Time Yields a New Lower Bound for Random 3-SAT." In *Proceedings of the 32nd Annual ACM Symposium on Theory of Computing (STOC'00)*, 28–37. New York: ACM Press, 2000.
3. Achlioptas, D., and C. Moore. "The Asymptotic Order of the Random k-SAT Threshold." In *Proceedings of the 43rd Annual IEEE Symposium on Foundations of Computer Science (FOCS'02)*, 779–788. IEEE Computer Society, 2002.
4. Achlioptas, D., and Y. Peres. "The Threshold for Random k-SAT is $2^k \ln 2 - O(k)$." In *Proceedings of the 35th Annual ACM Symposium on Theory of Computing (STOC'03)*, 223–231. New York: ACM Press, 2003.
5. Achlioptas, D., and G. B. Sorkin. "Optimal Myopic Algorithms for Random 3-SAT." In *Proceedings of the 41st Annual Symposium on Foundations of Computer Science (FOCS'00)*, 590–600. IEEE Computer Society, 2000.

6. Achlioptas, D., C. Gomes, H. Kautz, and B. Selman. "Generating Satisfiable Problem Instances." In *Proceedings of the 17th National Conference on Artificial Intelligence (AAAI'00)*, 256–261. Menlo Park, CA: AAAI Press, 2000.

7. Achlioptas, D., A. Chtcherba, G. Istrate, and C. Moore. "The Phase Transition in 1-in-k SAT and NAE 3-SAT." In *Proceedings of the 12th Annual ACM-SIAM Symposium on Discrete Algorithms (SODA'01)*, 721–722. New York: ACM Press, 2001.

8. Achlioptas, D., L. M. Kirousis, E. Kranakis, and D. Krizanc. "Rigorous Results for Random $(2 + p)$-SAT." *Theor. Comp. Sci.* 265 (2001): 109–129.

9. Achlioptas, D., P. Beame, and M. Molloy. "A Sharp Threshold in Proof Complexity." In *Proceedings of the 33rd Annual ACM Symposium on Theory of Computing (STOC'01)*, 337–346. New York: ACM Press, 2001.

10. Agarwal, S., and A. Condon. "On Approximation Algorithms for Hierarchical MAX-SAT." *J. Algorithms* 26 (1998): 141–165.

11. Ajtai, M., J. Komlós, and E. Szemerédi. "Largest Random Component of a k-Cube." *Combinatorica* 2 (1982): 1–7.

12. Akahori, J. "Asymptotics of Hedging Errors in a Slightly Incomplete Discrete Market: A Noise-Sensitive Example." Preprint, 2002. http://www.ritsumei.ac.jp/se/~akahori/papers/pp/slightly_incomplete.pdf (accessed September 27, 2005).

13. Aldous, D. "The Harmonic Mean Formula for Probabilities of Unions: Applications to Sparse Random Graphs." *Discrete Math.* 76 (1989): 167–176.

14. Alekhnovich, M., and E. Ben-Sasson. "Linear Upper Bounds for Random Walk on Small Density Random 3-CNFs." In *Proceedings of the 44th Annual Symposium on Foundations of Computer Science (FOCS'03)*, 352–361. IEEE Computer Society, 2003.

15. Aleksiejuk, A., J. A. Holyst, and D. Stauffer. "Ferromagnetic Phase Transition in Barabási–Albert Networks." *Physica A* 310 (2002): 260–266.

16. Alexander, S., and P. Pincus. "Phase Transitions of Some Fully Frustrated Models." *J. Phys. A: Math. Gen.* 13 (1980): 263–273.

17. Alon, N., and J. Spencer. *The Probabilistic Method.* New York: John Wiley & Sons, 1992.

18. Alon, N., I. Dinur, E. Friedgut, and B. Sudakov. "Graph Products, Fourier Analysis and Spectral Techniques." *Geom. Funct. Anal.* 14 (2004): 913–940.

19. Alur, R., and T. Henzinger. "A Really Temporal Logic." *J. ACM* 41 (1994): 181–204.

20. Arora, S., and S. Safra, "Probabilistic Checking of Proofs: A New Characterization of NP." *J. ACM* 45 (1998): 70–122.

21. Arora, S., C. Lund, R. Motwani, M. Sudan, and M. Szegedy. "Proof Verification and Intractability of Approximation Problems." *J. ACM* 45 (1998): 501–555.

22. Arrow, K. "A Difficulty in the Theory of Social Welfare." *J. Pol. Econ.* 58 (1950): 328–346.

23. Ausiello, G., P. Crescenzi, G. Gambosi, V. Kann, A. Marchetti-Spaccamela, and M. Protasi. *Complexity and Approximation: Combinatorial Optimization Problems and Their Approximability Properties*. Berlin: Springer-Verlag, 1999.

24. Bak, P., C. Tang, and K. Wiesenfeld. "Self-Organized Criticality: An Explanation of the 1/f Noise." *Phys. Rev. Lett.* 59 (1987): 381–384.

25. Baker, B. S. "Approximation Algorithms for NP-Complete Problems on Planar Graphs." *J. ACM* 41 (1994): 153–180.

26. Bálcazar, J. L., A. Lozano, and J. Toran. "The Complexity of Algorithmic Problems for Succinct Instances." In *Computer Science*, ed. R. Baeza-Yates, 351–377. New York: Plenum Press, 1992.

27. Baptista, L., and J. P. Marques-Silva. "Using Randomization and Learning to Solve Hard Real-World Instances of Satisfiability." In *Proceedings of the Sixth International Conference on Principles and Practice of Constraint Programming (CP 2000)*, ed. R. Dechter, 489–494. Lecture Notes in Computer Science, vol. 1894. Berlin: Springer-Verlag, 2000.

28. Barabási, A. L., and R. Albert. "Emergence of Scaling in Random Networks." *Science* 286 (1999): 509–512.

29. Barabási, A.-L., and H. E. Stanley. *Fractal Concepts in Surface Growth*. Cambridge, UK: Cambridge University Press, 1995.

30. Barabási, A. L., R. Albert, and H. Jeong. "Mean-Field Theory for Scale-Free Random Networks." *Physica A* 272 (1999): 173–187.

31. Barahona, F. "On the Computational Complexity of Ising Spin Glass Models." *J. Phys. A* 15 (1982): 3241–3253.

32. Barahona, F., R. Maynard, R. Rammal, and J. P. Uhry. "Morphology of Ground States of Two-Dimensional Frustration Model." *J. Phys. A: Math. Gen.* 15 (1982): 673–699.

33. Barg, A. "Complexity Issues in Coding Theory." In *Handbook of Coding Theory*, vol. 1, ed. V. S. Pless and W. C. Huffman, 649–754. Amsterdam: Elsevier Science, 1998.

34. Barrat, A., and M. Weigt. "On the Properties of Small-World Network Models." *Eur. Phys. J. B* 13 (2000): 547–560.

35. Barthe, F., P. Cattiaux, C. Roberto. "Interpolation Inequalities between Exponential and Gaussian, Orlicz Hypercontractivity and Applications to Isoperimetry." *Revista Mat. Iberoamericana* (2005): to appear.

36. Barthel, W., A. Hartmann, and M. Weigt. "Solving Satisfiability Problems by Fluctuations: An Approximate Description of the Dynamics of Stochastic Local Search Algorithms." *Phys. Rev. E* 67 (2003): 066104.

37. Bauer, M., and O. Golinelli. "Core Percolation in Random Graphs: A Critical Phenomena Analysis." *Eur. Phys. J. B* 24 (2001): 339–352.

38. Bauke, H., S. Mertens, and A. Engel. "Phase Transition in Multiprocessor Scheduling." *Phys. Rev. Lett.* 90 (2003): 158701.

39. Beame, P., R. Karp, T. Pitassi, and M. Saks. "The Efficiency of Resolution and Davis-Putnam Procedures." *SIAM J. Comp.* 31 (2002): 1048–1075.

Preliminary version in *Proceedings of the 30th Annual ACM Symposium on Theory of Computing (STOC'98)*, 561–571. New York: ACM Press, 1998.

40. Beckner, W. "Inequalities in Fourier Analysis." *Ann. Math.* 102 (1975): 159–182.

41. Beeri, C., and P. A. Bernstein. "Computational Problems Related to the Design of Normal Form Relational Schemas." In *ACM Trans. Database Systems (TODS'79)*, 30–59. New York: ACM Press, 1979.

42. Beigel, R., and W. I. Gasarch. "On the Complexity of Finding the Chromatic Number of Recursive Graphs: Parts I and II." *Ann. Pure. Appl. Logic* 45 (1989): 1–38; 227–247.

43. Bellare, M., O. Goldreich, and M. Sudan. "Free Bits, PCPs, and Non-approximability—Towards Tight Results." *SIAM J. Comp.* 27 (1998): 804–915.

44. Benjamini, I., G. Kalai, and O. Schramm. "First Passage Percolation has Sublinear Distance Variance." *Ann. Probab.* 31 (2003): 1970–1978.

45. Benjamini, I., G. Kalai, and O. Schramm. "Noise Sensitivity of Boolean Functions and Applications to Percolation." *Publ. I.H.E.S.* 90 (1999): 5–43.

46. Ben-Or, M., and N. Linial. "Collective Coin Flipping." In *Randomness and Computation*, ed. S. Micali, 91–115. New York: Academic Press, 1990.

47. Berg, B. A., U. E. Hansmann, and T. Celik. "Ground-State Properties of the Three-Dimensional Ising Spin Glass." *Phys. Rev. B* 50 (1994): 16444–16452.

48. Berger, R. "The Undecidability of the Domino Problem." *Mem. Amer. Math. Soc.* 66 (1966).

49. Berlekamp, E. R., R. J. McEliece, and H. C. A. van Tilborg. "On the Inherent Intractability of Certain Coding Problems." *IEEE Trans. Inf. Theory* 24 (1978): 384–386.

50. Bernstein, A. J. "Maximally Connected Arrays on the n-Cube." *SIAM J. Appl. Math.* 15 (1967): 1485–1489.

51. Bianconi, G. "Mean Field Solution of the Ising Model on a Barabási–Albert Network." *Phys. Lett. A* 303 (2002): 166–168.

52. Bieche, L., J. P. Uhry, R. Maynard, and R. Rammal. "On the Ground States of the Frustration Model of a Spin Glass by a Matching Method of Graph Theory." *J. Phys. A: Math Gen.* 13 (1980): 2553–2576.

53. Binder, K., and D. W. Herrmann. *Monte Carlo Simulation in Statistical Physics.* Berlin: Springer-Verlag, 1988

54. Birkhoff, G. *Lattice Theory*, 3d ed. Providence, RI: American Mathematics Society, 1966.

55. Biroli, G., and R. Monasson. "From Inherent Structures to Pure States: Some Simple Remarks and Examples." *Europhys. Lett.* 50 (2000): 155–161.

56. Biroli, G., R. Monasson, and M. Weigt. "A Variational Description of the Ground State Structure in Random Satisfiability Problems." *Eur. Phys. J. B* 14 (2000): 551–568.

57. Boettcher, S. Personal communication.

58. Boettcher, S., and M. Grigni. "Jamming Model for the Extremal Optimization Heuristic." *J. Phys. A* 35 (2002): 1109–1123.

59. Boettcher, S., and A. G. Percus. "Extremal Optimization at the Phase Transition of the Three-Coloring Problem." *Phys. Rev. E* 69 (2004): 066703.

60. Boettcher, S., and A. G. Percus. "Nature's Way of Optimizing." *Art. Intel.* 119 (2000): 275–286.

61. Boettcher, S., and A. G. Percus. "Optimization with Extremal Dynamics." *Phys. Rev. Lett.* 86 (2001): 5211–5214.

62. Bollobás, B. *Random Graphs*, 2d ed. Cambridge, UK: Cambridge University Press, 2001.

63. Bollobás, B., and S. Rasmussen. "First Cycles in Random Directed Graph Processes." *Discrete Math.* 75 (1989): 55–68.

64. Bollobás, B., and O. Riordan. "The Critical Probability for Random Voronoi Percolation in the Plane is 1/2." arXiv.org E-print Archive, Cornell University Library, 2004. http://arxiv.org/abs/math.PR/0410336 (accessed September 27, 2005).

65. Bollobás, B., and O. Riordan. "A Short Proof of the Harris-Kesten Theorem." arXiv.org E-print Archive, Cornell University Library, 2004. http://arxiv.org/abs/math.PR/0410359 (accessed September 27, 2005).

66. Bollobás, B., and A. Thomason. "Threshold Functions." *Combinatorica* 7 (1987): 35–38.

67. Bollobás, B., C. Borgs, J. T. Chayes, J. H. Kim, and D. B. Wilson. "The Scaling Window of the 2-SAT Transition." *Rand. Struct. & Algorithms* 18 (2001): 201–256.

68. Bonami, A. "Etude des Coefficients Fourier des Fonctiones de $L^p(G)$." *Ann. Inst. Fourier* 20 (1970): 335–402.

69. Boppana, R. "The Average Sensitivity of Bounded Depth Circuits." *Inf. Proc. Lett.* 63 (1997): 257–261.

70. Boppana, R. "Threshold Functions and Bounded Depth Monotone Circuits." In *Proceedings of the 16th Annual ACM Symposium on Theory of Computing (STOC'84)*, 475–479. New York: ACM Press, 1984.

71. Borgs, C., J. Chayes, and B. Pittel. "Phase Transition and Scaling Window for the Integer Partitioning Problem." *Rand. Struct. & Algorithms* 19 (2001): 247–288.

72. Borgs, C., J. T. Chayes, S. Mertens, and B. Pittel. "Phase Diagram for the Constrained Integer Partitioning Problem." *Rand. Struct. & Algorithms* 24 (2004): 315–380.

73. Bourgain, J. "On Sharp Thresholds of Monotone Properties, Appendix to E. Friedgut. "Sharp Thresholds of Graphs Properties, and the k-SAT Problem."' *J. Amer. Math. Soc.* 12 (1999): 1017–1054.

74. Bourgain, J., and G. Kalai. "Influences of Variables and Threshold Intervals under Group Symmetries." *Geom. Funct. Anal.* 7 (1997): 438–461.

75. Bourgain, J., J. Kahn, G. Kalai, Y. Katznelson, and N. Linial. "The Influence of Variables in Product Spaces." *Israel J. Math.* 77 (1992): 55–64.

76. Boyer, M., G. Brassard, P. Høyer, and A. Tapp. "Tight Bounds on Quantum Searching." In *Proceedings of the Workshop on Physics and Computation (PhysComp'96)*, 36–43. Cambridge, MA: New England Complex Systems Institute, 1996.

77. Brassard, G., P. Høyer, and A. Tapp. "Quantum Counting." In *Proceedings of the 25th International Colloquium on Automata, Languages, and Programming (ICALP'98)*, ed. K. Larsen, 820–831. Berlin: Springer, 1998.

78. Braunstein, A., R. Mulet, A. Pagnani, M. Weigt, and R. Zecchina. "Polynomial Iterative Algorithms for Coloring and Analyzing Random Graphs." *Phys. Rev. E* 68 (2003): 036702.

79. Braunstein, A., M. Mézard, and R. Zecchina. "Survey Propagation: An Algorithm for Satisfiability." *Rand. Struct. & Algorithms* 27 (2005): 201–226.

80. Broadbent, S. R., and J. M. Hammersley. "Percolation Processes: I. Crystals and Mazes." *Proc. Cambridge Phil. Soc.* 53 (1957): 629–641.

81. Broder, A. Z., A. Frieze, and E. Upfal. "On the Satisfiability and Maximum Satisfiability of Random 3-CNF Formulas." In *Proceedings of the 4th Annual ACM-SIAM Symposium on Discrete Algorithms (SODA'93)*, 322–330. New York: ACM Press, 1993.

82. Büchi, J. R. "Turing Machines and the Entscheidungsproblem." *Math. Ann.* 148 (1962): 201–213.

83. Burgers, M. *The Nonlinear Diffusion Equation*. Boston: Riedel, 1974.

84. Cech, T. R. "RNA As An Enzyme." *Sci. Am.* 255(5) (1986): 64–75.

85. Cech, T. R. "Self-Splicing RNA: Implications for Evolution." *Intl. Rev. Cytology* 93 (1985): 3–22.

86. Cerf, N. J., L. K. Grover, and C. P. Williams. "Nested Quantum Search and NP-Complete Problems." *Applicable Algebra in Engineering, Communication and Computing* 10 (2000): 311–338.

87. Chandy, K. M., and J. Misra. "Asynchronous Distributed Simulation via a Sequence of Parallel Computations." *Comm. ACM* 24 (1981): 198–206.

88. Chao, M. T., and J. Franco. "Probabilistic Analysis of a Generalization of the Unit-Clause Literal Selection Heuristics for the k-Satisfiability Problem." *Inf. Sci.* 51 (1990): 289–314.

89. Chao, M. T., and J. Franco. "Probabilistic Analysis of Two Heuristics for the 3-Satisfiability Problem." *SIAM J. Comp.* 15 (1986): 1106–1118.

90. Chayes, J. T., L. Chayes, D. S. Fisher, and T. Spencer. "Finite-Size Scaling and Correlation Length for Disordered Systems." *Phys. Rev. Lett.* 57 (1986): 2999–3002.

91. Cheeseman, P., B. Kanefsky, and W. M. Taylor. "Where the *Really* Hard Problems Are." In *Proceedings of the Twelfth International Joint Conference on Artificial Intelligence (IJCAI'91)*, ed. J. Mylopoulos and R. Rediter, 331–337. San Mateo, CA: Morgan Kaufmann, 1991.
Morgan Kaufmann, 1991.

92. Cheung, H.-F., and W. L. McMillan. "Equilibrium Properties of a Two-Dimensional Random Ising Model with a Continuous Distribution of Interactions." *J. Phys. C: Solid State Phys.* 16 (1983): 7027–7032.

93. Chung, S.-Y., G. D. Forney, Jr., T. J. Richardson, and R. Urbanke. "On the Design of Low-Density Parity-Check Codes within 0.0045 dB of the Shannon Limit." *IEEE Comm. Lett.* 5 (2001): 58–60.

94. Chvátal, V., and B. Reed. "Mick Gets Some (the Odds are on His Side)." In *Proceedings of the 32nd Annual IEEE Symposium on Foundations of Computer Science (FOCS'92)*, 620–627. IEEE Computer Society, 1992.

95. Chvátal, V., and E. Szemerédi. "Many Hard Examples for Resolution." *J. ACM* 35 (1988): 759–768.

96. Coarfa, C., D. D. Demopoulos, A. San Miguel Aguirre, D. Subramanian, and M. Y. Vardi. "Random 3-SAT: The Plot Thickens." In *Proceedings of the Sixth International Conference on Principles and Practice of Constraint Programming (CP 2000)*, ed. R. Dechter, 143–159. Lecture Notes in Computer Science, vol. 1894. Berlin: Springer-Verlag, 2000.

97. Cocco, S., and R. Monasson. "Analysis of the Computational Complexity of Solving Random Satisfiability Problems using Branch and Bound Search Algorithms." *Eur. Phys. J. B* 22 (2001): 505–531.

98. Cocco, S., and R. Monasson. "Exponentially Hard Problems are Sometimes Polynomial, A Large Deviation Analysis of Search Algorithms for the Random Satisfiability Problem, and Its Application to Stop-and-Restart Resolutions." *Phys. Rev. E* 66 (2002): 037101.

99. Cocco, S., and R. Monasson. "Restarts and Exponential Acceleration of the Davis-Putnam-Loveland-Logemann Algorithm. A Large Deviation Analysis of the Generalized Unit Clause Heuristic for Random 3-SAT." *Ann. Math. & Art. Intel.* 43 (2005): 153–172.

100. Cocco, S., and R. Monasson. "Trajectories in Phase Diagrams, Growth Processes and Computational Complexity: How Search Algorithms Solve the 3-Satisfiability Problem." *Phys. Rev. Lett.* 86 (2001): 1654–1657.

101. Cocco, S., O. Dubois, J. Mandler, and R. Monasson. "Rigorous Decimation-Based Construction of Ground Pure States for Spin Glass Models on Random Lattices." *Phys. Rev. Lett.* 90 (2003): 047205.

102. Coffman, E., and G. S. Lueker. *Probabilistic Analysis of Packing and Partitioning Algorithms.* New York: John Wiley & Sons, 1991.

103. Cohen, E., and N. Megiddo. "Recognizing Properties of Dynamic Graphs." In *Applied Geometry and Discrete Mathematics, The Victor Klee Festschrift*, ed. P. Gritzmann and B. Strumfels, 135–146. New York: ACM, 1991.

104. Cohen, E., and N. Megiddo. "Strongly Polynomial-time and NC Algorithms for Detecting Cycles in Dynamic Graphs." *J. ACM* 40 (1993): 791–830.

105. Computer Industry Almanac Inc. Home Page. http://www.c-i-a.com/ (accessed September 27, 2005).

106. Condon, A. "Approximate Solutions to Problems in PSPACE." *ACM SIGACT News* 26(2) (1995): 4–13.

107. Condon, A., J. Feigenbaum, C. Lund, and P. Shor. "Probabilistically Checkable Debate Systems and Approximation Algorithms for PSPACE-Hard Functions." *Chicago J. Theor. Comp. Sci.* (1995): Article no. 4.

108. Condon, A., J. Feigenbaum, C. Lund, and P. Shor. "Random Debaters and the Hardness of Approximating Stochastic Functions." *SIAM J. Comp.* 26 (1997): 369–400.

109. Cook, S. "The Complexity of Theorem-Proving Procedures." In *Proceedings of the 3rd Annual ACM Symposium on Theory of Computing (STOC'71)*, 151–158. New York: ACM Press, 1971.

110. Crawford, J. M., and L. D. Auton. "Experimental Results on the Crossover Point in Random 3-SAT." *Art. Intel.* 81 (1996): 31–57.

111. Creignou, N. "A Dichotomy Theorem for Maximum Generalized Satisfiability Problems." *J. Comp. Syst. Sci.* 51 (1995): 511–522.

112. Creignou, N., and H. Daudé. "Satisfiability Threshold for Random XOR-CNF Formulae." *Discrete Appl. Math.* 96/97 (1999): 41–53.

113. Creignou, N., S. Khanna, and M. Sudan. "Complexity Classifications of Boolean Constraint Satisfaction Problems." *SIGACT News* 32(4) (2001): 24–33.

114. Cugliandolo, L. F. "Dynamics of Glassy Systems." Lecture notes, Les Houches, July 2002. arXiv.org E-print Archive, Cornell University Library, 2002. http://arxiv.org/abs/cond-mat/0210312 (accessed September 27, 2005).

115. Dantzig, G., W. Blattner, and M. Rao. "Finding a Cycle in a Graph with Minimum Cost to Time Ratio with Application to a Ship Routing Problem." In *Theory of Graphs: International Symposium*, ed. P. Rosenstieh, 77–84. New York: Gordon and Breach, 1967.

116. Darling, R. W. R., and J. R. Norris. "Structure of Large Random Hypergraphs." *Ann. Appl. Prob.* 15 (2005): 125–152.

117. Davis, M., and H. Putnam. "A Computing Procedure for Quantification Theory." *J. ACM* 7 (1960): 201–215.

118. Davis, M., G. Logemann, and D. Loveland. "A Machine Program for Theorem Proving." *Comm. ACM* 5 (1962): 394–397.

119. Dean, D. S. "Metastable States of Spin Glasses on Random Thin Graphs." *Eur. Phys. J. B* 15 (2000): 493–498.

120. Derrida, B., Y. Pomeau, G. Toulouse, and J. Vannimenus. "Fully Frustrated Simple Cubic Lattices and the Overblocking Effect." *J. Physique* 40 (1979): 617–626.

121. De Simone, C., M. Diehl, M. Jünger, P. Mutzel, G. Reinelt, and G. Rinaldi. "Exact Ground States of Ising Spin Glasses: New Experimental Results with a Branch-and-Cut Algorithm." *J. Stat. Phys.* 80 (1995): 487–496.

122. De Simone, C., M. Diehl, M. Jünger, P. Mutzel, G. Reinelt, and G. Rinaldi. "Exact Ground States of Two-Dimensional ±J Ising Spin Glasses." *J. Stat. Phys.* 84 (1996): 1363–1371.

123. Deutsch, D. "Quantum Theory, the Church-Turing Principle and the Universal Quantum Computer." *Proc. Roy. Soc. Lond. A* 400 (1985): 97–117.

124. Devlin, K. *The Millenium Problems: The Seven Greatest Unsolved Mathematical Puzzles of our Time.* New York: Basic Books, 2002.

125. Dinur, I., and S. Safra. "On the Hardness of Approximating Minimum Vertex Cover." *Ann. Math.* 162 (2005): 439–486. Preliminary version in *Proceedings of the 34th Annual ACM Symposium on Theory of Computing (STOC'02)*, 33–42. New York: ACM Press, 2002.

126. DiVincenzo, D. P. "Quantum Computation." *Science* 270 (1995): 255–261.

127. Dorogovtsev, S. N., A. V. Goltsev, and J. F. F. Mendes. "Ising Model on Networks with an Arbitrary Distribution Of Connections." *Phys. Rev. E* 66 (2002): 016104.

128. Dowling, W. F., and J. H. Gallier. "Linear-Time Algorithms for Testing the Satisfiability of Propositional Horn Formulae." *J. Logic Programming* 1 (1984): 267–284.

129. Dubois, O. "Upper Bounds on the Satisfiability Threshold." *Theor. Comp. Sci.* 265 (2001): 187–197.

130. Dubois, O., and Y. Boufkhad. "A General Upper Bound for the Satisfiability Threshold of Random r-SAT Formulae." *J. Algorithms* 24 (1997): 395–420.

131. Dubois, O., and J. Mandler. "The 3-XORSAT Threshold." In *Proceedings of the 43rd Annual IEEE Symposium on Foundations of Computer Science (FOCS'02)*, 769–778. IEEE Computer Society, 2002.

132. Dubois, O., Y. Boufkhad, and J. Mandler. "Typical Random 3-SAT Formulae and the Satisfiability Threshold." In *Proceedings of the 11th Annual ACM-SIAM Symposium on Discrete Algorithms (SODA'00)*, 126–127. New York: ACM Press, 2000.

133. Dubois, O., R. Monasson, B. Selman, and R. Zecchina, eds. "Phase Transitions in Combinatorial Problems." *Theor. Comp. Sci.* 265 (2001): 1.

134. Dubois, O., Y. Boufkhad, and J. Mandler. "Typical Random 3-SAT Formulae and the Satisfiability Threshold." Report TR03-007, Electronic Colloquium on Computational Complexity, 2003.

135. Duchet, P. "Hypergraphs." In *Handbook of Combinatorics*, ed. R. Graham, M. Grötschel, and L. Lovász. Amsterdam: Elsevier Science, 1995.

136. Dyer, M. E., A. M. Frieze, and S. Suen. "Ordering Clone Libraries in Computational Biology." *J. Comp. Biol.* 2 (1995): 207–218.

137. Eastman, W. L. "Linear Programming with Pattern Constraints." Ph.D. thesis, Computation Laboratory, Harvard University, Cambridge, MA, 1958.

138. Edelkamp, S., and R. E. Korf. "The Branching Factor of Regular Search Spaces." In *Proceedings of the 15th National Conference on Artificial Intelligence (AAAI'98)*, 299–304. Menlo Park, CA: AAAI Press, 1998.

139. Edmonds, J. "Maximum Matchings and a Polyhedron with 0,1-Vertices." *J. Res. Nat'l. Bureau of Standards (Section B)* 69B (1965): 125–130.

140. Edmonds, J. "Paths, Trees and Flowers." *Canad. J. Math.* 17 (1965): 449–467.

141. Edwards, S. F., and P. W. Anderson. "Theory of Spin Glasses." *J. Phys. F: Metal Phys.* 5 (1975): 965–974.

142. Ein-Dor, L., and R. Monasson. "The Dynamics of Proving Uncolorability of Large Random Graphs: I. Symmetric Coloring Heuristic." *J. Phys. A: Math. Gen.* 36 (2003) 11055-11067.

143. Ellington, A. D. "Aptamers Achieve the Desired Recognition." *Curr. Biol.* 4 (1994): 427-429.

144. Ellington, A. D., and J. W. Szostak. "*In Vitro* Selection of RNA Molecules that Bind Specific Ligands." *Nature* 346 (1990): 818-822.

145. El Maftouhi, M., and W. Fernandez de la Vega. "On Random 3-SAT." *Comb. Prob. & Comp.* 4 (1995): 190-195.

146. Engel, A., and C. van den Broeck. *Statistical Mechanics of Learning.* Cambridge, UK: Cambridge University Press, 2001.

147. Erdős, P., and A. Rényi. "On the Evolution of Random Graphs." *Publ. Math. Inst. Hung. Acad. Sci.* 5 (1960): 17-61.

148. Erdős, P., and A. Rényi. "On Random Graphs I." *Publ. Math. Debrecen* 6 (1959): 290-297.

149. Fabrikant, A., and T. Hogg. "Graph Coloring with Quantum Heuristics." In *Proceedings of the 18th National Conference on Artificial Intelligence (AAAI'02)*, 22-27. Menlo Park, CA: AAAI Press, 2002.

150. Falik, D., and A. Samorodnitsky. "A Combinatorial Proof for a Theorem of Kahn, Kalai, and Linial and Some Applications." Preprint, 2005.

151. Farhi, E., J. Goldstone, S. Gutmann, J. Lapan, A. Lundgren, and D. Preda. "A Quantum Adiabatic Evolution Algorithm Applied to Random Instances of an NP-Complete Problem." *Science* 292 (2001): 472-476.

152. Feddersen, T., and W. Pesendorfer. "Convicting the Innocent: The Inferiority of Unanimous Jury Verdicts under Strategic Voting." *Amer. Pol. Sci. Rev.* 92 (1998): 23-35.

153. Feige, U. "A Threshold of ln n for Approximating Set Cover." *J. ACM* 45 (1998): 634-652.

154. Feigenbaum, J. "Games, Complexity Classes and Approximation Algorithms." Invited talk at the *International Congress on Mathematics,* Berlin, 1998.

155. Felderman, R. E., and L. Kleinrock. "Bounds and Approximations for Self-Initiating Distributed Simulation without Lookahead." *ACM Trans. Model. Comp. Simul.* 1 (1991): 386-406.

156. Fernandez de la Vega, W. "On Random 2-SAT." Unpublished manuscript, 1992.

157. Ferreira, F., and J. Fontanari. "Probabilistic Analysis of the Number Partitioning Problem." *J. Phys. A* 31 (1998): 3417-3428.

158. Feynman, R. P. *Feynman Lectures on Computation.* Reading, MA: Addison-Wesley, 1996.

159. Fisher, M. E. "The Theory of Equilibrium Critical Phenomena." *Rep. Prog. Phys.* 30 (1967): 615-730.

160. Flamm, C., I. L. Hofacker, and P. F. Stadler. "RNA *In Silico:* The Computational Biology of RNA Secondary Structures." *Adv. Compl. Syst.* 2 (1999): 65-90.

161. Fleischer, L. "Faster Algorithms for Quickest Transshipment Problem with Zero Transit Time." *SIAM J. Optimization* 12 (2001): 18–35.

162. Foltin, G., K. Oerding, Z. Rácz, R. L. Workman, and R. K. P. Zia. "Width Distribution for Random-Walk Interfaces." *Phys. Rev. E* 50 (1994): R639–R642.

163. Ford, L. R., and D. R. Fulkerson. "Constructing Maximal Dynamic Flows from Static Flows." *Oper. Res.* 6 (1958): 419–433.

164. Ford, L. R., and D. R. Fulkerson. *Flows in Networks*, 419–433. Princeton, NJ: Princeton University Press, 1958.

165. Forst, C. V., C. M. Reidys, and J. Weber. "Evolutionary Dynamics and Optimization: Neutral Networks as Model Landscapes for RNA Secondary Structure Landscapes." In *Advances in Artificial Life, Third European Conference on Artificial Life (ECAL 1995)*, ed. F. Morán, A. Moreno, J. J. Merelo Guervós, and P. Chacón, 128–147. Lecture Notes in Artificial Intelligence, vol. 929. New York: Springer Verlag, 1995.

166. Franco, J. "Probabilistic Analysis of the Pure Literal Heuristic for the Satisfiability Problem." *Ann. Oper. Res.* 1 (1984): 273–289.

167. Franco, J., and M. Paull. "Probabilistic Analysis of the Davis-Putnam Procedure for Solving the Satisfiability Problem." *Discrete Appl. Math.* 5 (1983): 77–87.

168. Frank, J., P. Cheeseman, and J. Stutz. "When Gravity Fails: Local Search Topology." *J. Art. Intel. Res.* 7 (1997): 249–281.

169. Franz, S., M. Leone, A. Montanari, and F. Ricci-Tersenghi. "The Dynamic Phase Transition for Decoding Algorithms." *Phys. Rev. E* 66 (2002): 046120.

170. Franz, S., M. Leone, F. Ricci-Tersenghi, and R. Zecchina. "Exact Solutions for Diluted Spin Glasses and Optimization Problems." *Phys. Rev. Lett.* 87 (2001): 127209.

171. Fraser, S. M. Private communication, 1997.

172. Freedman, M. "k-SAT on Groups and Undecidability." In *Proceedings of the 30th Annual ACM Symposium on Theory of Computing (STOC'98)*, 572–576. New York: ACM Press, 1998.

173. Freedman, M. "Limits, Logic and Computation." *PNAS* 95 (1998): 95–97.

174. Freuder, Eugene C., and Richard J. Wallace. "Partial Constraint Satisfaction." *Art. Intel.* 58 (1992): 21–70.

175. Friedgut, E. "Boolean Functions with Low Average Sensitivity Depend on Few Coordinates." *Combinatorica* 18 (1998): 27–35.

176. Friedgut, E. "Hunting for Sharp Thresholds." *Rand. Struct. & Algorithms.* 26(1-2) (2005): 37–51.

177. Friedgut, E. "Sharp Thresholds of Graph Properties, and the k-SAT Problem." *J. Amer. Math. Soc.* 12 (1999): 1017–1054.

178. Friedgut, E., and G. Kalai. "Every Monotone Graph Property Has a Sharp Threshold." *Proc. Amer. Math. Soc.* 124 (1996): 2993–3002.

179. Friedgut, E., J. Kahn, and A. Wigderson. "Computing Graph Properties by Randomized Subcube Partitions." In *Randomization and Approximation*

Techniques in Computer Science, 6th International Workshop (RANDOM 2002), 105–113. Berlin: Springer-Verlag, 2002.

180. Frieze, A. M. "On the Independence Number of Random Graphs." *Discrete Math.* 81 (1990): 171.

181. Frieze, A. M., and B. V. Halldórsson. "Optimal Sequencing by Hybridization in Rounds." *J. Comp. Biol.* 9 (2002): 355–369.

182. Frieze, A., and S. Suen. "Analysis of Two Simple Heuristics on a Random Instance of k-SAT." *J. Algorithms* 20 (1996): 312–335.

183. Frieze, A., and N. C. Wormald. "Random k-SAT: A Tight Threshold for Moderately Growing k." In *Proceedings of the 5th International Symposium on the Theory and Applications of Satisfiability Testing (SAT 2002)*, 1–6. University of Cincinnati, 2002. Available at http://gauss.ececs.uc.edu/Conferences/SAT2002/sat2002list.html (accessed September 27, 2005).

184. Frieze, A. M., F. P. Preparata, and E. Upfal. "Optimal Reconstruction of a Sequence from Its Probes." *J. Comp. Biol.* 6 (1999): 361–368.

185. Fu, Y. "The Use and Abuse of Statistical Mechanics in Computational Complexity." In *Lectures in the Sciences of Complexity*, ed. D. L. Stein, 815–826. Santa Fe Institute Studies in the Sciences of Complexity Series, vol. 1. Reading, MA: Addison-Wesley, 1989.

186. Fujimoto, R. "Parallel Discrete Event Simulation." *Commun. ACM* 33 (1990): 30–53.

187. Galambos, J. *The Asymptotic Theory of Extreme Order Statistics*. Malabar, Florida: Robert E. Krieger Publishing Co., 1987.

188. Gallager, R. G. *Low Density Parity-Check Codes*. Cambridge, MA: MIT Press, 1963.

189. Galperin, H. "Succinct Representation of Graphs." Ph.D. thesis, Princeton University, 1982.

190. Galperin, H., and A. Wigderson. "Succinct Representation of Graphs." *Information and Control* 56 (1983): 183–198.

191. Garey, M. R., and D. S. Johnson. *Computers and Intractability. A Guide to the Theory of NP-Completeness*. New York: W.H. Freeman, 1997.

192. Gasper, G., and M. Rahman. *Basic Hypergeometric Series*, 2d ed. Encyclopedia of Mathematics and Its Applications, vol. 96. Cambridge, UK: Cambridge University Press, 2004.

193. Gazmuri, P. G. "Independent Sets in Random Sparse Graphs." *Networks* 14 (1984): 367–377.

194. Gent, I. P., and T. Walsh. "Phase Transitions and Annealed Theories: Number Partitioning as a Case Study." In *Proceedings of ECAI-96*, ed. W. Wahlster, 170–174. New York: John Wiley & Sons, 1996.

195. Gent, I. P., E. MacIntyre, P. Prosser, and T. Walsh. "The Constrainedness of Search." In *Proceedings of the 13th National Conference on Artificial Intelligence (AAAI'96)*, 246–252. Menlo Park, CA: AAAI Press, 1996.

196. Gent, I., H. van Maaren, and T. Walsh, eds. *SAT2000: Highlights of Satisfiability Research in the Year 2000.* Frontiers in Artificial Intelligence and Applications, vol. 63. Amsterdam: IOS Press, 2000.

197. Gitterman, M. "Small-World Phenomena in Physics: The Ising Model." *J. Phys. A* 33 (2000): 8373–8381.

198. Global Grid Forum. Home Page. http://www.gridforum.org (accessed September 27, 2005).

199. Goemans, M. X., and D. P. Williamson. ".878 Approximation Algorithms for MAX CUT and MAX 2SAT." In *Proceedings of the 26th Annual ACM Symposium on Theory of Computing (STOC'94)*, 422–431. New York: ACM Press, 1994.

200. Goemans, M. X., and D. P. Williamson. "Improved Approximation Algorithms for Maximum Cut and Satisfiability Problems using Semidefinite Programming." *J. ACM* 42 (1995): 1115–1145.

201. Goerdt, A. "A Threshold for Unsatisfiability." In *Proceedings of the 17th Symposium on Mathematical Foundations of Computer Science (MFCS'92)*, ed. I. M. Havel and V. Koubek, 264–274. Lecture Notes in Computer Science, vol. 629. Berlin: Springer-Verlag, 1992.

202. Goerdt, A. "A Threshold for Unsatisfiability." *J. Comp. Syst. Sci.* 53 (1996): 469–486.

203. Goldberg, A. "On the Complexity of the Satisfiability Problem." Courant Computer Science Report No. 16, New York University, 1979.

204. Goldberg, A., P. W. Purdom, and C. A. Brown. "Average Time Analyses of Simplified Davis-Putnam Procedures." *Inf. Proc. Lett.* 15 (1982): 72–75.

205. Gomes, C. P., and B. Selman. "Algorithm Portfolio Design: Theory vs. Practice." In *Proceedings of the 13th Conference on Uncertainty in Artificial Intelligence (UAI'97)*, ed. D. Geiger and P. Shenoy, 190–197. San Francisco: Morgan Kaufmann, 1997.

206. Grädel, E. "Domino Games and Complexity." *SIAM J. Comp.* 19 (1990): 787–804.

207. Graham, B. T., and G. R. Grimmett. "Influence and Sharp Threshold Theorems for Monotonic Measures." arXiv.org E-print Archive, Cornell University Library, 2005. http://www.arxiv.org/abs/math.PR/0505057 (accessed September 27, 2005).

208. Graham, R., O. Patashnik, and D. E. Knuth. *Concrete Mathematics: A Foundation for Computer Science*, 2d ed. Reading, MA: Addison-Wesley, 1994.

209. Greenberg, A. G., B. D. Lubachevsky, D. M. Nicol, and P. E. Wright. "Efficient Massively Parallel Simulation of Dynamic Channel Assignment Schemes for Wireless Cellular Communications." In *Proceedings of the Eighth Workshop on Parallel and Distributed Simulation (PADS'94)*, 187–197. New York: ACM Press, 1994.

210. Grimmett, G. *Percolation.* Berlin: Springer-Verlag, 1989.

211. Gropengiesser, U. "The Ground-State Energy of the $\pm J$-Spin Glass—A Comparison of Various Biologically Motivated Algorithms." *J. Stat. Phys.* 79 (1995): 1005–1012.

212. Gross, L. "Hypercontractivity, Logarithmic Sobolev Inequalities and Applications: A Survey of Surveys." Preprint, 2005.

213. Gross, L. "Logarithmic Sobolev Inequalities." *Amer. J. Math.* 97 (1975): 1061–1083.

214. Grötschel, M., M. Jünger, and G. Reinelt. "Calculating Exact Ground States of Spin Glasses: A Polyhedral Approach." In *Heidelberg Colloquium on Glassy Dynamics*, ed. J. L. van Hemmen and I. Morgenstern, 325–353. Lecture Notes in Physics, vol. 275. Berlin: Springer-Verlag, 1987.

215. Grover, L. K. "Quantum Mechanics Helps in Searching for a Needle in a Haystack." *Phys. Rev. Lett.* 79 (1997): 325–328.

216. Grunbaum, B., and G. Shephard. *Tilings and Patterns*. New York: Freeman and Company, 1986.

217. Grüner, W., R. Giegerich, D. Strothmann, C. M. Reidys, J. Weber, I. L. Hofacker, P. F. Stadler, and P. K. Schuster. "Analysis of RNA Sequence Structure Maps by Exhaustive Enumeration I. Neutral Networks." *Chemical Monthly* 127 (1996): 355–374.

218. Grüner, W., R. Giegerich, D. Strothmann, C. M. Reidys, J. Weber, I. L. Hofacker, P. F. Stadler, and P. K. Schuster. "Analysis of RNA Sequence Structure Maps by Exhaustive Enumeration II. Structures of Neutral Networks and Shape Space Covering." *Chemical Monthly* 127 (1996): 375–389.

219. Gu, J., R. Puri, and B. Du. "Satisfiability Problems in VLSI Engineering." Unpublished manuscript, 1996.

220. Gu, J., P. W. Purdom, J. Franco, and B. W. Wah. "Algorithms for Satisfiability (SAT) Problem: A Survey." In *Satisfiability Problem: Theory and Applications*, ed. D. Du, J. Gu, and P. M. Pardalos, 19–151. DIMACS Series on Discrete Mathematics and Theoretical Computer Science, vol. 35. American Mathematical Society, 1997.

221. Häggström, O. "Zero-Temperature Dynamics for the Ferromagnetic Ising Model on Random Graph." *Physica A* 310 (2002): 275–284.

222. Häggström, O., G. Kalai, and E. Mossel. "A Law of Large Numbers for Weighted Majority." Discussion Paper No. 363, Center for Rationality and Interactive Decision Theory, Hebrew University, Jerusalem. http://www.ratio.huji.ac.il/dp363.pdf (accessed September 27, 2005).

223. Hajiaghayi, M., and G. B. Sorkin. "The Satisfiability Threshold of Random 3-SAT is at least 3.52." arXiv.org E-print Archive, Cornell University Library, 2003. http://arXiv.org/abs/math.CO/0310193 (accessed September 27, 2005).

224. Harel, D. "Recurring Dominos: Making Highly Undecidable Highly Understandable." *Ann. Discrete Math.* 24 (1985): 51–72.

225. Harper, L. H. "Optimal Numberings and Isoperimetric Problems on Graphs." *J. Comb. Theor.* 1 (1966): 385–393.

226. Hart, S. "A Note on the Edges of the n-Cube." *Discrete Math.* 14 (1976): 157–163.

227. Hartmanis, J., and R. Stearns. "On the Computational Complexity of Algorithms." *Trans. Am. Math. Soc.* 117 (1965): 285–306.

228. Hartmann, A. K. "Ground-State Clusters of Two-, Three-, and Four-Dimensional $\pm J$ Ising Spin Glasses." *Phys. Rev. E* 63 (2001): 016106.

229. Hartmann, A. K. "Ground-State Landscape of $\pm J$ Ising Spin Glasses." *Eur. Phys. J. B* 8 (1999): 619–626.

230. Hartmann, A. K., and H. Rieger. *Optimization Algorithms in Physics.* Berlin: Wiley-VHC Verlag, 2002.

231. Hartwig, A., F. Daske, and S. Kobe. "A Recursive Branch-and-Bound Algorithm for the Exact Ground State of Ising Spin-Glass Models." *Comp. Phys. Commun.* 32 (1984): 133–138.

232. Håstad, J. "Almost Optimal Lower Bounds for Small Depth Circuits." In *Randomness and Computation,* ed. S. Micali, 143–170. Advances in Computing Research, vol. 5. JAI Press, 1989.

233. Håstad, J. "Clique is Hard to Approximate within n to the Power $1 - \epsilon$." *Acta Mathematica* 182 (1999): 105–142.

234. Håstad, J. "A Slight Sharpening of LMN." *J. Comp. Syst. Sci.* 63 (2001): 498–508.

235. Håstad, J. "Some Optimal Inapproximability Results." *J. ACM* 48 (2001): 798–859.

236. Hayes, B. "Computing Science: Can't Get No Satisfaction." *Amer. Sci.* 85(2) (1997): 108–112.

237. Hayes, B. "Computing Science: The Easiest Hard Problem." *Amer. Sci.* 90(2) (2002): 113–117.

238. Hed, G., A. K. Hartmann, D. Stauffer, and E. Domany. "Spin Domains Generate Hierarchical Ground State Structure in $J = \pm 1$ Spin Glasses." *Phys. Rev. Lett.* 86 (2001): 3148–3151.

239. Henschen, L., and L. Wos. "Unit Refutations and Horn Sets." *J. ACM* 21 (1974): 590–605.

240. Hirst, T., and D. Harel. "Taking it to the Limit: On Infinite Variants of NP-Complete Problems." *J. Comp. Syst. Sci.* 53 (1996): 180–193.

241. Hofacker, I. L., P. K. Schuster, and P. F. Stadler. "Combinatorics of RNA Secondary Structures." *Discrete Appl. Math.* 88 (1998): 207–237.

242. Hofacker, I. L., W. Fontana, L. S. Stadler, P. F. Bonhoeffer, M. Tacker, and P. K. Schuster. Vienna RNA Package. http://www.tbi.univie.ac.at/~ivo/RNA/ (accessed September 27, 2005).

243. Höfting, F., and E. Wanke. "Minimum Cost Paths in Dynamic Graphs." *SIAM J. Comp.* 24 (1995): 1051–1067.

244. Höfting, F., and E. Wanke. "Polynomial Time Analysis of Toroidal Dynamic Graphs." *J. Algorithms* 34 (2000): 14–39.

245. Höfting, F., T. Lengauer, and E. Wanke. "Processing of Hierarchically Defined Graphs and Graph Families." In *Data Structures and Efficient Algorithms,* ed.

B. Monien and T. Ottmann, 44–69. Lecture Notes in Computer Science, vol. 594. Berlin: Springer-Verlag, 1992.

246. Hogg, T. "Adiabatic Quantum Computing for Random Satisfiability Problems." *Phys. Rev. A* 67 (2003): 022314.

247. Hogg, T. "Exploiting Problem Structure as a Search Heuristic." *Intl. J. Mod. Phys. C* 9 (1998): 13–29.

248. Hogg, T. "A Framework for Structured Quantum Search." *Physica D* 120 (1998): 102–116.

249. Hogg, T. "Highly Structured Searches with Quantum Computers." *Phys. Rev. Lett.* 80 (1998): 2473–2476.

250. Hogg, T. "Quantum Search Heuristics." *Phys. Rev. A* 61 (2000): 052311.

251. Hogg, T. "Single-Step Quantum Search using Problem Structure." *Intl. J. Mod. Phys. C* 11 (2000): 739–773.

252. Hogg, T, and D. Portnov. "Quantum Optimization." *Inf. Sci.* 128 (2000): 181–197.

253. Hogg, T., and C. P. Williams. "The Hardest Constraint Problems: A Double Phase Transition." *Art. Intel.* 69 (1994): 359–377.

254. Hogg, T., B. A. Huberman, and C. Williams, eds. "Frontiers in Problem Solving: Phase Transitions and Complexity." *Art. Intel.* 81 (1996): 1–15.

255. Hogg, T., C. Mochon, E. Rieffel, and W. Polak. "Tools for Quantum Algorithms." *Intl. J. Mod. Phys. C* 10 (1999): 1347–1361.

256. Hong, H., B. J. Kim, and M. Y. Choi. "Comment on 'Ising Model on a Small World Network.'" *Phys. Rev. E* 66 (2002): 018101.

257. Hong, H., M. Y. Choi, and B. J. Kim. "Phase Ordering on Small-World Networks with Nearest-Neighbor Edges." *Phys. Rev. E* 65 (2002): 047104.

258. Hoos, H. H. "A Mixture-Model for the Behavior of SLS Algorithms for SAT." In *Proceedings of the 18th National Conference on Artificial Intelligence (AAAI'02)*, 661–667. Menlo Park, CA: AAAI Press, 2002.

259. Hoos, H. H., and T. Stützle. "Local Search Algorithms for SAT: An Empirical Evaluation." *J. Automated Reasoning* 24 (2000): 421–481.

260. Hoppe, B., and E. Tardos. "The Quickest Transshipment Problem." *Math. Oper. Res.* 25 (2000): 36–62.

261. Horvitz, E., Y. Ruan, C. P. Gomes, H. Kautz, B. Selman, and D. M. Chickering. "A Bayesian Approach to Tackling Hard Computational Problems." In *Proceedings of the 17th Conference on Uncertainty in Artificial Intelligence (UAI'01)*, ed. J. Breeze and D. Koller, 235–244. San Francisco: Morgan Kaufmann, 2001.

262. Houdayer, J., and O. C. Martin. "Hierarchical Approach for Computing Spin Glass Ground States." *Phys. Rev. E* 64 (2001): 056704.

263. Howell, J. A., T. F. Smith, and M. S. Waterman. "Computation of Generating Functions for Biological Molecules." *SIAM J. Appl. Math.* 39 (1980): 119–133.

264. Huberman, B. A., and T. Hogg. "Phase Transitions in Artificial Intelligence Systems." *Art. Intel.* 33 (1987): 155–171.

265. Huberman, B. A., R. M. Lukose, and T. Hogg. "An Economics Approach to Hard Computational Problems." *Science* 275 (1997): 51–54.

266. Hunt, H., III, M. Marathe, V. Radhakrishnan, S. Ravi, D. Rosenkrantz, and R. Stearns. "NC-approximation Schemes for NP- and PSPACE-hard Problems for Geometric Graphs." *J. Algorithms* 26 (1998): 238–274.

267. Hunt, H., III, R. Stearns, and M. Marathe. "Relational Representability, Local Reductions and the Complexity of Generalized Satisfiability Problem." Technical Report No. LA-UR-00-6108, Los Alamos National Laboratory, Los Alamos, NM, 2000.

268. Hunt, H., III, R. Stearns and M. Marathe. "Strongly Local Reductions and the Complexity/Efficient Approximability of Algebra and Optimization on Abstract Algebraic Structures." In *Proceedings of the Seventeenth International Joint Conference on Artificial Intelligence (IJCAI'01)*, ed. B. Nebel, 183–191. San Mateo, CA: Morgan Kaufmann, 2001.

269. Hunt, H., III, M. Marathe. V. Radhakrishnan, S. Ravi, D. Rosenkrantz, and R. Stearns. "Parallel Approximation Schemes for a Class of Planar and Near Planar Combinatorial Problems." *Information and Computation* 173 (2002): 40–63.

270. Istrate, G. "Computational Complexity and Phase Transitions." In *Proceedings of the 15th IEEE Annual Conference on Computational Complexity (CCC 2000)*, 104–115. IEEE Computer Society, 2000.

271. Istrate, G. "The Phase Transition in Random Horn Satisfiability and Its Algorithmic Implications." *Rand. Struct. & Algorithms* 20 (2002): 483–506.

272. Istrate, G. "On the Satisfiability of Random k-Horn Formulae." In *Graphs, Morphisms and Statistical Physics*, ed. J. Nesetril and P. Winkler, 113–136. DIMACS Series in Discrete Mathematics and Theoretical Computer Science, vol. 63. American Mathematical Society, 2004.

273. Istrate, G., S. Boettcher, and A. G. Percus. "Spines of Random Constraint Satisfication Problems: Definition and Connection with Computational Complexity." *Ann. Math. & Art. Intel.* 44 (2004): 353–372.

274. Iwano, K., and K. Steiglitz. "Optimization of One-Bit Full Adders Embedded in Regular Structures." *IEEE Trans. Acoustics, Speech, and Signal Processing* 34 (1986): 1289-1300. Reprinted in *Computer Arithmetic*, ed. E. Swartzlander, 193–204. Los Alamitos, CA: IEEE Computer Society, 1990.

275. Iwano, K., and K. Steiglitz. "Planarity Testing of Doubly Connected Dynamic Infinite Graphs." *Networks* 18 (1988): 205–222.

276. Iwano, K., and K. Steiglitz. "Testing for Cycles in Infinite Graphs with Dynamic Structure." In *Proceedings of the 19th Annual ACM Symposium on Theory of Computing (STOC'87)*, 46–53. New York: ACM Press, 1987.

277. Janson, S. "New Versions of Suen's Correlation Inequality." *Rand. Struct. & Algorithms* 13 (1998): 467–483.

278. Janson, S., Y. C. Stamatiou, and M. Vamvakari. "Bounding the Unsatisfiability Threshold of Random 3-SAT." *Rand. Struct. & Algorithms* 17 (2000): 103–116.

279. Janson, S., T. Luczak, and A. Rucinski. *Random Graphs.* New York: John Wiley & Sons, 2000.

280. Jefferson, D. R. "Virtual Time." *ACM Trans. Prog. Lang. Syst.* 7 (1985): 404–425.

281. Johnson, D. S., C. R. Aragon, L. A. McGeoch, and C. Schevon. "Optimization by Simulated Annealing: An Experimental Evaluation; Part II, Graph Coloring and Number Partitioning." *Oper. Res.* 39 (1991): 378–406.

282. Johnson, D. S., L. A. McGeoch, and E. E. Rothberg. "Asymptotic Experimental Analysis for the Held-Karp Traveling Salesman Bound." In *Proceedings of the 7th Annual ACM-SIAM Symposium on Discrete Algorithms (SODA'96)*, 341–350. New York: ACM Press, 1996.

283. Jones, N. D., and W. T. Laaser. "Complete Problems for Deterministic Polynomial Time." *Theor. Comp. Sci.* 3 (1977): 105–117.

284. Joyce, G. F. "Directed Molecular Evolution." *Sci. Am.* 267(6) (1992): 90–97.

285. Kadanoff, L. P., W. Götze, D. Hamblen, R. Hecht, E. A. S. Lewis, V. V. Palciauskas, M. Rayl, J. Swift, D. Aspnes, and J. Kane. "Static Phenomena near Critical Points: Theory and Experiment." *Rev. Mod. Phys.* 39 (1967): 395–431.

286. Kahn, J., and G. Kalai. "A Discrete Isoperimetric Conjecture with Probabilistic Applications." Preprint (2005).

287. Kahn, J., G. Kalai, and N. Linial. "The Influence of Variables on Boolean Functions. In *Proceedings of the 29th Annual IEEE Symposium on Foundations of Computer Science (FOCS'88)*, 68–80. IEEE Computer Society, 1988.

288. Kalai, G. "A Fourier-Theoretic Perspective for the Condorcet Paradox and Arrow's Theorem." *Adv. in Appl. Math.* 29 (2002): 412–426.

289. Kalai, G. "Noise Sensitivity and Chaos in Social Choice Theory." Discussion Paper No. 399. Center for Rationality and Interactive Decision Theory, Hebrew University, Jerusalem. http://ratio.huji.ac.il/dp/dp399.pdf (accessed September 27, 2005).

290. Kalai, G. "Social Choice and Threshold Phenomena." Discussion Paper No. 279. Center for Rationality and Interactive Decision Theory, Hebrew University, Jerusalem. http://ratio.huji.ac.il/dp/dp279.pdf (accessed September 27, 2005).

291. Kalai, G. "Social Indeterminacy." *Econometrica* 72 (2004): 1565–1581.

292. Kamath, A., R. Motwani, K. Palem, and P. Spirakis. "Tail Bounds for Occupancy and the Satisfiability Threshold Conjecture." *Rand. Struct. & Algorithms* 7 (1995): 59–80.

293. Kaporis, A. C., L. M. Kirousis, Y. C. Stamatiou, M. Vamvakari, and M. Zito. "The Unsatisfiability Threshold Revisited." Paper presented at LICS 2001 Workshop on Theory and Applications of Satisfiability Testing." *Elec. Notes Discrete Math.* 9 (2001): Paper no. 7.

294. Kaporis, A. C., L. M. Kirousis, and E. G. Lalas. "The Probabilistic Analysis of a Greedy Satisfiability Algorithm." In *Proceedings of the 10th Annual Eu-*

ropean Symposium on Algorithms, ed. R. H. Möhring and R. Raman, 574–585. Lecture Notes in Computer Science, vol. 2461. Berlin: Springer-Verlag, 2002.

295. Kaporis, A. C., L. M. Kirousis, and E. G. Lalas. "Selecting Complementary Pairs of Literals." *Elec. Notes Discrete Math.* 16 (2003): 47–70.

296. Kardar, M., G. Parisi, and Y.-C. Zhang. "Dynamic Scaling of Growing Interfaces." *Phys. Rev. Lett.* 56 (1986): 889–892.

297. Karmarkar, N., and R. M. Karp. "The Differencing Method of Set Partitioning." Technical Report UCB/CSD 81/113, Computer Science Division, University of California, Berkeley, 1982.

298. Karmarkar, N., R. M. Karp, G. S. Lueker, and A. M. Odlyzko. "Probabilistic Analysis of Optimum Partitioning." *J. Appl. Prob.* 23 (1986): 626–645.

299. Karp, R. M. "Reducibility among Combinatorial Problems." In *Complexity of Computer Computations*, ed. R. E. Miller and J. W. Thatcher, 85–103. New York: Plenum Press, 1972.

300. Karp, R. M. "The Transitive Closure of a Random Digraph." *Rand. Struct. & Algorithms* 1 (1990): 73–93.

301. Karp, R. M., and J. Pearl. "Searching for an Optimal Path in a Tree with Random Costs." *Art. Intel.* 21 (1983): 99–116.

302. Karp, R. M., R. E. Miller, and S. Winograd. "The Organization of Computations for Uniform Recurrence Equations." *J. ACM* 14 (1967): 563–590.

303. Kautz, H., Y. Ruan, D. Achlioptas, C. Gomes, B. Selman, and M. Stickel. "Balance and Filtering in Structured Satisfiable Problems." In *Proceedings of the Seventeenth International Joint Conference on Artificial Intelligence (IJCAI'01)*, ed. B. Nebel, 351–358. San Mateo, CA: Morgan Kaufmann, 2001. Available at http://www.cs.cornell.edu/gomes/new-papers.htm (accessed September 27, 2005).

304. Kautz, H., E. Horvitz, Y. Ruan, C. Gomes, and B. Selman. "Dynamic Restart Policies." In *Proceedings of the 18th National Conference on Artificial Intelligence (AAAI'02)*, 674–681. Menlo Park, CA: AAAI Press, 2002.

305. Kempe, D., J. Kleinberg, and A. Kumar. "Connectivity and Inference Problems for Temporal Networks." *J. Comp. Syst. Sci.* 64 (2002): 820–842.

306. Kesten, H. "The Critical Probability of Bond Percolation on the Square Lattice Equals 1/2." *Comm. Math. Phys.* 74 (1980): 41–59.

307. Kesten, H. "Scaling Relations for 2D-Percolation." *Comm. Math. Phys.* 109 (1987): 109–156.

308. Kesten, H., and Y. Zhang. "Strict Inequalities for Some Critical Exponents in 2D-Percolation." *J. Stat. Phys.* 46 (1987): 1031–1055.

309. Khanna, S., and R. Motwani. "Towards a Syntactic Characterization of PTAS." In *Proceedings of the 28th Annual ACM Symposium on Theory of Computing (STOC'96)*, 329–337. New York: ACM Press, 1996.

310. Khanna, S., M. Sudan, L. Trevisan, and D. Williamson. "The Approximability of Constraint Satisfaction Problems." *SIAM J. Comp.* 30 (2001): 1863–1920.

311. Khintchine, A. "Über dyadische Brüche." *Math. Z.* 18 (1923): 109–116.

312. Khot, S. "On the Power of Unique 2-Prover 1-Round Games." In *Proceedings of the 34th Annual ACM Symposium on Theory of Computing (STOC'02)*, 767–775. New York: ACM Press, 2002.

313. Khot, S., and O. Regev. "Vertex Cover Might be Hard to Approximate to Within $2 - \varepsilon$." In *Proceedings of the 18th IEEE Annual Conference on Computational Complexity (CCC'03)*, 379–386. IEEE Computer Society, 2003.

314. Khot, S., and N. Vishnoi. "The Unique Games Conjecture, Integrality Gap for Cut Problems and Embeddability of Negative Type Metrics into ℓ_1." In *Proceedings of the 46th Annual IEEE Symposium on Foundations of Computer Science (FOCS'05)*, to appear. IEEE Computer Society, 2005.

315. Khot, S., G. Kindler, E. Mossel, and R. O'Donnell. "Optimal Inapproximability Results for MAX-CUT and Other 2-Variable CSPs?" In *Proceedings of the 45th Annual IEEE Symposium on Foundations of Computer Science (FOCS'04)*, 146–154. IEEE Computer Society, 2004.

316. Kim, B. J., H. Hong, P. Holme, G. S. Jeon, P. Minnhagen, and M. Y. Choi. "XY Model in Small-World Networks." *Phys. Rev. E* 64 (2001): 056135.

317. Kimura, M. *The Neutral Theory of Molecular Evolution.* Cambridge, UK: Cambridge University Press, 1983.

318. Kirkpatrick, S. "Rough Times Ahead." *Science* 299 (2003): 668–669.

319. Kirkpatrick, S., and B. Selman. "Critical Behavior in the Satisfiability of Random Boolean Expressions." *Science* 264 (1994): 1297–1301.

320. Kirkpatrick, S., C. D. Gelatt, and M. P. Vecchi. "Optimization by Simulated Annealing." *Science* 220 (1983): 671–680.

321. Kirkpatrick, S., G. Györgi, N. Tishby, and L. Troyansky. "The Statistical Mechanics of k-Satisfaction." In *Advances in Neural Information Processing Systems*, ed. J. D. Cowan, G. Tesauro, and J. Alspector, vol. 6, 439–446. San Mateo, CA: Morgan Kaufmann Publishers, 1993.

322. Kirousis, L. M., E. Kranakis, and D. Krizanc. "An Upper Bound for a Basic Hypergeometric Series." Technical Report TR-96-07, Carleton University, School of Computer Science, Canada, 1996. Available at http://www.scs.carleton.ca/research/tech_reports/1996 (accessed September 27, 2005).

323. Kirousis, L. M., E. Kranakis, and D. Krizanc. "A Better Upper Bound for the Unsatisfiability Threshold." In *Satisfiability Problem: Theory and Applications*, ed. D. Du, J. Gu, and P. M. Pardalos, 643–648. DIMACS Series in Discrete Mathematics and Theoretical Computer Science, vol. 35. American Mathematical Society, 1997.

324. Kirousis, L. M., E. Kranakis, D. Krizanc, and Y. C. Stamatiou. "Approximating the Unsatisfiability Threshold of Random Formulas." *Rand. Struct. & Algorithms* 12 (1998): 253–269.

325. Kirousis, L. M., Y. C. Stamatiou, and M. Vamvakari. "Upper Bounds and Asymptotics for the q-Binomial Coefficients." *Stud. Appl. Math.* 107 (2001): 43–62.

326. Kleinberg, J. "Navigation in a Small World." *Nature* 406 (2000): 845.

327. Klotz, T. "Zur Phasenraumstruktur in ungeordneten $\pm I$ Ising-Spinsystemen." Thesis, Technische Universität Dresden, 1996

328. Klotz, T., and S. Kobe. "Cluster Structures in the Configuration Space and Relaxation in 3d $\pm I$ Ising Spin-Glass Model." In *Hayashibara Forum '95, Int. Symp. Coherent Approaches to Fluctuations*, ed. by M. Suzuki and N. Kawashima, 192–195. Singapore: World Scientific, 1996.

329. Klotz, T., and S. Kobe. "Exact Low-Energy Landscape and Relaxation Phenomena in Ising Spin Glasses." *Acta Phys. Slovaca* 44 (1994): 347–356.

330. Klotz, T., and S. Kobe. "'Valley Structure' in the Phase Space of a Finite 3D Ising Spin Glass with $\pm I$ Interactions." *J. Phys. A: Math. Gen.* 27 (1994): L95–L100.

331. Klug, S. J., and M. Famulok. "All You Wanted to Know about SELEX." *Mol. Biol. Rep.* 20 (1994): 97–107.

332. Knuth, D. E. *Stable Marriage and its Relation to Other Combinatorial Problems: An Introduction to the Mathematical Analysis of Algorithms.* CRM Proceedings & Lecture Notes, vol. 10. American Mathematical Society, 1997. First French edition: Les Presses de l'Université de Montréal, 1976.

333. Knuth, D. E., R. Motwani, and B. Pittel. "Stable Husbands." *Rand. Struct. & Algorithms* 1 (1990): 1–14.

334. Kobe, S. "Ground State of an Ising Antiferromagnet with Dense Random Packing Structure." In *Amorphous Magnetism II*, ed. R. A. Levy and R. Hasegawa, 529–534. New York and London: Plenum Press, 1977.

335. Kobe, S., and K. Handrich. "Correlation Function and Misfit in a Computer-Simulated Two-Dimensional Amorphous Ising Antiferromagnet." *Phys. Stat. Sol.* 73(b) (1976): K65–K67.

336. Kobe, S., and A. Hartwig. "Exact Ground State of Finite Amorphous Ising Systems." *Comp. Phys. Commun.* 16 (1978): 1–4.

337. Kobe, S., and T. Klotz. "Frustration: How It Can be Measured." *Phys. Rev. E* 52 (1995): 5660–5663.

338. Kodialam, M., and J. B. Orlin. "Recognizing Strong Connectivity in Dynamic Graphs and Its Relation to Integer Programming." In *Proceedings of the 2nd Annual ACM-SIAM Symposium on Discrete Algorithms (SODA '91)*, 131–135. New York: ACM Press, 1991.

339. Kolaitis, P., and T. Raffill. "In Search of a Phase Transition in the AC-Matching Problem." In *Proceedings of the Seventh International Conference on Principles and Practice of Constraint Programming (CP 2001)*, ed. T. Walsh, 433–450. Berlin: Springer-Verlag, 2001.

340. Kolakowska, A. K., M. A. Novotny, and G. Korniss. "Algorithmic Scalability in Globally Constrained Conservative Parallel Discrete Event Simulations of Asynchronous Systems." *Phys. Rev. E* 67 (2003): 046703.

341. Korf, R. E. "A Complete Anytime Algorithm for Number Partitioning." *Art. Intel.* 106 (1998): 181–203.

342. Korniss, G., M. A. Novotny, and P. A. Rikvold. "Parallelization of a Dynamic Monte Carlo Algorithm: A Partially Rejection-Free Conservative Approach." *J. Comp. Phys.* 153 (1999): 488–508.

343. Korniss, G., Z. Toroczkai, M. A. Novotny, and P. A. Rikvold. "From Massively Parallel Algorithms and Fluctuating Time Horizons to Nonequilibrium Surface Growth." *Phys. Rev. Lett.* 84 (2000): 1351–1354.

344. Korniss, G., M. A. Novotny, Z. Toroczkai, and P. A. Rikvold. "Non-Equilibrium Surface Growth and Scalability of Parallel Algorithms for Large Asynchronous Systems." In *Computer Simulated Studies in Condensed Matter Physics XIII*, ed. D. P. Landau, S. P. Lewis and H.-B. Schüttler, 183–188. Berlin: Springer-Verlag, 2001.

345. Korniss, G., M. A. Novotny, P. A. Rikvold, H. Guclu, and Z. Toroczkai. "Going Through Rough Times: From Non-Equilibrium Surface Growth to Algorithmic Scalability." *Mat. Res. Soc. Symp. Proc.* 700 (2002): 297–308.

346. Korniss, G., M. A. Novotny, H. Guclu, Z. Toroczkai, and P. A. Rikvold. "Suppressing Roughness of Virtual Times in Parallel Discrete-Event Simulations." *Science* 299 (2003): 677–679.

347. Kosaraju, K. R., and G. F. Sullivan. "Detecting Cycles in Dynamic Graphs in Polynomial Time." In *Proceedings of the 27th Annual IEEE Symposium on Foundations of Computer Science (FOCS'88)*, 398–406. IEEE Computer Society, 1988.

348. Krauth, W., and M. Mézard. "The Cavity Method and the Travelling-Salesman Problem." *Europhys. Lett.* 8 (1989): 213–218.

349. Krawczyk, J., and S. Kobe. "Low-Temperature Dynamics of Spin Glasses: Walking in the Energy Landscape." *Physica A* 315 (2002): 302–307.

350. Krug, J., and P. Meakin. "Universal Finite-Size Effects in the Rate of Growth Processes." *J. Phys. A* 23 (1990): L987–L994.

351. Kschischang, F. R., B. J. Frey, and H.-A. Loeliger. "Factor Graphs and the Sum-Product Algorithm." *IEEE Trans. Inf. Theory* 47 (2001): 498–519.

352. Ladner, R. "On the Structure of Polynomial Time Reducibility." *J. ACM* 22 (1975): 155–171.

353. Lancaster, D. "Two Combinatorial Models with Identical Statics Yet Different Dynamics." *J. Phys. A: Math. Gen.* 37 (2003): 1125–1143. Cond-mat/0310743.

354. Land, A. H., and A. G. Doig. "An Automatic Method for Solving Discrete Programming Problems." *Econometrica* 28 (1960): 497–520.

355. Lawler, G., O. Schramm, and W. Werner. "One-Arm Exponent for Critical 2D Percolation." *Elec. J. Prob.* 7 (2002): Paper no. 2.

356. Lebrecht, W., and E. E. Vogel. "Ground-State Properties of Finite Square and Triangular Ising Lattices with Mixed Exchange Interactions." *Phys. Rev. B* 49 (1994): 6018–6027.

357. Ledoux, M. *The Concentration of Measure Phenomenon*, Mathematical Surveys and Monographs, 89. Providence, RI: American Mathematical Society, 2001.

358. Lengauer, C. "Loop Parallelization in the Polytope Model." In *Proceedings of the 4th International Conference on Concurrency Theory*, 398–416. Lecture Notes in Computer Science, vol. 715. Berlin: Springer-Verlag, 1993.

359. Leone, M., A. Vázquez, A. Vespignani, and R. Zecchina. "Ferromagnetic Ordering in Graphs with Arbitrary Degree Distribution." *Eur. Phys. J. B* 28 (2002): 191–197.

360. Levin, L. "Universal Sequential Search Problems." *Problems of Information Transmission* 9(3) (1973): 265–266.

361. Liebmann, R. *Statistical Mechanics of Periodic Frustrated Ising Systems.* Lecture Notes in Physics, vol. 251. Berlin: Springer-Verlag, 1986

362. Linial, N., Y. Mansour, and N. Nisan. "Constant Depth Circuits, Fourier Transform, and Learnability." *J. Assoc. Comp. Mach.* 40 (1993): 607–620.

363. Little, J. D. C., K. G. Murty, D. W. Sweeney, and C. Karel. "An Algorithm for the Traveling-Salesman Problem." *Oper. Res.* 11 (1963): 972–989.

364. Loomis, L., and H. Whitney. "An Inequality Related to the Isoperimetric Inequality." *Bull. Amer. Math. Soc.* 55 (1949): 961–962.

365. Lubachevsky, B. D. "Efficient Parallel Simulations of Asynchronous Cellular Arrays." *Complex Systems* 1 (1987): 1099–1123.

366. Lubachevsky, B. D. "Efficient Parallel Simulations of Dynamic Ising Spin Systems." *J. Comp. Phys.* 75 (1988): 103–122.

367. Lubachevsky, B. D. "Fast Simulation of Multicomponent Dynamic Systems." *Bell Labs Tech. J.* 5 (2000): 134–156.

368. Luby, M., and W. Ertel. "Optimal Parallelization of Las Vegas Algorithms." In *STACS '94: 11th Annual Symposium on Theoretical Aspects of Computer Science*, ed. P. Enjalbert, E. Mayr, and K.W. Wagner, 463–475. Lecture Notes in Computer Science, vol. 775. Berlin: Springer-Verlag, 1994.

369. Luby, M., A. Sinclair, and D. Zuckerman. "Optimal Speedup of Las Vegas Algorithms." *Inf. Proc. Lett.* 47 (1993): 173–180.

370. Lueker, G. S. "Exponentially Small Bounds on the Expected Optimum of the Partition and Subset Sum Problems." *Rand. Struct. & Algorithms* 12 (1998): 51–62.

371. Lynch, J. F. "On the Threshold of Chaos in Random Boolean Cellular Automata." *Rand. Struct. & Algorithms* 6 (1995): 239–260.

372. Lynch, J. F. "A Phase Transition in Random Boolean Networks." In *Artificial Life IV*, ed. Rodney A. Brooks and Pattie Maes, 236–245. Cambridge, MA: MIT Press, 1994.

373. MacLane, S., and G. Birkhoff. *Algebra.* New York: Macmillan, 1967.

374. Majumdar, S. N., and P. L. Krapivsky. "Extreme Value Statistics and Traveling Fronts: Application to Computer Science." *Phys. Rev. E* 65 (2002): 036127.

375. Makowsky, J. A. "Why Horn Formulas Matter in Computer Science: Initial Structures and Generic Examples." *J. Comp. Syst. Sci.* 34 (1987): 266–292.

376. Marathe, M.V. "Towards a Predictive Computational Complexity Theory." In *Proceedings of the 29th International Colloquium on Automata Languages*

and Programming, ed. P. Widmayer, F. T. Ruiz, R. Morales, M. Hennessy, S. Eidenbenz, and R. Conejo, 22–31. Lecture Notes in Computer Science, vol. 2380. Berlin: Springer-Verlag, 2002.

377. Marathe, M. V., H. B. Hunt III, R. E. Stearns, and V. Radhakrishnan. "A Dichotomy Theorem for Hierarchically and 1-Dimensional Periodically Specified Satisfiability Problems with Applications." Unpublished manuscript, 1996.

378. Marathe, M. V., H. B. Hunt III, R. E. Stearns, and V. Radhakrishnan. "Complexity of Hierarchically and 1-Dimensional Dynamically Specified Problems I: Hardness Results." In *Satisfiability Problem: Theory and Application*, ed. D. Du., J. Gu, and P. M. Pardalos, 225–259. DIMACS Series on Discrete Mathematics and Theoretical Computer Science, vol. 35. American Mathematical Society, 1997.

379. Marathe, M. V., H. B. Hunt III, R. E. Stearns, and V. Radhakrishnan. "Approximation Algorithms for PSPACE-Hard Hierarchically and Dynamically Specified Problems." *SIAM J. Comp.* 27 (1998): 1237–1261.

380. Marathe, M. V., H. B. Hunt III, D. J. Rosenkrantz, and R. E. Stearns. "Theory of Dynamically Specified Problems: Complexity and Approximability." In *Proceedings of the 13th IEEE Annual Conference on Computational Complexity*, 106–119. Washington, DC: IEEE Computer Society, 1998.

381. Margulis, G. "Probabilistic Characteristics of Graphs with Large Connectivity (in Russian)." *Probl. Pered. Inf.* 10 (1974): 101–108.

382. Martin, O. C., R. Monasson, and R. Zecchina. "Statistical Mechanics Methods and Phase Transitions in Optimization Problems." *Theor. Comp. Sci.* 265 (2001): 3–67.

383. Maurer, S. M., T. Hogg, and B. A. Huberman. "Portfolios of Quantum Algorithms." *Phys. Rev. Lett.* 87 (2001): 257901.

384. McAllester, D., B. Selman, and H. Kautz. "Evidence for Invariants in Local Search." In *Proceedings of the 14th National Conference on Artificial Intelligence (AAAI'97)*, 321–326. Menlo Park, CA: AAAI Press, 1997.

385. McKane, A. J., M. Droz, J. Vannimenus, and D. Wolf, eds. *Scale Invariance, Interfaces and Non-Equilibrium Dynamics*. NATO ASI Series B, vol. 344. New York: Plenum, 1995.

386. Melin, R., J. C. Angles d'Auriac, P. Chandra, and B. Douçot. "Glassy Behavior in the Ferromagnetic Ising Model on a Cayley Tree." *J. Phys. A* 29 (1996): 5773–5804.

387. Merkle, R. C., and M. E. Hellman. "Hiding Informations and Signatures in Trapdoor Knapsacks." *IEEE Trans. Inf. Theory* 24 (1978): 525–530.

388. Mertens, S. "Computational Complexity for Physicists." *Comp. Sci. & Eng.* 4(3) (2002): 31–47.

389. Mertens, S. "Phase Transition in the Number Partitioning Problem" *Phys. Rev. Lett.* 81 (1998): 4281–4284.

390. Mertens, S. "A Physicist's Approach to Number Partitioning." *Theor. Comp. Sci.* 265 (2001): 79–108.

391. Mertens, S. "Random Costs in Combinatorial Optimization." *Phys. Rev. Lett.* 84 (2000): 1347–1350.

392. Mézard, M., and G. Parisi. "The Cavity Method at Zero Temperature." *J. Stat. Phys.* 111 (2003): 1–34.

393. Mézard, M., and G. Parisi. "A Replica Analysis of the Travelling Salesman Problem." *J. Physique* 47 (1986): 1285–1296.

394. Mézard, M., and G. Parisi. "Replicas and Optimization." *J. Physique Lett.* 46 (1985): L771–L778.

395. Mézard, M., and R. Zecchina. "Random K-Satisfiability Problem: From an Analytic Solution to an Efficient Algorithm." *Phys. Rev. E* 66 (2002): 056126.

396. Mézard, M., G. Parisi, and M. A. Virasoro. *Spin Glass Theory and Beyond.* Singapore: World Scientific, 1987.

397. Mézard, M., G. Parisi, and R. Zecchina. "Analytic and Algorithmic Solutions of Random Satisfiability Problems." *Science* 297 (2002): 812–815.

398. Mézard, M., F. Ricci-Tersenghi, and R. Zecchina. "Two Solutions to Diluted p-Spin Models and XORSAT Problems." *J. Stat. Phys.* 111 (2003): 505–533.

399. Mills, D. R., R. L. Peterson, and S. Spiegelman. "An Extracellular Darwinian Experiment with a Self-Duplicating Nucleic Acid Molecule." *PNAS* 58 (1967): 217–224.

400. Mitchell, D., B. Selman, and H. Levesque. "Hard and Easy Distributions of SAT Problems." In *Proceedings of the Tenth National Conference on Artificial Intelligence (AAAI'92)*, 440–446. Menlo Park, CA: AAAI Press, 1992.

401. Molloy, M. "Models for Random Constraint Satisfaction Problems." *SIAM J. Comp.* 32 (2003): 935–949.

402. Molloy, M. Private communication, February 26, 2001.

403. Molloy, M. "Thresholds for Colourability and Satisfiability in Random Graphs and Boolean Formulae." In *Surveys in Combinatorics, 2001*, ed. J. Hirschfeld, 166–200. LMS Lecture Note Series 288. Cambridge, UK: Cambridge University Press, 2001.

404. Monasson, R. "Structural Glass Transition and the Entropy of the Metastable States." *Phys. Rev. Lett.* 75 (1995): 2847–2850.

405. Monasson, R., and R. Zecchina. "Statistical Mechanics of the Random K-Satisfiability Model." *Phys. Rev. E* 56 (1997): 1357–1370.

406. Monasson, R., R. Zecchina, S. Kirkpatrick, B. Selman, and L. Troyansky. "Determining Computational Complexity from Characteristic 'Phase Transitions'." *Nature* 400 (1999):133–137.

407. Monasson, R., R. Zecchina, S. Kirkpatrick, B. Selman, and L. Troyansky. "2+p-SAT: Relation of Typical-Case Complexity to the Nature of the Phase Transition." *Rand. Struct. & Algorithms* 15 (1999): 414–435.

408. Montanari, A., and R. Zecchina. "Optimizing Searches via Rare Events. *Phys. Rev. Lett.* 88 (2002): 178701.

409. Moore, C. Personal communication, 2003.

410. Mossel, E., R. O'Donnell, and F. Oleszkiewicz. "Noise Stability of Functions with Low Influences: Invariance and Optimality." arXiv.org E-print Archive,

Cornell University Library, 2005. http://www.arxiv.org/abs/math/0503503 (accessed September 27, 2005).

411. Motwani, R., and P. Raghavan. *Randomized Algorithms*. Cambridge, UK: Cambridge University Press, 2000.

412. Mulet, R., A. Pagnani, M. Weigt, and R. Zecchina. "Coloring Random Graphs." *Phys. Rev. Lett.* 89 (2002): 268701.

413. Nelson, E. "The Free Markov Field." *J. Functional Analysis* 12 (1973): 211–227.

414. Netcraft Ltd. "Web Server Survey." Netcraft web server surveys from 2005. http://news.netcraft.com/archives/web_server_survey.html (accessed September 27, 2005).

415. Newman, M. E. J., and D. J. Watts. "Renormalization Group Analysis of the Small-World Network Model." *Phys. Lett. A* 263 (1999): 341–346.

416. Nicol, D. M., and R. M. Fujimoto. "Parallel Simulation Today." *Ann. Oper. Res.* 53 (1994): 249–286.

417. Odlyzko, A. M. "Asymptotic Enumeration Methods." In *Handbook of Combinatorics*, ed. R. L. Graham, M. Grötschel, and L. Lovász, vol. 2, 1063–1229. Amsterdam: Elsevier Science, 1995.

418. O'Donnell, R., M. Saks, O. Schramm, and R. Servedio. "Every Decision Tree has an Influential Variable." In *Proceedings of the 46th Annual IEEE Symposium on Foundations of Computer Science (FOCS'05)*, to appear. IEEE Computer Society, 2005.

419. Orlin, J. B. "The Complexity of Dynamic/Dynamic Languages and Optimization Problems." Sloan Working paper 1679-86. Alfred P. Sloan School of Management, MIT, Cambridge, MA, July 1985. Preliminary version in *Proceedings of the 13th Annual ACM Symposium on Theory of Computing (STOC'78)*, 218-227. New York: ACM Press, 1981.

420. Orlin, J. B. "Minimum Convex Cost Dynamic Network Flows." *Math. Oper. Res.* 9 (1984): 190–207.

421. Orlin, J. B. "Some Problems on Dynamic/Dynamic Graphs." In *Progress in Combinatorial Optimization*, ed. W. R. Pullerybank, 273-293. Toronto: Academic Press, 1984.

422. Palmer, R. G., and J. Adler. "Ground States for Large Samples of Two-Dimensional Ising Spin Glasses." *Int. J. Mod. Phys. C* 10 (1999): 667–675.

423. Papadimitriou, C. H. *Computational Complexity*. Reading, MA: Addison-Wesley, 1994.

424. Papadimitriou, C. H. "On Selecting a Satisfying Truth Assignment." In *Proceedings of the 32nd Annual IEEE Symposium on Foundations of Computer Science* (1991): 163–169. IEEE Computer Society, 1991.

425. Papadimitriou, C. H., and K. Steiglitz. *Combinatorial Optimization: Algorithms and Complexity*. Englewood Cliffs, NJ: Prentice-Hall, 1982.

426. Papadimitriou, C. H., and M. Yannakakis. "A Note on Succinct Representations of Graphs." *Information and Control* 71 (1986): 181–185.

427. Parisi, G. "Constraint Optimization and Statistical Mechanics." In *The Physics of Complex Systems*, ed. F. Mallamace and H. E. Stanley, 205–228. International School of Physics Enrico Fermi, vol. 134. Amsterdam: IOS Press, 2004.

428. Parisi, G. "Some Remarks on the Survey Decimation Algorithm for K-Satisfiability." arXiv.org E-print Archive, Cornell University Library, 2003. http://arxiv.org/abs/cs/0301015 (accessed September 27, 2005).

429. Parisi, G. "On the Survey-Propagation Equations for the Random K-Satisfiability Problem." arXiv.org E-print Archive, Cornell University Library, 2002. http://arxiv.org/abs/cs.CC/0212009 (accessed September 27, 2005).

430. Parkes, A. J. "Scaling Properties of Pure Random Walk on Random 3-SAT." In *Proceedings of the Eighth International Conference on Principles and Practice of Constraint Programming (CP 2002)*, ed. P. von Hentenryek, 708–713. Lecture Notes in Computer Science, vol. 2470. Berlin: Springer-Verlag, 2002.

431. Pearl, J. *Probabilistic Reasoning in Intelligent Systems: Network of Plausible Inference.* San Francisco: Morgan Kaufmann, 1988.

432. Pemberton, J. C., and W. Zhang. "Epsilon-Transformation: Exploiting Phase Transitions to Solve Combinatorial Optimization Problems." *Art. Intel.* 81 (1996): 297–325.

433. Pennock, D. M., and Q. F. Stout. "Exploiting a Theory of Phase Transitions in Three-Satisfiability Problems." In *Proceedings of the 13th National Conference on Artificial Intelligence (AAAI'96)*, 253–258. Menlo Park, CA: AAAI Press, 1996.

434. Percus, A. G., and O. C. Martin. "The Stochastic Traveling Salesman Problem: Finite Size Scaling and the Cavity Prediction." *J. Stat. Phys.* 94 (1999): 739–758.

435. Peres, Y. "Noise Stability of Weighted Majority." arXiv.org E-print Archive, Cornell University Library, 2005. http://arxiv.org/abs/math.PR/0412377 (accessed September 27, 2005).

436. Pittel, B., J. Spencer, and N. Wormald. "Sudden Emergence of a Giant k-Core in a Random Graph." *J. Comb. Theor. B* 67 (1996): 111–151.

437. Powell, W., P. Jaillet, and A. Odoni. "Stochastic and Dynamic Networks and Routing." In *Handbook of Operations Research and Management Science: Network and Routing*, ed. M. Ball, T. Magnanti, C. Monma, and G. Nemhauser, 141–295. Amsterdam: Elsevier, 1999.

438. Raz, R. "A Parallel Repetition Theorem." *SIAM J. Comp.* 27 (1998): 763–803.

439. Raz, R., and S. Safra. "A Sub-Constant Error-Probability Low-Degree Test, and A Sub-Constant Error-Probability PCP Characterization of NP." In *Proceedings of the 29th Annual ACM Symposium on Theory of Computing (STOC'97)*, 475–484. New York: ACM Press, 1997.

440. Reidys, C. M. "Distances in Random Induced Subgraphs of Generalized n-Cubes." *Comb. Prob. & Comp.* 11 (2002): 599–605.

441. Reidys, C. M. "Random Induced Subgraphs of Generalized n-Cubes." *Adv. Appl. Math.* 19 (1997): 360–377.

442. Reidys, C. M., and P. F. Stadler. "Bio-Molecular Shapes and Algebraic Structures." *Comp. & Chem.* 20 (1996): 85–94.

443. Reidys, C. M., P. F. Stadler, and P. Schuster. "Generic Properties of Combinatory Maps: Neutral Networks of RNA Secondary Structures." *Bull. Math. Biol.* 59 (1997): 339–397.

444. Reinelt, G. *The Travelling Salesman. Computational Solutions for TSP Applications.* Lecture Notes in Computer Science, vol. 840. Berlin: Springer-Verlag, 1994.

445. Ricci-Tersenghi, F., M. Weigt, and R. Zecchina. "Simplest Random K-Satisfiability Problem." *Phys. Rev. E* 63 (2001): 026702.

446. Richardson, T., and R. Urbanke. "An Introduction to the Analysis of Iterative Coding Systems." In *Codes, Systems, and Graphical Models*, ed. B. Marcus and J. Rosenthal, 1–37. New York: Springer, 2001.

447. Robinson, C. *Dynamical Systems Stability, Symbolic Dynamics, and Chaos*, 2d ed. Boca Raton, FL: CRC Press, 1999.

448. Ruml, W., J. Ngo, J. Marks, and S. Shieber. "Easily Searched Encodings for Number Partitioning." *J. Opt. Theory Appl.* 89 (1996): 251–291.

449. Russo, L. "An Approximate Zero-One Law." *Zeitschrift für Wahrscheinlichkeitstheorie und Verwandte Gebiete* 61 (1982): 129–139.

450. Russo, L. "A Note on Percolation." *Zeitschrift für Wahrscheinlichkeitstheorie und Verwandte Gebiete* 43 (1978): 39–48.

451. Samet, Y. "Equilibria with Information Aggregation in Sharp Threshold Voting Rules." M.Sc. thesis, Hebrew University of Jerusalem, 2004.

452. Sassanfar, M., and J. W. Szostak. "An RNA Motif that Binds ATP." *Nature* 364 (1993): 550–553.

453. Sato, H., and R. Kikuchi. Dilute Antiferromagnetic Systems in FCC and BCC Lattices." In *AIP Conf. Proc. No. 18, Part 1, Magnetism and Magnetic Materials*, ed. C. D. Graham, Jr., and J. J. Rhyne, 605–609. New York: AIP, 1974.

454. Savelsberg, M., and P. van Emde Boas. "Bounded Tiling, An Alternative to SATISFIABILITY." In *Proceedings of the 2nd Frege Conference*, ed. G. Wechsung, 354–363. Berlin: Akademie Verlag, 1984.

455. Schaefer, T. "The Complexity of Satisfiability Problems." In *Proceedings of the 10th Annual ACM Symposium on Theory of Computing (STOC'78)*, 216–226. New York: ACM Press, 1978.

456. Schmitt, W. R., and M. S. Waterman. "Plane Trees and RNA Secondary Structure." *Discrete Appl. Math.* 51 (1994): 317–323.

457. Schoneveld, A. "Parallel Complex Systems Simulation." Ph.D. thesis, Universiteit van Amsterdam, 1999.

458. Schöning, U. "A Probabilistic Algorithm for k-SAT Based on Limited Local Search and Restart. *Algorithmica* 32 (2002): 615–623.

459. Schramm, O., and J. Steif. "Quantitative Noise Sensitivity and Exceptional Times for Percolation." arXiv.org E-print Archive, Cornell University Library, 2005. http://arxiv.org/abs/math/0504586 (accessed September 27, 2005).

460. Schramm, O., and B. Tsirelson. "Trees, Not Cubes: Hypercontractivity, Cosiness, and Noise Stability." *Elec. Comm. Prob.* 4 (1999): 39–49.

461. Schultes, E., and P. Bartels. "One Sequence, Two Ribozymes: Implications for the Emergence of New Ribozyme Folds." *Science* 289 (2000): 448–452.

462. Schuster, P. K. "Landscapes and Molecular Evolution." *Physica D* 107 (1997): 351–365.

463. Schuster, P. K., and P. F. Stadler. "Landscapes: Complex Optimization Problems and Biopolymer Structures." *Comp. & Chem.* 18 (1994): 295–314.

464. Schuster, P. K., W. Fontana, P. F. Stadler, and I. L. Hofacker. "From Sequences to Shapes and Back: A Case Study in RNA Secondary Structures." *Proc. Roy. Soc. Lond. B* 255 (1994): 279–284.

465. Schuster, P. K., J. Weber, W. Grüner, and C. M. Reidys. "Molecular Evolutionary Biology: From Concepts to Technology." In *Physics of Biological Systems: From Molecules to Species*, ed. H. Flyvbjerg, J. Hertz, M. H. Jensen, K. Sneppen, and O. G. Mouritsen, 283–306. Berlin: Springer, 1997.

466. Selman, B., and S. Kirkpatrick. "Critical Behavior in the Computational Cost of Satisfiability Testing." *Art. Intel.* 81 (1996): 273–295.

467. Selman, B., H. Levesque, and D. Mitchell. "A New Method for Solving Hard Satisfiability Problems." In *Proceedings of the 10th National Conference on Artificial Intelligence (AAAI'92)*, 440–446. Menlo Park, CA: AAAI Press, 1992.

468. Selman, B., H. Kautz, and B. Cohen. "Noise Strategies for Improving Local Search." In *Proceedings of the 12th National Conference on Artificial Intelligence (AAAI'94)*, 337–343. Menlo Park, CA: AAAI Press, 1994.

469. Selman, B., D. G. Mitchell, and H. J. Levesque. "Generating Hard Satisfiability Problems." *Art. Intel.* 81 (1996): 17–29.

470. Selman, B., H. Kautz, and B. Cohen. "Local Search Strategies for Satisfiability Testing." In *Cliques, Coloring, and Satisfiability: Second DIMACS Implementation Challenge*, ed. D. S. Johnson and M. A. Trick, 521–531. DIMACS Series in Discrete Mathematics and Theoretical Computer Science, vol. 26. American Mathematical Society, 1996.

471. Semerjian, G., and L. F. Cugliandolo. "Cluster Expansions in Dilute Systems: Applications to Satisfiability Problems and Spin Glasses." *Phys. Rev. E* 64 (2001): 036115.

472. Semerjian, G., and R. Monasson. "Relaxation and Metastability in a Local Search Procedure for the Random Satisfiability Problem." *Phys. Rev. E* 67 (2003): 066103.

473. SETI@home. Home Page. http://setiathome.ssl.berkeley.edu/ (accessed September 27, 2005).

474. Sherrington, D., and S. Kirkpatrick. "Solvable Model of a Spin-Glass." *Phys. Rev. Lett.* 35 (1975): 1792–1796.

475. Simon, J. C., J. Carlier, O. Dubois, and O. Moulines. "Étude statistique de l'existence de solutions de problèmes SAT, application aux systèmes-experts." *C.R. Acad. Sci. Paris. Sér. I Math.* 302 (1986): 283–286.

476. Sistla, A. P., and E. Clark. "Complexity of Linear Temporal Logics." *J. ACM* 32 (1985): 733–749.

477. Slaney, J., and T. Walsh. "Backbones in Optimization and Approximation." In *Proceedings of the Seventeenth International Joint Conference on Artificial Intelligence (IJCAI'01)*, ed. B. Nebel, 254–259. San Francisco: Morgan Kaufmann, 2001.

478. Sloot, P. M. A., B. J. Overeinder, and A. Schoneveld. "Self-Organized Criticality in Simulated Correlated Systems." *Comp. Phys. Commun.* 142 (2001): 76–81.

479. Smirnov, S. "Critical Percolation in the Plane: Conformal Invariance, Cardy's Formula, Scaling Limits." *C. R. Acad. Sci. Paris Sér. I Math.* 333 (2001): 239–244.

480. Snedecor, G. W., and W. G. Cochran. *Statistical Methods*, 6th ed. Ames, Iowa: Iowa State University Press, 1967.

481. Spielman, D. A. "The Complexity of Error-Correcting Codes." In *Proceedings of the 11th International Symposium on Fundamentals of Computation Theory (FCT 1997)*, ed. B. S. Chlebus and L. Czaja, 67–84. Lecture Notes in Computer Science, vol. 1279. Berlin: Springer-Verlag, 1997.

482. Spirin, V., P. L. Krapivsky, and S. Redner. "Freezing in Ising Ferromagnets." *Phys. Rev. E* 65 (2001): 016119.

483. Stadler, P. F., W. Hordijk, and J. F. Fontanari. "Phase Transition and Landscape Statistics of the Number Partitioning Problem." *Phys. Rev. E* 67 (2003): 056701.

484. Stauffer, D. "Frustration and Simulation." *Physics World* 23. May 1999.

485. Stauffer, D., and A. Aharony. *Introduction to Percolation Theory*. London: Taylor and Francis, 1994.

486. Steane, A. "Quantum Computing." *Rep. Prog. Phys.* 61 (1998): 117–173.

487. Stearns, R. E. "It's Time to Reconsider Time." *Comm. ACM* 37 (1994): 95–99.

488. Stearns, R. E., and H. B. Hunt III. "Power Indices and Easier Hard Problems." *Math. Syst. Theory* 23 (1990): 209–225.

489. Steffen, M., W. van Dam, T. Hogg, G. Breyta, and I. Chuang. "Experimental Implementation of an Adiabatic Quantum Optimization Algorithm." *Phys. Rev. Lett.* 90 (2003): 067903.

490. Stone, H. S., and P. Sipala. "The Average Complexity of Depth-First Search with Backtracking and Cutoff." *IBM J. Res. & Dev.* 30 (1986): 242–258.

491. Strogatz, S. H. "Exploring Complex Networks." *Nature* 410 (2001): 268–276.

492. Suen, W. C. "A Correlation Inequality and a Poisson Limit Theorem for Nonoverlapping Balanced Subgraphs of a Random Graph." *Rand. Struct. & Algorithms* 1 (1990): 231–242.

493. Svenson, P. "Freezing in Random Graph Ferromagnets." *Phys. Rev. E* 64 (2001): 036122.

494. Svenson, P., and M. G. Nordhal. "Relaxation in Graph Coloring and Satisfiability Problems." *Phys. Rev. E* 59 (1999): 3983–3999.

495. Talagrand, M. "On Boundaries and Influences." *Combinatorica* 17 (1997): 275–285.

496. Talagrand, M. "Concentration and Influences." *Israel J. Math.* 111 (1999): 275–284.

497. Talagrand, M. "Concentration of Measure and Isoperimetric Inequalities in Product Spaces." *Publ. I.H.E.S.* 81 (1995): 73–205.

498. Talagrand, M. "Isoperimetry, Logarithmic Sobolev Inequalities on the Discrete Cube, and Margulis' Graph Connectivity Theorem." *Geom. & Funct. Anal.* 3 (1993): 295–314.

499. Talagrand, M. "On Russo's Approximate Zero-One Law." *Ann. Prob.* 22 (1994): 1576–1587.

500. Tina: a Beowulf Supercomputer. Institute of Theoretical Physics, Otto-von-Guericke-Universität, Magdeburg. http://tina.nat.uni-magdeburg.de (accessed September 27, 2005).

501. Toroczkai, Z., G. Korniss, S. Das Sarma, and R. K. P. Zia. "Extremal-Point Densities of Interface Fluctuations." *Phys. Rev. E* 62 (2000): 276–294.

502. Toulouse, G. "Theory of the Frustration Effect in Spin Glasses: I." *Comm. Physics* 2 (1977): 115–119.

503. Trotter, H. F. "Approximation of Semi-Groups of Operators." *Pacific J. Math.* 8 (1958): 887–919.

504. Tsai, L.-H. "Asymptotic Analysis of an Algorithm for Balanced Parallel Processor Scheduling." *SIAM J. Comp.* 21 (1992): 59–64.

505. Tsirelson, B. "Scaling Limit, Noise, Stability." In *Lectures on Probability Theory and Statistics*, 1–106. Lecture Notes in Mathematics, 1840. Berlin: Springer, 2004.

506. Tsirelson, B., and A. Vershik. "Examples of Nonlinear Continuous Tensor Products of Measure Spaces and Non-Fock Factorizations." *Rev. Math. Phys.* 10 (1998): 81–145.

507. Tuerk, C., and L. Gold. "Systematic Evolution of Ligands by Exponential Enrichment: RNA Ligands to Bacteriophage T4 DNA Polymerase." *Science* 249 (1990): 505–510.

508. van Dam, W., M. Mosca, and U. Vazirani. "How Powerful is Adiabatic Quantum Computation?" In *Proceedings of the 42nd Annual IEEE Symposium on Foundations of Computer Science (FOCS'01)*, 279–287. IEEE Computer Society, 2001.

509. van Emde Boas, P. "Dominoes are Forever." In *Proceedings of the 1st GTI Workshop*, 76–95. Paderborn, 1983.

510. Vardi, M. Y., and P. Wolper. "An Automata-Theoretic Approach to Automatic Program Verification (preliminary report)." In *Proceedings of the 1st IEEE Symposium on Logic in Computer Science (LICS'86)*, 332–344. Washington, DC: IEEE Computer Society, 1986.

511. Vogel, E. E., J. Cartes, S. Contreras, W. Lebrecht, and J. Villegas. "Ground-State Properties of Finite Square and Triangular Ising Lattices with Mixed Exchange Interactions." *Phys. Rev. B* 49 (1994): 6018–6027.

512. Vogel, E. E., J. Cartes, P. Vargas, D. Altbir, S. Kobe, T. Klotz, and M. Nogala. "Hysteresis in $\pm J$ Ising Square Lattices." *Phys. Rev. B* 59 (1999): 3325–3328

513. Vogel, E. E., A. J. Ramirez-Pastor, and F. Nieto. "Detailed Structure of Configuration Space and Its Importance on Ergodic Separation of $\pm J$ Ising Lattices." *Physica A* 310 (2002): 384–396.

514. Walsh, T. "The Constrainedness Knife-Edge." In *Proceedings of the 15th National Conference on Artificial Intelligence (AAAI'98)*, 406–411. Menlo Park, CA: AAAI Press, 1998.

515. Wang, H. "Proving Theorems by Pattern Recognition II." *Bell System Tech. J.* 40 (1961): 1–41.

516. Wanke, E. "Paths and Cycles in Finite Dynamic Graphs." In *Proceedings of the 20th Symposium on Mathematical Foundations of Computer Science (MFCS'93)*, ed. A. M. Borzyszkowski and S. Sokolowski, 751–760. Lecture Notes in Computer Science, vol. 711. Berlin: Springer-Verlag, 1993.

517. Wannier, G. H. "Antiferromagnetism. The Triangular Ising Net." *Phys. Rev.* 79 (1950): 357–364.

518. Waterman, M. S. "Combinatorics of RNA Hairpins and Cloverleaves." *Stud. Appl. Math.* 60 (1978): 91–96.

519. Watts, D.J., and S. H. Strogatz. "Collective Dynamics of 'Small-World' Networks." *Nature* 393 (1998): 440–442.

520. Weber, J. *Dynamics on Neutral Evolution.* Ph.D. thesis, Friedrich Schiller University, Jena, 1997.

521. Weigt, M. "Dynamics of Heuristic Optimization Algorithms on Random Graphs." *Eur. Phys. J. B* 28 (2002): 369–381.

522. Weigt, M., and A. K. Hartmann. "Minimal Vertex Covers on Finite-Connectivity Random Graphs: A Hard-Sphere Lattice-Gas Picture." *Phys. Rev. E* 63 (2001): 056127.

523. Weigt, M., and A. K. Hartmann. "The Number of Guards Needed by a Museum: A Phase Transition in Vertex Covering of Random Graphs." *Phys. Rev. Lett.* 84 (2000): 6118–6121.

524. Weigt, M., and A. K. Hartmann. "Typical Solution Time for a Vertex-Covering Algorithm on Finite-Connectivity Random Graphs." *Phys. Rev. Lett.* 86 (2001): 1658–1661.

525. West, D. *Introduction to Graph Theory.* Princeton, NJ: Prentice Hall, 1994.

526. Westheimer, F. H. "Polyribonucleic Acids as Enzymes." *Nature* 319 (1986): 534–536.

527. Wilson, D. B. "On the Critical Exponents of Random k-SAT." *Rand. Struct. & Algorithms.* 21 (2002): 182–195.

528. Wolfram, S. *Theory and Applications of Cellular Automata.* Singapore: World Scientific, 1987.

529. Wormald, N. C. "Differential Equations for Random Processes and Random Graphs." *Ann. Appl. Prob.* 5 (1995): 1217–1235.

530. Wright, S. "Random Drift and the Shifting Balance Theory of Evolution." In *Mathematical Topics in Population Genetics*, ed. K. Kojima, 1–31. Berlin: Springer-Verlag, 1970.

531. Wright, S. "The Roles of Mutation, Inbreeding, Crossbreeeding and Selection in Evolution." In *International Proceedings of the Sixth International Congress on Genetics*, ed. D. F. Jones, vol. 1, 356–366. Menosha, WI: Brooklyn Botanic Garden, 1932. On-line facsimile version available from Electronic Scholarly Publishing. http://www.esp.org/books/6th-congress/facsimile (accessed September 27, 2005).

532. Yakir, B. "The Differencing Algorithm LDM for Partitioning: A Proof of a Conjecture of Karmarkar and Karp." *Math. Oper. Res.* 21 (1996): 85–99.

533. Yedidia, J. S., W. T. Freeman and Y. Weiss. "Generalized Belief Propagation." In *Advances in Neural Information Processing Systems (NIPS)*, ed. T. K. Leen, T. G. Dietterich, and V. Tresp, vol. 13, 689–695. Cambridge, MA: MIT Press, 2001.

534. Young, A. P. "Spin Glasses: A Computational Challenge for the 21st Century." *Comp. Phys. Commun.* 146 (2002): 107–112.

535. Young, H. P. "Condorcet's Theory of Voting." *Amer. Pol. Sci. Rev.* 82 (1988): 1231–1244.

536. Zecchina, R. Survey Based Algorithm for Random Satisfiability. http://www.ictp.trieste.it/~zecchina/SP (accessed September 27, 2005).

537. Zhan, Z. F., L. W. Lee, and J.-S. Wang. "A New Approach to the Study of the Ground-State Properties of 2D Ising Spin Glass." *Physica A* 285 (2000): 239–247.

538. Zito, M. "Randomised Techniques in Combinatorial Algorithmics." Ph.D. thesis, Department of Computer Science, University of Warwick, November 1999.

Index

Printed in the United States
By Bookmasters